Ergebnisse der Mathematik und ihrer Grenzgebiete. 3. Folge/A Series of Modern Surveys in Mathematics

Volume 77

Ergebnisse der Mathematik und ihrer Grenzgebiete, now in its third sequence, aims to provide reports, at a high level, on important topics of mathematical research. Each book is designed as a reliable reference covering a significant area of advanced mathematics, spelling out related open questions, and incorporating a comprehensive, up-to-date bibliography.

Michael W. Davis

Infinite Group Actions on Polyhedra

Michael W. Davis
Department of Mathematics
The Ohio State University
Columbus, OH, USA

ISSN 0071-1136 ISSN 2197-5655 (electronic)
Ergebnisse der Mathematik und ihrer Grenzgebiete. 3. Folge/A Series of Modern Surveys in Mathematics
ISBN 978-3-031-48442-1 ISBN 978-3-031-48443-8 (eBook)
https://doi.org/10.1007/978-3-031-48443-8

Mathematics Subject Classification: 20F65, 20F67, 20F36, 20F55, 52B11, 57M07, 57Q15, 57K32, 57P05

This Springer imprint is published by the registered company Springer Nature Switzerland AG
The registered company address is: Gewerbestrasse 11, 6330 Cham, Switzerland

Paper in this product is recyclable.

Preface

In the spring of 2018 Misha Gromov asked me if I wanted to write a book, *Infinite Group Actions on Polyhedra*, for Springer's Ergebnisse der Mathematik series and I agreed. This is the result. When I first mentioned the proposed title to Tadeusz Januszkiewicz, he responded "Well, that includes just about everything." Indeed, the title encompasses much of the rapidly growing field of geometric group theory. Since I have not kept up with many of the recent advances in the field, the project has proved to be more difficult than I originally envisioned. In the end, I have tried to focus on the subjects I understood best: nonpositive curvature, Coxeter groups, Artin groups, groups acting on buildings, simple complexes of groups, and hyperbolization techniques. My one attempt at covering a more recent topic is Chap. 5 on the Haglund–Wise theory of special cube complexes and its application by Agol to the theory of hyperbolic 3-manifolds.

The main ideas in geometric group theory originated in Gromov's extended essay, *Hyperbolic Groups* [136]. The key idea of a nonpositively curved polyhedron appeared there as did the basic notion of a hyperbolic group. Many of the topics in this book can also be found in the text by Bridson and Haefliger, *Metric Spaces of Non-positive Curvature* [35]. There is also considerable overlap with my earlier book, *The Geometry and Topology of Coxeter Groups* [82]. The book in hand should be regarded as a companion text to the two aforementioned books. An important feature of nonpositive curvature is that the universal cover of a nonpositively curved polyhedron is CAT(0) and hence, is contractible. The main subject of this book is on infinite groups acting on contractible polyhedra. Here my focus has been on describing techniques for constructing examples.

This book is intended for graduate students and researchers in geometric group theory.

Because of my many typing mistakes, proofreading became a difficult time-consuming task. I thank Jingyin Huang, Conchita Martinez, Grigori Avramidi, Kevin Schreve, and Boris Okun for their comments and considerable help with proofreading. Frédéric Haglund made a number of helpful suggestions, both mathematical and typographical, concerning Chap. 5. I also thank my old friend Elliott Stein for spending a good deal of time and effort proofreading the entire manuscript and making many substantial suggestions. My thanks also go to my student, Katherine Goldman, for doing the figures and illustrations. I am grateful to the editor of the series, Remi Lodh, who provided a good deal of help and encouragement.

Columbus, OH, USA
December, 2023

Michael W. Davis

Contents

Acronyms

CAT	Comparison of Alexandrov and Toponogov
CG$\mathcal{R}$	Category for a complex of groups
FC	the flag complex condition
GHS	Generalized homology sphere
HNN	Higman–Neumann–Neumann
NPC	Nonpositively curved
RAAG	Right-angled Artin group
RACG	Right-angled Coxeter group
RACS	Right-angled Coxeter system

Part I
Introduction

Chapter 1
Overview

Mathematics is only a story of groups.
(H. Poincaré)
The theory of groups can be summarily defined as a theory of symmetry, indistinguishability and homogeneity.
(J. Tits)[1]

For over 170 years, the connection between group theory and the fields of geometry and topology has held a central position in mathematics—the connection is that the symmetries of a geometric object form a group. For example, the full symmetry group of a regular n-simplex is the symmetric group on $n + 1$ letters. Hence, any finite group can be realized as a group of symmetries of a regular simplex. As another example, any finitely generated group G acts on its Cayley graph as a group of symmetries of the graph. The pantheon of great mathematicians who worked on symmetry groups during the second half of the nineteenth century includes Möbius, Schläfli, Jordan, Cayley, Lie, Klein, and Poincaré. Around 1850 Möbius determined the full symmetry groups of 3-dimensional regular polytopes (or equivalently of tessellations of $\mathbb{S}^2$ by spherical triangles). This was extended to higher dimensions by Schläfli. Möbius' results were also extended to triangular tessellations of the euclidean and hyperbolic planes by Riemann and Schwarz and later to polygonal tessellations of the hyperbolic plane by Poincaré and Klein. In this early work the group was a discrete group of symmetries of a geometric object, namely, a polytope or a tessellation of a space by polytopes. Such discrete symmetry groups are the main focus of this book. In the nineteenth century research also began on "continuous" symmetry groups with work of Lie and Klein on Klein's Erlangen Program. This program identified a geometry with its group of symmetries. Discrete subgroups of Lie groups acting an appropriate geometries also play a role in this volume. For the first half of the twentieth century, we mention only the work Weyl, E. Cartan, Chevalley, and Coxeter. The first three made important contributions to continuous

[1] I found both of these quotes in a paper of Étienne Ghys.

M. W. Davis, *Infinite Group Actions on Polyhedra*, Ergebnisse der Mathematik und ihrer Grenzgebiete. 3. Folge / A Series of Modern Surveys in Mathematics 77, https://doi.org/10.1007/978-3-031-48443-8_1

symmetry groups. Cartan's theory of symmetric spaces and their isometry groups (i.e., semisimple Lie groups), as well as the theory of locally symmetric spaces (i.e., quotients of symmetric spaces by a discrete subgroups of isometries of cofinite volume) stands out. In the 1930s Coxeter [66, 67] classified irreducible discrete reflection groups on the n-sphere and on n-dimensional Euclidean space. He showed that each was associated with a symmetry group of a regular tessellation or with symmetry group of a root system. The theory of Coxeter groups developed out of this work. The first half of the twentieth century also saw the introduction into group theory of the notion of a "presentation" of a group via generators and relations. This led to the introduction of topological methods in group theory and to the field of combinatorial group theory. For the second half of the twentieth century, we only mention Tits, Mostow, Thurston, and Gromov. Tits was responsible for the theory of buildings (cf. Sect. 4.4). In particular, his theory of "spherical buildings" provides polyhedra on which finite simple groups of Lie type (also called "Chevalley groups") act as symmetry groups. Mostow's Rigidity Theorem showed that in many cases, a locally symmetric space of finite volume is determined, up to isometry and homothety, by the discrete subgroup of isometries of the symmetric space. Thurston's Geometrization Conjecture, in particular his work on putting hyperbolic structures on 3-manifolds, vastly expanded our view of the connection between group theory, geometry and topology. Beginning in the late 1970s Gromov introduced some of the main ideas in geometric group theory: the basic idea was to study "coarse" properties of a group G (i.e., properties of the Cayley graph of G which are invariant under quasi-isometries). A early success appeared in Gromov [135] where it is proved that any group of polynomial growth is virtually a subgroup of nilpotent Lie group. An important idea of Gromov [136] was the notion of a "hyperbolic group," meaning a group that is coarsely negatively curved.

In this book the terms "cell complex" and "polyhedron" are used synonymously. The cells are usually convex polytopes. We are primarily concerned with polyhedra whose universal covers are contractible. The group of deck transformations is a symmetry group of such a universal cover. Since we are implicitly assuming that symmetries preserve the cell structure, these symmetry groups will be discrete subgroups of the group of homeomorphisms of the underlying topological space. As basic examples, the reader can think of tessellations of euclidean space or hyperbolic space. However, such tessellations only give a tiny fraction of the possibilities. For example, the universal cover of any connected graph is a tree, i.e., it is a 1-dimensional contractible polyhedron, usually not a manifold. The reasons for studying polyhedra with contractible universal covers are rooted in topology.

In the mid twentieth century Alexsandrov and Busemann introduced notions of what it means for certain singular metric spaces (geodesic spaces) to be "nonpositively curved" (abbreviated as NPC). These notions give a geometric method for proving that the universal cover of a connected polyhedron is contractible: if a polyhedron admits a NPC polyhedral metric, then its universal cover is contractible. This method was used by Gromov [136] to produce many new examples of nonpositively curved polyhedra. The description of such examples is a principal focus of this book. Although such polyhedra need not be manifolds, the case in which they are is particularly interesting. In Sect. 3.2 and Chap. 6 we construct NPC metrics on polyhedra which happen to be topological manifolds. These manifolds need not admit any nonpositively curved Riemannian metric. For example, the universal cover of such an NPC polyhedral manifold, although contractible, need not be homeomorphic to euclidean space. By the Cartan–Hadamard Theorem such a manifold cannot admit any NPC Riemannian metric.

There are two methods for constructing actions of discrete groups on polyhedra. One method involves the fundamental group and covering space theory. The other method involves the theory of complexes of groups.

Covering Spaces If a space X admits a universal covering space (i.e., if X is semi-locally 1-connected), then its fundamental group acts as the group of deck transformations on its universal cover $\widetilde{X}$. Given a presentation for a group π, there is a standard construction of a 2-dimensional, connected CW complex X with $\pi_1(X) = \pi$: start with a wedge of circles, one for each generator and then attach a 2-cell for each relation. It follows from van Kampen's Theorem that $\pi_1(X) = \pi$. One can continue to attach cells to X of dimension greater than 2 to kill the higher homotopy groups, $\pi_i(X)$, for all $i \geq 2$. The result is a CW complex with fundamental group π and with all higher homotopy groups equal to 0. We denote such a space by $B\pi$ and call it the *classifying space* of π. The CW complex $B\pi$ is characterized by the fact that its universal cover, $E\pi$, is contractible. Standard arguments in homotopy theory show that $B\pi$ is unique up to homotopy equivalence, e.g., see [146, Theorem 1B.8]. Thus, algebraic topological properties of $B\pi$ (such as its cohomology groups) are properties of the group π. The space $B\pi$ is also called an "Eilenberg-MacLane complex" $K(\pi, 1)$. Any such space in which all higher homotopy groups vanish is said to be *aspherical*.

Simple Complexes of Groups A 1-dimensional complex of groups is a *graph of groups* as defined in Serre's book [207]. The theory of graphs of groups inspired several people, notably, Haefliger [138], to define a higher dimensional version called a "complex of groups." Roughly, the data for a complex of groups consists of a cell complex X, a family of subcomplexes and an assignment of a group, called the "local group" to each subcomplex in the family. This data defines a group, called the "fundamental group" of the complex of groups, as well as a new cell complex Y, called the "universal cover of the complex of groups," on which the "fundamental group of the complex of groups" acts. The local groups are cell stabilizers. For example, if each local group is trivial, then the fundamental group of the complex of groups is the ordinary fundamental group of the underlying CW complex and its universal cover is the ordinary universal cover. At the other extreme, there is the notion of a "simple" complex of groups. Simple complexes of groups arise from group actions on polyhedra with strict fundamental domains. This means that the family of local groups is a poset of groups and that the fundamental group is just the direct limit of this system of groups (at least in the case when the geometric realization of the underlying poset is simply connected). In the case of a graph of groups, the underlying 1-complex is a graph, the universal cover is a tree, and the local groups are the stabilizers of edges or vertices. When the underlying graph is a tree, the graph of groups is a simple complex of groups; so, its fundamental group is the direct limit of the poset of local groups. In dimensions greater than 1 the universal cover of a simple complex of groups need not be contractible. For example, if a triangle of groups comes from the action of a discrete reflection group on S^2, then its fundamental group is a finite Coxeter group and the universal cover is S^2.

CAT(**0**) **Spaces and Nonpositive Curvature** Geometric group theory essentially begins with Gromov's essay [136]. Nonpositive curvature is the main current in the ocean of geometric group theory. One of the innovations of [136] was to define the notion of a nonpositively curved (or a "locally CAT(0)") polyhedron. Gromov defines a geodesic space to be CAT(0) if each of its geodesic triangles is "thinner" than a corresponding comparison triangle in the euclidean plane. Any CAT(0) space is contractible. The reason is that there is a unique geodesic segment connecting any point to a base point: geodesic contraction then provides the contraction. A prototypical example of a CAT(0) space is the universal cover of a closed, nonpositively curved Riemannian manifold. In this book we will be concerned with "singular" geodesic spaces (e.g. spaces with polyhedral metrics) which need not be manifolds. Such a geodesic space is *nonpositively curved* if it is locally CAT(0). The universal cover of any NPC space can then be seen to be CAT(0) and hence, to be contractible. This is the main use of nonpositive curvature in geometric group theory—it provides a method for showing that a polyhedron is aspherical. Similarly, one can define what it means for a geodesic space to be CAT(-1) or CAT(1) by using comparison triangles in the hyperbolic plane $\mathbb{H}^2$ or the unit 2-sphere S^2, respectively. It is explained in Sect. 2.1 that a piecewise euclidean polyhedron X is nonpositively curved (abbreviated NPC) if and only if the link of any cell is CAT(1). (See Theorem 2.11.) This condition on a polyhedron X is called the "Link Condition." It means, for example, that a piecewise euclidean surface is nonpositively curved if and only if the cone angle at each of its vertices is $\geq 2\pi$.

Nonpositively Curved Cube Complexes A polyhedron X is a *cube complex*[2] if each of its k-dimensional cells is combinatorially equivalent to the unit k-dimensional cube in some euclidean space (possibly with some identifications of faces on its boundary). Cube complexes can be given a piecewise euclidean metric: just declare each cell to be locally isometric to the corresponding unit cube in euclidean space. For cube complexes, the Link Condition is equivalent to an easy combinatorial condition—the link of each cell must be a "flag complex." This is Gromov's Lemma from Sect. 2.2. (A simplicial complex is a *flag complex* if every complete subgraph of its 1-skeleton spans a simplex.) The study of nonpositively curved cube complexes is a major subject in this book, in fact, it is the primary topic in Chaps. 3 and 5.

[2] Recently, the use of the term "cube complex" instead of "cubical complex" has come into vogue.

Polyhedral Products and Their Universal Covers Polyhedral products provide a method for constructing NPC cube complexes. The data for a polyhedral product consists of a simplicial complex L, whose vertex set is denoted I, together with an I-tuple $\{(A_i, B_i)\}_{i\in I}$ of pairs of spaces. The *polyhedral product* $\{(A_i, B_i)\}^L$ is then defined to be a certain subspace of the product $\prod_{i\in I} A_i$ that depends on which subsets of I are vertex sets of simplices in L (see Definition 3.1). A prototypical example is the case when each A_i is the interval $[-1, 1]$ and B_i is its boundary $\{\pm 1\}$. The polyhedral product $\{([-1, 1], \partial[-1, 1])\}^L$ is denoted by P_L in Sect. 3.1. It is a subcomplex of the cube $[-1, 1]^I$. It follows from Gromov's Lemma that P_L is NPC if and only if L is a flag complex (usually we assume this). Hence, its universal cover $\tilde{P}_L$ is a CAT(0) cube complex, called the "Davis–Moussong complex." Let $\mathbf{C}_2 = \{\pm 1\}$ denote the cyclic group of order 2 acting on $[-1, 1]$ by reflection across 0. Then $(\mathbf{C}_2)^I$ acts on $[-1, 1]^I$ via reflections across the coordinate hyperplanes. The fundamental domain for the $(\mathbf{C}_2)^I$-action is the polyhedral product, where each $(A_i, B_i) = ([0, 1], 1)$. This fundamental domain is denoted K. It is homeomorphic to the cone on the barycentric subdivision of L. The subcomplex P_L is stable under the $(\mathbf{C}_2)^I$-action. The group of all lifts of elements of $(\mathbf{C}_2)^I$ to the universal cover $\tilde{P}_L$ is denoted by W_L and called the *right-angled Coxeter group* (abbreviated as RACG) associated to L. The Davis–Moussong complex together with its W_L-action is one of the principal examples in this book.

More generally, given any I-tuple of discrete groups $(G_i)_{i\in I}$, take $(A_i, B_i) = (\text{Cone}(G_i), G_i)$ and form the polyhedral product $Z_L := \{(\text{Cone}(G_i), G_i)\}^L$. The direct sum (i.e., the weak direct product), $\sum G_i$ acts on Z_L. The group of all lifts of elements of $\sum G_i$ to the universal cover $\tilde{Z}_L$ is called the *graph product* of the G_i. It is denoted by $\prod_{L^1} G_i$. As before, if L is a flag complex, then $\tilde{Z}_L$ is a CAT(0) cube complex. It is the basic example of a *right-angled building* (abbreviated RAB). If each G_i is infinite cyclic, then the graph product $\prod_{L^1} \mathbb{Z}$ is denoted A_L and called the *right-angled Artin group* (abbreviated RAAG) associated to the 1-skeleton of L. The associated RAB, $\tilde{Z}_L$, is called the *Deligne complex* for A_L in Chap. 4.

Polyhedral products also provide a method for constructing the classifying space for a graph product $\Gamma = \prod_{L^1} G_i$ from the classifying spaces of the factors. In fact, in the case $A_i = BG_i$ and B_i is a base point $*_i$, we get that $B\Gamma$ is homotopy equivalent to the polyhedral product:

$$B\Gamma = \{(BG_i, *_i)\}^L,$$

provided that L is a flag complex. (See Proposition 3.27 in Sect. 3.1.2 and Theorem A.19 in Appendix A.4.) In the case of the infinite cyclic group, $\mathbb{Z}$, the classifying space $B\mathbb{Z}$ is S^1. So, the classifying space BA_L for the right-angled Artin group A_L is the polyhedral product, $\{(S^1, *)\}^L$, which is a subcomplex of the torus $\mathbb{T}^I$ (that is, the I-fold product of copies of S^1).

The $K(\pi, 1)$-Question Suppose $\mathcal{G} = \{G_\sigma\}_{\sigma\in\mathcal{Q}}$ is a simple complex of groups with local groups G_σ indexed by a poset $\mathcal{Q}$. Let Γ be the fundamental group of the complex of groups $\mathcal{G}$ and let $Y(\mathcal{G})$ be the universal cover of its development. One can glue together the corresponding classifying spaces BG_σ to get a new space $B\mathcal{G}$. It follows from van Kampen's Theorem that the ordinary fundamental group of $B\mathcal{G}$ is Γ. The $K(\pi, 1)$-Question for $\mathcal{G}$ asks: Is $B\mathcal{G}$ homotopy equivalent to $B\Gamma$? In other words, if we use a simple complex of groups to glue together a poset $\{BG_\sigma\}$ of aspherical complexes, is the result a $K(\pi, 1)$? In fact this question was a principal motivation for the early development of the theory of complexes of groups by Gersten and Stallings (cf. [209]). Indeed, they showed that for nonpositively curved triangles of groups, the answer is affirmative. In general, it turns out that the answer is affirmative if and only if the universal cover $Y(\mathcal{G})$ of $\mathcal{G}$ is contractible. In particular, this is true if $Y(\mathcal{G})$ is CAT(0). Therefore, it follows from the previous paragraph that the $K(\pi, 1)$-Question has an affirmative answer for the simple complex of groups corresponding to RACGs, RAAGs and RABs (cf. Sect. 4.1.3).

Right-Angled Reflection Groups on Manifolds In Sect. 3.2 we use polyhedral products to start the discussion of right-angled reflection groups on aspherical manifolds and their universal covers. If L is a piecewise linear triangulation of S^{n-1}, then the fundamental chamber K $(= K(W, S))$ is an n-disk, P_L is an n-manifold, the action of $(\mathbf{C}_2)^I$ as a reflection group on P_L is locally linear and the cubical structure on P_L is dual to its cellulation by chambers. If the simplicial complex L is only required to be a generalized homology $(n - 1)$-sphere, then P_L is a homology n-manifold that is NPC. When L is simply connected, P_L actually is a topological manifold. When L is not a PL-sphere, we can get examples where the CAT(0) manifold $\tilde{P}$ is not simply connected at infinity and hence, is not homeomorphic to euclidean n-space. (These topics are discussed in Sect. 3.2, as well as in my earlier book, [82].) In Sect. 3.2 we also take an extended detour into some related topics: the theory of real quasi-toric manifolds (or "small covers" of the fundamental orbifold K), as well as the theory of higher-dimensional Haken manifolds. In Sect. 3.2.6 we explain the Reflection Group Trick. The main consequence of this trick is that given an finite aspherical complex B, there is a closed aspherical manifold M which retracts onto it. Actually, $\pi_1(M)$ is a torsion-free subgroup of finite index in a reflection group which has as its fundamental chamber a thickening of B. Projection onto the fundamental chamber is a retraction that leads to a retraction on the level of groups, $\pi_1(M) \to \pi_1(B)$. From this one sees that many pathological properties which are for fundamental groups of aspherical polyhedra are also possible for fundamental groups of aspherical manifolds. For example such groups need not be residually finite.

Chapter 4 The heart of this book is *Chapter 4: Coxeter groups, Artin groups, buildings*. It describes the main classes of examples of group actions on polyhedra in which we are interested. Each of these three classes is based on the theory of Coxeter systems. In contrast to the right-angled examples discussed above, these examples need not be cube complexes. However, each has an associated simple complex of groups which is often (conjecturally always is) nonpositively curved.

Coxeter Groups The theory of (not necessarily right-angled) Coxeter groups is essentially synonymous with the theory of groups generated by reflections. If a group W is a discrete group generated by isometric reflections on a simply connected manifold M of constant curvature, then there is a natural strict fundamental domain for the W-action on M, namely, the closure of any component of the complement of the union of reflecting hyperplanes in M. If we choose one such component and denote its closure by K, then K is a convex polytope and the set of reflections S across its codimension-one faces is a set of generators for W (K is called a "fundamental chamber.") The pair (W, S) is a *Coxeter system* and W is a *Coxeter group*. The theory of Coxeter groups is the topic of my previous book [82]. It is shown in [82] how to reverse this procedure: given an "abstract" Coxeter system (W, S) one can build a CAT(0) cell complex $\Sigma(W, S)$, called the "Davis–Moussong complex," on which W acts as a proper, cocompact reflection group with a strict fundamental domain, denoted by $K(W, S)$. A key feature is that the strata of $K(W, S)$ correspond to the subsets J of S that generate finite subgroups of W; these are the so-called spherical subsets of S. It follows that the cells of $\Sigma(W, S)$ are "Coxeter zonotopes" corresponding to the spherical subsets of S. General, not necessarily right-angled, Coxeter systems are intimately connected to buildings, as well as to Artin groups. We discuss some of these connections in the next two paragraphs.

Buildings The theory of buildings was developed by Tits, cf. [216, 218]. A "building" $\mathcal{C}$ is a certain combinatorial object $\mathcal{C}$ consisting of a set of "chambers," $\mathrm{Ch}(\mathcal{C})$, that satisfies certain additional conditions. Each building has an associated Coxeter system (W, S) called its *type*. In the standard realization of a building each element of $\mathrm{Ch}(\mathcal{C})$ is homeomorphic to the fundamental chamber $K(W, S)$ of the associated Coxeter system (W, S). The "standard realization" of $\mathcal{C}$ is the polyhedron $|\mathcal{C}|$ formed by gluing together copies of $K(W, S)$, one for each element of $\mathrm{Ch}(\mathcal{C})$, along codimension-one faces. The piecewise euclidean metric on $K(W, S)$ induces a piecewise euclidean metric on $|\mathcal{C}|$. The space $|\mathcal{C}|$ is a union of copies of $\Sigma(W, S)$; each such copy is called an "apartment." It follows from the axioms for buildings that each apartment is a totally geodesic subspace of $|\mathcal{C}|$ and that $|\mathcal{C}|$ is a CAT(0) space. Here is the picture to have in mind: in an apartment exactly 2 chambers meet along each codimension-one face of $K(W, S)$, while in a building the number of adjacent chambers along any codimension-one face is always ≥ 2. For example, any tree without terminal vertices is a building (whose type is the infinite dihedral group), the chambers are the edges, an apartment is an embedded geodesic line in the tree. Any product of trees is an example of a *right-angled building* (abbreviated as RAB). The cube complex associated to any graph product of groups is another example of a RAB. If $\mathcal{C}$ admits a group G of automorphisms that is transitive on $\mathrm{Ch}(\mathcal{C})$, then the isotropy subgroups of G give us a simple complex of groups over $K(W, S)$. The local groups are the parabolic subgroups G_J corresponding to the spherical subgroups J of S. The universal cover of this simple complex of groups is $|\mathcal{C}|$.

Artin Groups Associated to any Coxeter system (W, S) there is a presentation for an *Artin group* A $(= A_{(W,S)})$. The set of generators $\{a_s\}_{s \in S}$ in the presentation is bijective with the set of generators S for the Coxeter group, except that now the generator a_s corresponding to s will have infinite order rather than order 2. In the Coxeter group we had a relation $(st)^{m(s,t)} = 1$ for each edge $\{s, t\}$ of the 1-skeleton of L, while in the Artin group this relation is replaced by an "Artin relation" of the form:

$$\text{prod}(a_s, a_t; m(s,t)) = \text{prod}(a_t, a_s; m(t,s)),$$

where $\text{prod}(a, b; m)$ means the alternating word in a, b of length m starting with a.

As before we get a simple complex of groups over $K(W, S)$; the local groups are the spherical Artin groups. The fundamental group of this simple complex of groups is the Artin group A. Its universal cover (also called its "development") is the *Deligne complex*, denoted by $\Lambda(W, S)$. If A is a RAAG, then its Deligne complex is a RAB.

For any Coxeter system the Davis–Moussong complex, $\Sigma(W, S)$, is contractible. The same is true for the standard realization of any building. However, for the Deligne complex, $\Lambda(W, S)$, associated to an Artin group, the answer is not known. Conjecturally, as in the case for Coxeter groups and buildings, the Deligne complex is CAT(0). (See Conjecture 4.54 in Sect. 4.3.1.) As explained above, this would imply that the $K(\pi, 1)$-Question for the complex of spherical Artin subgroups $\{A_\sigma\}$ has a positive answer. This is the "$K(\pi, 1)$-Conjecture for Artin groups". It is the most important unsolved problem in this area. The answer is known in many cases, for example, when A is spherical type, is type FC, or when $\dim \Lambda(W, S) \leq 2$. (See Corollaries 4.59 and 4.57 in Sect. 4.3.1.) It is also known to be true when the associated Coxeter system is euclidean, cf. [188].

The General Theory of CAT(0) Cube Complexes In Chap. 5 we return to cube complexes. For any cube complex there are naturally defined "hyperplanes." These are certain totally geodesic immersed subcomplexes of codimension one. In a CAT(0) cube complex X, each hyperplane is embedded and separates X into two open "half-spaces;" furthermore, each such hyperplane is itself a CAT(0) cube complex.

Sageev [200] gave a definition of an abstract CAT(0) cube complex as a certain poset, the elements of which are called "half-spaces," which satisfies certain properties. This abstraction is called a "pocset," a "half-space system," or sometimes a "wall set." Given an abstract half-space system one can recover a geometric CAT(0) cube complex. (See [182, 197, 202].) Sageev used this construction in the following way: given a group G and a "sufficiently deep," codimension-one subgroup H, one can define a half-space system where the abstract half-spaces are certain subsets of G. Hence, one obtains a G-action on a CAT(0) cube complex.

The basic example to keep in mind is when G is the fundamental group of a manifold M and H is a π_1-injective codimension-one submanifold N that separates M into two pieces, M_1 and M_2, with respective fundamental groups, G_1 and G_2. The lifted hypersurfaces divide the universal cover $\tilde{M}$ into regions, each of which is a copy of the universal cover of either M_1 or M_2. In this way we get a decomposition of G as an amalgamated free product $G = G_1 *_H G_2$, where $H = \pi_1(N)$, so that G acts on the associated Bass-Serre tree with one vertex for each region tree with one vertex for each region (cf. Fig. 5.1 in Sect. 5.1.2). Sageev realized that if the π_1-injective hypersurface N is only required to be immersed, then we can still require the lifts of N to be embedded hypersurfaces that intersect transversely. The regions then correspond to most of the vertices of a CAT(0) cube complex. Thus, groups acting on such CAT(0) cube complexes are a generalizations of graphs of groups.

As examples, Niblo–Reeves [183] proved that every Coxeter group (not necessarily right-angled) acts properly on a CAT(0) cube complex. Also, Bergeron-Haglund-Wise [17] proved that the fundamental group of any "standard" arithmetic hyperbolic n-manifold acts on a CAT(0) cube complex.

Special Cube Complexes The theory of special cube complexes was developed by Haglund–Wise [143]. An NPC cube complex is "special" if certain types of intersections of hyperplanes are avoided—the main requirement is that hyperplanes with self-intersections are forbidden. The main result of [143] is that if X is an NPC special cube complex, then X isometrically immerses in the standard cube complex of some RAAG. (See Theorem 5.27.) Alternatively, X isometrically immerses in the Davis–Moussong complex for some RACG. It follows that $\pi_1(X)$ virtually embeds in some RAAG (i.e., that $\pi_1(X)$ has a subgroup of finite index which embeds in a RAAG). This is important because RAAGs have good separability properties which are inherited by subgroups. For example, any RAAG is residually finite and all of its "word quasi-convex subgroups" are virtual retracts. Using work of Kahn-Marković [162] and a result of Agol [3], one can show that the fundamental group of any hyperbolic 3-manifold is virtually special (i.e., has a subgroup of finite index that acts on a special cube complex). As a consequence of the resulting separability properties for hyperbolic 3-manifold groups, Agol [3] proved Thurston's remaining conjectures about 3-manifolds, e.g., every closed hyperbolic 3-manifold is virtually Haken. This also uses Agol's proof of a conjecture of Wise: if a word hyperbolic group acts properly and cocompactly on a CAT(0) cube complex, then it is virtually special.

Morse Theory on CAT(0) Polyhedra "Morse theory" on polyhedra is an important tool for studying the topology of CAT(0) polyhedral complexes. Given a real-valued function f on a CAT(0) polyhedron Y, one asks how the topology of the level sets $f^{-1}(t)$ changes as t crosses the value of f at a vertex. (Often one requires that the restriction of f to each cell be an affine map.) The answer is that the level set is changed by attaching a copy of the cone on the "ascending link" of f at v, denoted by $\mathrm{Lk}_{\uparrow}(v)$. It consists of the directions in $\mathrm{Lk}(v)$ in which f is increasing. The following two uses of Morse theory are explained in Chap. 7:

(i) The first use is to study the topology at infinity of Y.
(ii) Secondly, suppose G is a group that acts properly and cocompactly on a CAT(0) cube complex Y and that $\varphi : G \to \mathbb{Z}$ is an epiomorphism corresponding to a Morse function $f : Y \to \mathbb{R}$. In [23] Bestvina and Brady analyzed the finiteness properties of the group $\ker \varphi$ by studying the level sets $f^{-1}(t)$.

With regard to (i), suppose $f : Y \to [0, \infty)$ is the distance from a given vertex v_0, i.e., $f(y) = d(y, v_0)$. The level set $f^{-1}(t)$ is then the sphere of radius t about v_0. One can then prove a result of Brady-Meier [34]: if each ascending link $\mathrm{Lk}_{\uparrow}(v_0)$ is m-acyclic (resp. m-connected), then Y is m-acyclic (resp. m-connected) at infinity. (The relevant definitions can be found in Sects. 7.1.2 and 7.1.3.) Applications to the topology at infinity of RACGs and RAAGs are given in Sect. 7.1.3.

The main case of interest for (i) is when $G = A_L$, the RAAG corresponding to the flag complex L. The cube complex BA_L denotes the standard model for its classifying space as a subcomplex of $\mathbb{T}^I$; its universal cover is the CAT(0) cube complex $Y_L = EA_L$. The vertex set of Y_L can be identified with A_L. Let $\varphi : A_L \to \mathbb{Z}$ be the homomorphism which sends each Artin generator a_i to the element $1 \in \mathbb{Z}$. The homomorphism φ extends to a function $f : Y_L \to \mathbb{R}$ which is an affine map on each cube. The group $\ker(\varphi)$ acts freely and properly on any level set $f^{-1}(t)$. The link of each vertex in Y_L is the simplicial complex OL formed by "doubling each vertex." Each ascending link $\mathrm{Lk}_{\uparrow}(v)$ is a certain subcomplex of OL that can be identified with L. Bestvina–Brady prove that each level set $f^{-1}(t)$ is homotopy equivalent to Y_L if and only if each ascending link L is contractible and hence, that $f^{-1}(t)/\ker \varphi$ is a model for $B(\ker \varphi)$. Moreover, the inclusion $f^{-1}(t) \hookrightarrow Y_L$ induces an isomorphism on homology if and only if L is acyclic. The main result of [23] is that if the simplicial complex L is acyclic but is not simply connected, then the subgroup $\ker \varphi$ is type FP but not type F. (This means that $\ker \varphi$ has the homological properties required for $B(\ker \varphi)$ to be a finite complex except that it need not be finitely presented.) Interesting relatives of this result are explained in Sect. 7.4. First, if Q is a finite group of simplicial automorphisms of L, then the wreath graph product $A_L \rtimes Q$, acts properly on Y_L, i.e., Y_L is a cocompact model for $\underline{E}(A_L \rtimes Q)$, the universal space for proper $A_L \rtimes Q$-actions. Leary–Nucinkis [170] show that there are examples where $\ker \varphi$ is type F yet there is no cocompact model for $\underline{E}(\ker \varphi \rtimes Q)$. Other relatives of Bestvina–Brady groups are studied by Leary [169]. He shows that there are uncountably many groups of type FP which are not finitely presentable (and hence, are not type F).

Examples The principal theme of this book is the description of some general methods for constructing examples of group actions on contractible polyhedra. These methods include Coxeter groups, Artin groups, graph products and wreath graph products of groups, groups acting on buildings and the theory of complexes of groups. Another general method for constructing examples is the technique of hyperbolization, which is the subject of Chap. 6. It also is briefly discussed two paragraphs below. We begin by highlighting some other 2-dimensional examples.

- *Polygons of groups* (cf.Sect. 2.4.5). Some of the first work on complexes of groups, by Gersten and Stallings, was on triangles of groups.
- *Gromov polyhedra.* This term refers to 2-dimensional polygon complexes which are regular in the sense of Coxeter, i.e., their symmetry group is transitive on the set of 2-simplices in the barycentric subdivision of the polygon complex. So, Gromov polyhedra can be constructed from triangles of groups. (See Example 4.17.)
- *Burger–Mozes groups.* The product of two locally finite trees is a CAT(0) square complex. Burger–Mozes showed that such products of trees can have cocompact symmetry groups which, as abstract groups, are simple (cf. Sect. 2.4.4). In other words, there is a compact NPC square complex whose universal covers is a product of trees and whose fundamental group is simple. Although the Burger–Mozes groups were quite surprising when they were first discovered, nowadays it is known to be a quite common phenomenon for a building to have a discrete cocompact group of automorphisms which is an infinite simple group. (See the last paragraph of Sect. 4.4.8 and [52] for more details.)
- 1*-relator groups.* There are many groups with 2-dimensional models for their classifying spaces which are not NPC polyhedra. A 1*-relator group* is a group that has a presentation with just 1-relation, such that the relation is not a nontrivial power. It is a classical result that the presentation complex of any 1-relator group is aspherical. For example, the fundamental group of any aspherical surface is a 1-relator group (cf. [146, p. 51]). Another example of a 1-relator group is the *Baumslag–Solitar group* $BS(n, m)$ defined by the presentation,

$$BS(n, m) = \langle x, y \mid x^n y x^{-n} y^{-m} \rangle.$$

  Its presentation complex cannot be given a nonpositively curved structure unless it is a torus or Klein bottle. (The action of $BS(n, m)$ on the universal cover of the presentation complex cannot be by semisimple isometries since the relation implies that the elements x and y usually must have translation length equal to 0.)
- *Higman groups.* Let H be the group defined by the presentation:

$$H = \langle x_1, x_2, x_3, x_4 \mid x_i x_{i+1} x_i^{-1} = x_{i+1}^2; \quad i \in \mathbb{Z}/4 \rangle.$$

  We note that H is a square of groups. Each vertex group is isomorphic to the Baumslag–Solitar group $BS(1, 2)$ and each edge group is infinite cyclic. Higman [148] used this group to give the first proof of the existence of a finitely generated infinite simple group, namely, the quotient of H by a maximal normal subgroup. (See Corollary 2.46 in Sect. 2.4.3.)

Hyperbolization In [136] Gromov shows that nonpositively curved polyhedra and manifolds are quite common. For example, among closed surfaces only the sphere and the real projective plane fail to be aspherical. Some initial evidence demonstrating the genericity of asphericity was provided by the result of Kan–

Thurston [163]: any compact polyhedron P has an "asphericalization," $\mathcal{A}(P)$, together with a map $c : \mathcal{A}(P) \to P$ that induces an isomorphism on homology. So, the asphericity of a CW complex imposes no condition on its homology. One drawback of the Kan–Thurston procedure is that one cannot require such an asphericalization to take closed manifolds to closed manifolds—there is no aspherical 2-manifold to choose for $\mathcal{A}(S^2)$. However, Gromov noticed that if one weakens the requirement of being an isomorphism on homology to being a surjection on homology, then it is easy to produce such a procedure taking closed manifolds to closed manifolds; moreover, one can require that the asphericalization be nonpositively curved. Such a method is called a "hyperbolization" procedure.

Starting with a simplicial complex J, a *hyperbolization procedure* converts it into a nonpositively curved polyhedron $\mathcal{H}(J)$, usually a cube complex, by defining $\mathcal{H}(J)$ to be the union of hyperbolized simplices $\mathcal{H}(\sigma)$ where σ is a simplex of J. If J is an n-manifold, then so is $\mathcal{H}(J)$. There is a map $c : \mathcal{H}(J) \to J$. In many situations the map c induces a surjection on homology. The main use of hyperbolization has been to produce examples of nonpositively curved manifolds. In Sect. 6.2 we explain the following two corollaries to the existence of such procedures.

(i) Any manifold M^n is cobordant to an aspherical manifold $\mathcal{H}(M^n)$. Moreover, if two aspherical manifolds are cobordant, then the cobordism may be taken to be an aspherical $(n+1)$-manifold (cf. [136, pp. 117–118] and [96]).
(ii) There are aspherical topological n-manifolds ($n = 4$ or $n \geq 6$) that cannot be triangulated.

In Sect. 6.3.1 it also is explained how the Reflection Group Trick is essentially the same as a certain relative version of hyperbolization that is compatible with nonpositive curvature. Thus, for any nonpositively curved polyhedron B, there is a nonpositively manifold which retracts onto it. In Sect. 6.5 we explain a "strict" hyperbolization procedure which produces complexes and manifolds with curvature ≤ -1.

This book is far from being a comprehensive survey of the field. First of all, there is no discussion of the vast literature on cell complexes for the mapping class group or other outer automorphism groups such as $\mathrm{Out}(F_n)$. Because of this I have also not discussed many related recent concepts involving the term "hyperbolic," such as "acylindrically hyperbolic" or "hierarchically hyperbolic" groups (cf. [70]). Instead, I have focused on older topics that I better understand. As for nonpositive curvature, some major areas are completely absent from this book, e.g., the theory of systolic spaces and groups developed by Januszkiewicz–Świątkowski [160] (and others), as well as the relatively new theory of Helly spaces and groups (cf. Huang–Osajda [157]). Originally I had planned to include discussions of several minor topics concerning examples of nonpositively curved cube complexes, e.g., configuration spaces of graphs (cf. [2, 125, 213]), polygonal complexes with Platonic symmetry (cf. [212]) and alternating link complements (cf. [19], [35, Thm. 5.35, p. 220]; however, I could not find a good place to include them here.

Part II
Nonpositively Curved Cube Complexes

Chapter 2
Polyhedral Preliminaries

This chapter deals with cell complexes (also called "polyhedra") with metrics of piecewise constant curvature. In other words, each cell is required to be isometric to a convex polytope in a space of constant curvature κ. As $\kappa = -1, 0, +1$ such metrics are called, respectively, piecewise hyperbolic, piecewise euclidean, or piecewise spherical. A geodesic metric space is CAT(κ) if geodesic triangles in it satisfy Gromov's comparison inequality of Cartan, Aleksandrov and Toponogov with respect to triangles in a plane of constant curvature κ. The fundamental result is that a piecewise constant curvature polyhedron is locally CAT(κ) if and only if the Link Condition holds for each of its cells, i.e., if each such link is CAT(1). The link of a cell in a cube complex is a piecewise spherical complex with "all right" simplices. Gromov's Lemma asserts that such a link is CAT(1) if and only if it is a flag complex.

In Sect. 2.3 we state two standard results about group actions on CAT(0) spaces, the Bruhat–Tits Fixed Point Theorem and the Flat Torus Theorem. Some examples of nonpositively curved polygons of groups are explained in Sect. 2.4: Higman groups in Sect. 2.4.3, Burger–Mozes groups in Sect. 2.4.4 and nonpositively curved polygons of groups in Sect. 2.4.5.

2.1 Cell Complexes, Links

By a *cell complex* we will mean a space X formed by gluing together convex polytopes (in some euclidean space) via isometries between faces. The decomposition of X into polytopes ("cells") is part of the structure. The word "polyhedron" will

---

The original version of the chapter has been revised. A correction to this chapter can be found at https://doi.org/10.1007/978-3-031-48443-8_8

M. W. Davis, *Infinite Group Actions on Polyhedra*, Ergebnisse der Mathematik und ihrer Grenzgebiete. 3. Folge / A Series of Modern Surveys in Mathematics 77,
https://doi.org/10.1007/978-3-031-48443-8_2

be used synonymously with "cell complex." In the classical definition of a convex cell complex, one often adds the conditions that each cell is embedded and that the intersection of two cells is either empty or a common face of both. We call these two requirements the *classical conditions*. The first of these conditions rules out the possibility of a 1-gon, while the second rules out a 2-gon. (Even when X does not satisfy the classical conditions, the induced cell structure on its universal cover might well satisfy the classical conditions.) Two types of cell complexes are of particular interest: first, if each cell is required to be a simplex we have a "Δ-complex" in the sense of Hatcher [146, Section 2.1]: second, if each cell is a cube, we have a "cube complex." A Δ-complex or a cube complex can have "multiple faces;" for example, in a Δ-complex two different simplices can have the same vertex set. If a Δ-complex satisfies the classical conditions, then it is a *simplicial complex*.

Given a face F of a convex polytope P and a point x in its relative interior, the *normal cone* $N(F, P)$ is the set of all inward-pointing tangent vectors at x that are orthogonal to F. It is a convex polyhedral cone in the tangent space $T_x P$. The dimension of $N(F, P)$ is the codimension of F in P. The *link of F in P*, denoted $\mathrm{Lk}(F, P)$, is set of unit tangent vectors in $N(F, P)$. In other words, $\mathrm{Lk}(F, P)$ is the space of inward-pointing normal directions to F in P. It is a geodesically convex polytope in the unit sphere of $T_x P$. If $F < F'$, where F' is another face of P, then $\mathrm{Lk}(F, F')$ is identified with a face of $\mathrm{Lk}(F, P)$. Given a cell τ of a cell complex X, we can then glue together the corresponding links to form a cell complex:

$$\mathrm{Lk}(\tau, X) := \bigcup_{\substack{\sigma \\ \tau < \sigma}} \mathrm{Lk}(\tau, \sigma), \tag{2.1}$$

and call it the *link of τ in X*. (The above definition (2.1) needs to be slightly modified when there are self-gluings. For example, if P is the polytope corresponding to the cell σ and faces F_1, F_2 of P are identified to get a single cell τ of X, then $\mathrm{Lk}(\tau, \sigma)$ should be interpreted to be the disjoint union, $\mathrm{Lk}(F_1, P) \sqcup \mathrm{Lk}(F_2, P)$.)

A convex polytope P of dimension n is a *simple polytope* if the link of each of its k-dimensional faces is a spherical $(n-k-1)$-simplex. For example, a simplex is simple; so is a cube. An octahedron is not simple. If each cell of a convex cell complex X is a simple polytope, then each link of X is a Δ-complex. If each link is, in fact, a simplicial complex, then X is a *simple cell complex*. For example, if X is the square complex formed by doubling a square along its boundary to get a "pillow," then X is not simple since the link of any vertex is a cycle of length 2. On the other hand, if X if the square complex formed by identifying opposite edges of a square to get a 2-torus, then X is simple, since the link of a vertex is a cycle of length 4. If all cells of the cell complex X are simple polytopes and if X satisfies the classical conditions, then it is a simple cell complex.

If P an n-simplex and $F < P$ is a k-dimensional face, then $\mathrm{Lk}(F, P)$ is a spherical $(n-k-1)$-simplex. So, if X is a Δ-complex and σ is a cell in X, then $\mathrm{Lk}(\sigma, X)$ is a Δ-complex. Moreover, if X is a simplicial complex, then so is $\mathrm{Lk}(\sigma, X)$. Similarly, if F is a k-dimensional face of an n-cube P, then $\mathrm{Lk}(F, P)$ is an $(n-k-1)$-simplex. Hence, if X is a cube complex, then $\mathrm{Lk}(\sigma, X)$ is a Δ-complex and if X is a simple cube complex, then $\mathrm{Lk}(\sigma, X)$ is a simplicial complex.

Piecewise Constant Curvature Polyhedra In any simply connected space of constant curvature κ one can define the notion of a geodesically convex polytope. In dimension $n \geq 2$ the possibilities are: euclidean n-space, $\mathbb{E}^n$ for $\kappa = 0$, the n-sphere of curvature κ, $\mathbb{S}^n_\kappa$, for $\kappa > 0$, and hyperbolic n-space of curvature κ, $\mathbb{H}^n_\kappa$, for $\kappa < 0$. A cell complex X made by gluing together polytopes in a space of constant curvature κ is called a *piecewise constant curvature polyhedron*. For more details see [35, Ch. I.7] or [82, Appendix I.3]. (In [35] these are called M_κ polyhedral complexes.) We shall be interested only in the cases $\kappa = +1, 0$, or -1. Then X is *piecewise spherical*, *piecewise euclidean* or *piecewise hyperbolic*, respectively. If a piecewise constant curvature polyhedron X is connected, then it inherits a natural length metric as follows. The length of a geodesic segment in a polytope of constant curvature κ is defined by the usual formulas as in [35, Ch. I.2]. If $x, y \in X$, then $d(x, y)$ is defined to be the infimum of the lengths of all piecewise linear paths from x to y. If X is locally finite (or more generally if it has finitely many shapes of cells, cf. [35, Ch. I.7]), then X has the structure of a *geodesic space* meaning that the infimum $d(x, y)$ is always realized by a geodesic path, i.e., an isometric embedding $\gamma : [0, d] \to X$ with $\gamma(0) = x$, $\gamma(d) = y$. (For more details see [35, Ch. I.3, I.7].)

Cones Suppose L is a piecewise spherical polyhedron. Define the *euclidean cone* on L by

$$\mathrm{Cone}\, L := (L \times [0, \infty))/\sim, \tag{2.2}$$

where $L \times 0$ is identified to a point. If $(\theta, r) \in \mathbb{S}^{n-1} \times [0, \infty)$ are polar coordinates in $\mathbb{R}^n$, then by using the Law of Cosines one can write a formula for the euclidean distance on $\mathbb{R}^n$ (the cone on $\mathbb{S}^{n-1}$). By using $(\theta, r) \in L \times [0, \infty)$ as polar coordinates the same formula defines a metric on the euclidean cone, $\mathrm{Cone}(L)$. Similarly, for any real number κ and nonnegative real number r, define the *κ-cone of radius r on L*, by

$$\mathrm{Cone}^\kappa_r(L) := (L \times [0, r])/\sim, \tag{2.3}$$

where the metric is defined using polar coordinates and the Law of Cosines in the space of constant curvature κ. (See [35, p. 59] for more details.) If $\kappa > 0$, we require the radius of the cone to be $< \pi/\sqrt{\kappa}$.

Since the link of a face of any polytope in a constant curvature space is always a spherical polytope in $\mathbb{S}^n_1$, we see that the link of any cell of a piecewise constant curvature polyhedron is always a piecewise spherical polyhedron. Similarly, for any point $x \in X$, let $\mathrm{Lk}(x, X)$ be the union of all unit tangent vectors which point into a cell of X. (If x belongs to the relative interior of a k-cell σ, then $\mathrm{Lk}(x, \sigma) = \mathbb{S}^{k-1}$. When this is the case, it follows that $\mathrm{Lk}(x, X)$ can be identified with $\mathbb{S}^{k-1} * \mathrm{Lk}(\sigma, X)$, the k-fold suspension of $\mathrm{Lk}(\sigma, X)$.) Let $N_\varepsilon(x, X)$ (resp. $\partial N_\varepsilon(x, X)$) denote the ball (resp. sphere) of radius ε about x, i.e., $N_\varepsilon(x, X) = \{y \in X \mid d(x, y) \leq \varepsilon\}$ and

$\partial N_\varepsilon(x, X) = \{y \in X \mid d(x, y) = \varepsilon\}$. When X is locally finite or has finitely many shapes of cells, one sees $\partial N_\varepsilon(x, X)$ can be identified with $\mathrm{Lk}(x, X)$ and that

$$N_\varepsilon(x, X) \cong \mathrm{Cone}_\varepsilon(\mathrm{Lk}(x, X)), \tag{2.4}$$

where $\mathrm{Cone}_\varepsilon(A)$ stands for the *cone of radius* ε, that is, $A \times [0, \varepsilon]$ with $A \times 0$ collapsed to a point.

Formula (2.4) implies that $N_\varepsilon(x, X)$ is homeomorphic to the cone on $\mathrm{Lk}(x, X)$; in particular, $\mathrm{Lk}(x, X)$ determines the local topology of X at x. For example, X is a PL n-manifold if and only if $\mathrm{Lk}(x, X)$ is piecewise linearly homeomorphic to S^{n-1} for each $x \in X$.

2.1.1 *The* CAT(κ)*-inequality*

Using ideas of Alexandrov and Busemann, Gromov [136] defined what it means for a geodesic metric space X to have curvature bounded above by a real number κ (as before, usually κ will belong to $\{-1, 0, +1\}$). First, a geodesic triangle in X is said to satisfy the CAT(κ)*-inequality*, if the distance between any two points of the triangle is no greater than the distance between the corresponding points of a comparison triangle in the simply connected 2-manifold of constant curvature κ, i.e., in the hyperbolic plane, the euclidean plane, or the 2-sphere, as $\kappa = -1, 0$ or $+1$, respectively. A geodesic space X is CAT(κ) if each triangle satisfies the CAT(κ)-inequality. The CAT(κ)-inequality implies the CAT(κ')-inequality for $\kappa' > \kappa$. In particular, if a space is CAT(κ) with $\kappa \leq 0$, then it is CAT(0).

Remark 2.1 The terminology "CAT(κ)-inequality" is due to Gromov [136, p. 106]. The acronym CAT stands for "Comparison inequality of Alexandrov and Toponogov," although Gromov sometimes also has said that the "C" stands for "Cartan."

The uniqueness of the geodesic connecting two points in a CAT(0) space X follows immediately from the definitions. With a little work one can see that the constant map $X \to X$ which sends each point to a base point is homotopic to the identity via geodesic contraction. This gives the following.

Theorem 2.2 *If a complete geodesic space is* CAT(0), *then it is contractible.*

The space X has *curvature* $\leq \kappa$ if it satisfies the CAT(κ)-inequality locally. We say that X is *nonpositively curved* (abbreviated NPC) if this is true for $\kappa = 0$. (See [35].)

Theorem 2.3 (Globalization Theorem of Cartan–Hadamard, cf. [136, p. 119], [35, p. 193], or [4]) *Let X be a complete geodesic space of curvature $\leq \kappa$*

(i) Suppose $\kappa \leq 0$. Then X is CAT(κ) *if and only if it is simply connected.*
(ii) (Gromov [136, p. 122]). *Suppose $\kappa > 0$. Then X is* CAT(κ) *if and only if it has no closed geodesics of length $< 2\pi/\sqrt{\kappa}$.*

An immediate corollary to the first part of this theorem is the following.

Corollary 2.4 *Suppose a complete geodesic space X has curvature $\leq \kappa$ with $\kappa \leq 0$. Then its universal cover $\tilde{X}$ is* $\mathrm{CAT}(\kappa)$.

The relevance of nonpositive curvature to geometric group theory is explained by the next theorem, which combines Theorem 2.2 and Corollary 2.4.

Theorem 2.5 *If a complete geodesic space is NPC, then it is aspherical.*

Example 2.6 (Dimension One) Suppose X is a locally finite, connected graph. Assign a fixed length to each edge, then X is a constant curvature polyhedron for any curvature constant κ. Since each link is 0-dimensional, it is locally $\mathrm{CAT}(\kappa)$. If each edge of X is assigned a length of 1, then X is an NPC cube complex. If X is a tree with each edge having length 1, then, by Theorem 2.3 (i), X is a CAT(0) cube complex. We can also regard the metric on a graph X as giving it a piecewise spherical structure. In this case, by part (ii) of Theorem 2.3, X is CAT(1) if and only if it has no circuit (= closed geodesic) of length $< 2\pi$. In particular, if each edge has length $\pi/2$, then X is CAT(1) if and only if it has no circuits of length 3. i.e., if its girth is at least 4.

The Link Condition We turn next to the question of when a piecewise constant curvature polyhedron (with curvature constant κ) has curvature $\leq \kappa$. The answer is provided by the "Link Condition," defined in Definition 2.10 below. The proof of the Link Condition is a combination of three facts: first, the link of any cell in a piecewise constant curvature polyhedron X is naturally a piecewise spherical polyhedron; second, an open ball centered at a point $x \in X$ is isometric to an open neighborhood of the cone point in the "κ-cone" on a piecewise spherical polyhedron (essentially the link of x); and third, the κ-cone on a piecewise spherical polyhedron L is $\mathrm{CAT}(\kappa)$ if and only if L is CAT(1). Since graphs as in Example 2.6 occur as links in 2-dimensional piecewise constant curvature polyhedra, we get, via the Link Condition in Theorem 2.11 below, a condition for a 2-dimensional polyhedron to be locally $\mathrm{CAT}(\kappa)$.

The next lemma was pointed out earlier.

Lemma 2.7 *Let X be a piecewise constant curvature polyhedron, let τ be a cell of X and let x be a point in X. Then* $\mathrm{Lk}(\tau, X)$ *and* $\mathrm{Lk}(x, X)$ *are piecewise spherical polyhedra.*

The proof of the next lemma is straightforward.

Lemma 2.8 ([35, pp. 206–207]) $\mathrm{Cone}^{\kappa}_{r}(L)$ *is* $\mathrm{CAT}(\kappa)$ *if and only if L is* CAT(1).

Lemma 2.8 and Eq. (2.4) immediately imply the next lemma.

Lemma 2.9 ([35, pp. 206–207]) *If x is a point of a piecewise constant curvature polyhedron X, then for small enough ε the ball of radius ε centered at x is isometric to* $\mathrm{Cone}^{\kappa}_{\varepsilon}(\mathrm{Lk}(x, X))$. *Hence, X has curvature $\leq \kappa$ if and only if for all $x \in X$, each connected component of* $\mathrm{Lk}(x, X)$ *is* CAT(1).

Definition 2.10 A piecewise constant curvature polyhedron X satisfies the *Link Condition* if for each vertex v, each component of $\mathrm{Lk}(v, X)$ is CAT(1).

Theorem 2.11 (The Link Condition, cf. [35, Theorem 5.4, p. 206]) *A polyhedron X of piecewise constant curvature κ has curvature $\leq \kappa$ if and only if it satisfies the Link Condition.*

Theorem 2.11 is almost the same as Lemma 2.9. The only difference is that in Lemma 2.9 links are CAT(1) at all points $x \in X$, while in Theorem 2.11 this need only be true at vertices.

Example 2.12 (The Link Condition for 2-dimensional Piecewise Euclidean Polyhedra)

(a) Let X be a piecewise euclidean polyhedron of dimension 2. A circle is CAT(1) if and only if its length is $\geq 2\pi$. If X is a surface, then the link of any vertex v is a circle. Each edge of the link corresponds to a 2-cell containing v. With its natural spherical structure, the length of the edge is equal to the interior angle of the 2-cell at v. So, we get the following well-known folklore statement: the surface X is NPC if and only if the sum of the angles at each vertex is $\geq 2\pi$. If X is a general 2-dimensional piecewise euclidean polyhedron, then the link of each vertex is a graph and the Link Condition says that X is NPC if and only if for each vertex v, $\mathrm{Lk}(v, X)$ contains no circuit of length $< 2\pi$ (cf. Example 2.6).

(b) If X is a square complex, then since each interior angle in a square is $\pi/2$, the Link Condition says each link must be simple graph without 3-circuits. More generally, if each cell of X is a regular euclidean m-gon, then each interior angle is $(m-2)\pi/m$ and hence, this is the length of each edge in $\mathrm{Lk}(v, X)$. So, when $m \geq 6$, we see that X is NPC whenever each link is a simple graph (i.e., has no circuits of length 1 or 2.

Example 2.13 (Regular $2k$-gon Complexes, cf. Examples 4.15 and 4.17 of Sect. 4.2) Suppose L^1 is a simple graph and that k is an integer ≥ 2. In Sect. 4.2, by using the theory of Coxeter groups, we shall see that there is a simply connected 2-complex X_k so that each 2-cell is a regular euclidean $2k$-gon and so that the link of each vertex is L^1. Since the vertex angle of a regular euclidean $2k$-gon is $\pi - \pi/k$, L^1 should be given a piecewise spherical structure with each edge having length $\pi - \pi/k$. When $k = 2$, L^1 is CAT(1) if and only if it has no circuits of length 3, and when $k > 2$, it is always CAT(1). Thus, for $k > 2$ the 2-complex X_k is CAT(0) for any L^1, while for $k = 2$ it is CAT(0) provided L^1 has no circuits of length 3.

To prove Theorem 2.11 we need Lemmas 2.14 and 2.15 below. Before stating Lemma 2.14 we need the notion of a "spherical join." For $i = 1, 2$, suppose σ_i is a spherical polytope in $\mathbb{S}^{m_i}$. The join of these two polytopes, denoted $\sigma_1 * \sigma_2$, is naturally a spherical polytope in $\mathbb{S}^{m_1+m_2+1}$. The euclidean cones $\mathrm{Cone}^0(\sigma_i)$ are polyhedral cones in $\mathbb{R}^{m_i+1}$ and the product $\mathrm{Cone}^0(\sigma_1) \times \mathrm{Cone}^0(\sigma_2)$ is the polyhedral cone in $\mathbb{R}^{m_1+m_2+1}$ corresponding to the spherical polytope $\sigma = \sigma_1 * \sigma_2$. In other words, the link of the cone point of $\mathrm{Cone}^0(\sigma_1) \times \mathrm{Cone}^0(\sigma_2)$ is the spherical join $\sigma = \sigma_1 * \sigma_2$, where the spherical polytope $\sigma_1 * \sigma_2$ has its natural spherical metric. If, for $i = 1, 2$, L_i is a piecewise spherical polyhedron, then the *spherical join*, $L_1 * L_2$, is defined by

$$L_1 * L_2 := \bigcup_{(\sigma_1, \sigma_2)} \sigma_1 * \sigma_2, \tag{2.5}$$

(Here $(\sigma_1, \sigma_2) \in \mathcal{P}(L_1) \times \mathcal{P}(L_2)$, where $\mathcal{P}(L_i)$ means the poset of cells in L_i, including the empty cell, and either σ_1 or σ_2 is nonempty.)

The proof of the next lemma is easy.

Lemma 2.14 *Suppose L_1 and L_2 are piecewise spherical polyhedra. Then $L_1 * L_2$ is* CAT(1) *if and only if both L_1 and L_2 are* CAT(1).

Lemma 2.15 *If X is a constant curvature polyhedron, then each of the following conditions is equivalent to the Link Condition for X.*

(1) *For each $x \in X$,* $\mathrm{Lk}(x, X)$ *is* CAT(1).
(2) *For each cell τ of X,* $\mathrm{Lk}(\tau, X)$ *is* CAT(1).
(3) *For each cell τ of X,* $\mathrm{Lk}(\tau, X)$ *has no closed geodesic of length $\leq 2\pi$.*

Proof If x lies in the relative interior of a k-dimensional cell τ of X, then $\mathrm{Lk}(x, X)$ is isometric to the spherical join $\mathbb{S}^{k-1} * \mathrm{Lk}(\tau, X)$. By Lemma 2.14, either $\mathrm{Lk}(x, X)$ and $\mathrm{Lk}(\tau, X)$ are both CAT(1) or neither is CAT(1).

Proof of Theorem 2.11 Given a vertex $v \in X$, put $L = \mathrm{Lk}(v, X)$. If τ is a cell of X containing v, then there is a corresponding cell τ' in L (with $\dim \tau' = \dim \tau - 1$) and $\mathrm{Lk}(\tau, X)$ can be identified with $\mathrm{Lk}(\tau', L)$. If L is locally CAT(1), then $\mathrm{Lk}(\tau', L)$ is CAT(1). Hence, if the link of each vertex of X is CAT(1), then $\mathrm{Lk}(\tau, X)$ is CAT(1) for each cell τ of X and therefore, by Lemma 2.15, $\mathrm{Lk}(x.X)$ is CAT(1) for all points $x \in X$. By Lemma 2.9, this is equivalent to X having curvature $\leq \kappa$. Conversely, suppose $\mathrm{Lk}(v, X)$ is CAT(1) at some vertex $v \in X$. Then, by Lemmas 2.8 and 2.9, a conical neighborhood of v is CAT(κ).

2.1.2 Piecewise Hyperbolic Polyhedra

The material in this subsection will be used later in Sects. 4.2.4 and 4.2.3. (See Examples 4.16, 4.17, Proposition 4.20 and Theorem 4.23.)

We consider the possibility of deforming the metric on a piecewise euclidean cell complex X_e to a piecewise hyperbolic cell complex X_h. If X_e is NPC, then when will X_h be CAT(-1)? The problem is that since the angles change when we deform a euclidean polytope to one which is hyperbolic, the links in X_e might not remain CAT(1) in X_h. If P_e is a small euclidean polytope, say with all edge lengths $< \varepsilon$, then a small deformation will be a hyperbolic polytope P_h. The link of a face in P_h will be a small deformation of the link of the corresponding face of P_e. A precise formulation of this result can be found in [179]. However, after even a small deformation, the link of the cell in X_h may fail to be CAT(1). The reason is that if a closed geodesic in the link of a cell in a piecewise euclidean cell complex has length $= 2\pi$, then under a small deformation its length can become $< 2\pi$. For example, each angle of a regular convex 4-gon in $\mathbb{H}^2$ is $< \pi/2$. So, if a euclidean square complex has a vertex link containing a circuit with 4 edges, then after deformation this circuit will become a closed geodesic in the link of length $< 2\pi$; hence, the hyperbolic square complex will fail to be locally CAT(-1). The condition that the link has no such 4-circuits is the "no $\square$-condition" discussed below. More generally, given a compact 2-dimensional piecewise euclidean cell complex X, it can be deformed to a piecewise hyperbolic complex of curvature ≤ -1 provided that for each vertex v of X, each circuit of $\mathrm{Lk}(v, X)$ has length $> 2\pi$, cf. Example 2.12. (It may be necessary to first rescale the metric on X so that all its edge lengths are small.)

Definition 2.16 A piecewise spherical complex L is *large* if it is CAT(1). The cell complex L is *extra large* if the length of the shortest closed geodesic in L is strictly greater than 2π and if the same holds true for $\mathrm{Lk}(\sigma, L)$ for any cell σ in L.

Lemma 2.17 (Moussong [179, Lemma 5.11]) *Suppose L is a piecewise spherical cell complex with finitely many cells. If L is extra large, then any sufficiently small deformation of it will be* CAT(1).

A corollary of this lemma is the following.

Proposition 2.18 (Moussong [179]) *Suppose X is a NPC piecewise euclidean cell complex with finitely many shapes of cells so that for each cell σ of X, $\mathrm{Lk}(\sigma, X)$ is extra large. Then, after rescaling the metric, X can be deformed to a piecewise hyperbolic cell complex with curvature ≤ -1.*

2.1.3 *Quasi-Isometries, Word Hyperbolic Groups*

Definition 2.19 Given metric spaces X and Y, a *quasi-isometric embedding* from X to Y is a (not necessarily continuous) map $f : X \to Y$ so that there are constants $\lambda \geq 1$ and $C \geq 0$ such that

$$\frac{1}{\lambda} d_X(x, y) - C \leq d_Y(f(x), f(y)) \leq \lambda d_X(x, y) + C. \tag{2.6}$$

If, in addition, f is a quasi-surjection, i.e., if there is an $\varepsilon > 0$ so that each point of Y lies within ε of $f(X)$, then f is a *quasi-isometry*.

If G is a finitely generated group with finite set of generators S, then the Cayley graph $\mathrm{Cay}(G, S)$ is a metric space where each edge has length 1. The restriction of this metric to G (the vertex set of $\mathrm{Cay}(G, S)$) is called the *word metric* on G with respect to S. The quasi-isometry class of the word metric is independent of the choice of generating set S. One of the principal goals of geometric group theory is to understand group actions up to quasi-isometry.

Definition 2.20 A finitely generated subgroup H in a finitely generated group G is *quasiconvex* (or *word-quasiconvex*) if there is a constant C so that any geodesic segment in $\mathrm{Cay}(G, S)$ that connects two points of H is contained in the C-neighborhood of H.

If H is a quasiconvex subgroup of G, then the inclusion $H \hookrightarrow G$ is a quasi-isometric embedding. (N.B. in general, this notion depends on the choice of the set of generators for H)

An important chapter in geometric group theory begins with Gromov's work on hyperbolic groups in [136]. He first defines what it means for a metric space to be hyperbolic—his definition encapsulates the notion of having negative curvature in the large. The definition is easiest to state for a geodesic metric space. A triangle in a geodesic metric space X is *δ-thin* if the distance from any point x on one side of the triangle to the union of the other two sides is $\leq \delta$. The geodesic space X is *hyperbolic* if there is a constant $\delta > 0$ so that every triangle in X is δ-thin. (See [35, pp. 408–409], [136].) A finitely generated group G with a finite set of generators S is *word hyperbolic* if its Cayley graph $\mathrm{Cay}(G, S)$ is a hyperbolic metric space. Gromov [136] proves that the property of being hyperbolic is a quasi-isometry invariant.

The action of a group on a metric space X by isometries is *geometric* if the action is proper and cocompact. The Fundamental Lemma of Geometric Group Theory states that if a group G acts geometrically on two connected metric spaces X and X', then they are quasi-isometric; in particular, any such X is quasi-isometric to $\mathrm{Cay}(G, S)$. Since the hyperbolic plane $\mathbb{H}^2$ is a hyperbolic metric space, any $\mathrm{CAT}(-1)$-space is a hyperbolic metric space; hence, any group which acts properly on a $\mathrm{CAT}(-1)$-space is word hyperbolic. The above discussion is summarized in the next lemma.

Lemma 2.21 *If a group G acts geometrically on a piecewise hyperbolic,* $\mathrm{CAT}(-1)$ *cell complex X, then G is word hyperbolic.*

If G is word hyperbolic, then a finitely generated subgroup H is word-quasiconvex if and only if it is quasi-isometrically embedded in G. (This is because if a geodesic space is hyperbolic, then the a quasi-geodesic segment between two points lies within some bounded neighborhood of an actual geodesic.) Moreover, when H is a word-quasiconvex subgroup, then H also is word hyperbolic.

In a word hyperbolic group, normal quasiconvex subgroups are rare. For example, the following theorem is proved in the 1989 seminar notes from MSRI [6].

Theorem 2.22 (cf. [6]) *Suppose* $1 \to H \to G \to Q \to 1$ *is a short exact sequence of finitely generated groups. If* G *is word hyperbolic and* H *is a word-quasiconvex subgroup of* G*, then either* H *is finite or* Q *is finite.*

This means, for example, that for a word hyperbolic G the kernel of a nontrivial homomorphism $G \to \mathbb{Z}$ can never be word-quasiconvex.

2.2 Flag Complexes and Gromov's Lemma

A Δ-complex L is a *flag complex* if it is a simplicial complex and if it satisfies the following: any finite set of vertices that are pairwise connected by edges of L, spans a simplex of L. In other words, any clique in the 1-skeleton L^1 spans a simplex of L. (A *clique* in a simplicial graph Γ is a finite set of vertices of Γ such that the induced subgraph is a complete graph.) Here is a slick version of this definition: a simplicial complex L is a flag complex if and only if every non-simplex contains a non-edge (a set of vertices is a "non-simplex" if it is not the vertex set of a simplex).

Remark 2.23 (Comments on Terminology) Combinatorialists use the term "clique complex" instead of "flag complex." The *clique complex* of a simplicial graph Γ is the simplicial complex whose simplices are the cliques in Γ. So, a simplicial complex L is a flag complex if and only if L is equal to the clique complex determined by L^1. Instead of "flag complex" Gromov [136] says that L satisfies "the no Δ-condition" (pronounced "the no triangle condition").

A simplicial graph is a *circuit of length* m if it is isomorphic to the boundary complex of an m-gon. A circuit Γ of length m is a flag complex if and only if $m > 3$. More generally, a simplicial graph is a flag complex if and only if it contains no circuits of length 3. A natural way in which flag complexes arise is as order complexes of partially ordered sets.

Example 2.24 (Barycentric Subdivisions) If X is a classical convex cell complex, then the order complex of its poset of cells is a simplicial complex. The geometric realization of the order complex is the barycentric subdivision of X. It follows that the requirement of being a flag complex puts no restriction on the topological type of X.

The next lemma is left as an exercise for the reader.

Lemma 2.25 (Some Properties of Flag Complexes)

(i) *A full subcomplex of a flag complex is a flag complex.*
(ii) *The join of two flag complexes is a flag complex.*
(iii) *The link of a simplex in a flag complex is a flag complex.*

All Right Simplices and Simplicial Complexes The following terminology is due to Moussong.

Definition 2.26 A spherical n-simplex is *all right* if it is isometric to the simplex in $\mathbb{S}^n$ spanned by the standard orthonormal basis for $\mathbb{R}^{n+1}$. Similarly, a piecewise spherical simplicial complex is *all right* if each of its simplices is all right.

In other words, a spherical n-simplex is all right if it is isometric to the intersection of $\mathbb{S}^n$ with the standard orthant in $\mathbb{R}^{n+1}$ defined by $x_i \geq 0$, for $1 \leq i \leq n+1$.

We note that the length of every edge in an all right simplex is $\pi/2$ and that each of its dihedral angles is also $\pi/2$. The next lemma, which is a simple observation, shows the connection between cube complexes and all right simplicial complexes.

Lemma 2.27 *The link of any k-dimensional face in a regular euclidean cube $\square^n$ is an all right $(n-k-1)$-simplex. Hence, in any piecewise euclidean cube complex satisfying the classical conditions each of its links is all right.*

One of the key lemmas in this book is the following.

Lemma 2.28 (Gromov's Lemma, cf. [136, p. 122], [35, p. 211] or [82, Lemma I.6.1]) *Suppose L is a finite dimensional all right, piecewise spherical simplicial complex (or Δ-complex). Then L is* CAT(1) *if and only if it is a flag complex.*

Example 2.29 (An All Right Simplicial Complex that Is Not CAT(1)**)** Start with the triangulation of $\mathbb{S}^2$ as the boundary complex of an octahedron in which each 2-simplex is all right. Let L be the complement of the interior of one of the 2-simplices Δ. Of course, L is not a flag complex since it has an empty triangle, namely $\partial\Delta$. Note that $\partial\Delta$ is a short closed geodesic in L (its length is $3\pi/2$). (To see that $\partial\Delta$ is a closed geodesic, it suffices to show that it is a local geodesic. This follows from the fact that at each vertex of $\partial\Delta$ the link is a circular arc of length $3\pi/2$.)

Combining Gromov's Lemma with the observation in Example 2.24 about barycentric subdivisions, we get the following well-known result of Berestovskii.

Corollary 2.30 (Berestovskii) *Any finite dimensional polyhedron can be given a piecewise spherical* CAT(1) *metric.*

The form in which Lemma 2.28 will most often be used is given in the following corollary. This corollary will also be referred to as "Gromov's Lemma."

Corollary 2.31 (Gromov's Lemma for Cube Complexes) *A piecewise euclidean cube complex is NPC if and only if the link of each vertex is a flag complex.*

If L' is a simplicial complex with 1-skeleton a simplicial graph Γ, then the *flag completion* of L' is the clique complex of Γ. (This clique complex is also said to be the *flag complex determined by* Γ.) Similarly, for cube complexes we have the following definition which we shall need in Sect. 5.2.

Definition 2.32 A simple cube complex X is *completable* if its 2-skeleton is isomorphic to a subcomplex of the 2-skeleton of an NPC cube complex. The *completion* of a completable cube complex X is the smallest NPC cube complex whose 2-skeleton is the 2-skeleton of X.

(The link of a vertex in the completion of a cube complex is the flag complex completion of the link in the cube complex.)

Proof of Gromov's Lemma 2.28 If L has curvature ≤ 1 and is not a flag complex, then an argument similar to that in Example 2.29 shows that L has a short closed geodesic. It follows by induction on dimension that if L is not a flag complex, then at least one of its links has a short closed geodesic of length $3\pi/2$.

In an all right simplicial complex the closed star of a vertex v is the closed ball $B_{\pi/2}(v)$. An abbreviated version of Gromov's argument goes as follows. Suppose L is a flag complex and that γ is a closed geodesic which intersects the interior of $B_{\pi/2}(v)$ in a geodesic arc γ'. We claim that $l(\gamma') = \pi$. (This is proved by developing the surface determined by v and γ' onto the northern hemisphere of $\mathbb{S}^2$.) Hence, if $l(\gamma) < 2\pi$, then it cannot intersect two disjoint open stars about vertices. So, if V denotes the set of vertices v such that γ intersects the interior of $B_{\pi/2}(v)$, then any two element of V are connected by an edge. Since L is a flag complex, V must span a simplex of L. But an all right spherical simplex does not contain a closed geodesic. So, $l(\gamma) \geq 2\pi$.

The No □-condition When is an all right flag complex L extra large as in Definition 2.16? Since all edge lengths equal $\pi/2$, all closed geodesic of length 2π must correspond to an "empty 4-circuit" in L, i.e., a 4-circuit in L so that there is no other edge of L connecting a pair of diagonally opposite edges in the 4-circuit. So, the answer is simple: L cannot contain any empty 4-circuit. (In [136] this is called the *no □-condition.*) Proposition 2.18 then yields the following.

Lemma 2.33 *A piecewise euclidean structure on a cube complex can be deformed to a piecewise hyperbolic structure with curvature ≤ -1 if and only if the link of each vertex is a flag complex satisfying the no □-condition.*

2.3 Group Actions on CAT(0) Spaces

The Bruhat–Tits Fixed Point Theorem. The following fixed point theorem is due to Bruhat–Tits. Its proof goes back to Cartan.

Theorem 2.34 (Bruhat–Tits Fixed Point Theorem, cf. [1, p. 558]) *Suppose G is a group of isometries of a complete* CAT(0) *space X and that G has a bounded orbit on X. (Note that this hypothesis holds whenever the group G is compact.) Then the fixed set X^G is a nonempty convex subset of X.*

The proof is based on the fact that in a complete CAT(0) space any bounded subset B has a unique *circumcenter*, that is, the center of a circumscribed sphere containing B. If B is G-stable, then so is its circumcenter, i.e., the circumcenter of any bounded orbit is a fixed point. Uniqueness of geodesics in a CAT(0) space implies that any nonempty fixed point set is convex. A corollary to the Bruhat–Tits Fixed Point Theorem is the following.

Corollary 2.35 *Suppose a discrete group G acts properly and cocompactly on a* CAT(0) *space X. Then X is a model for the universal proper G-space, $\underline{E}G$ (cf. Definition 7.8).*

The Flat Torus Theorem If g is an isometry of a CAT(0) space, then its *translation length* $\tau(g)$ is the infimum of $\{d(x, gx) \mid x \in X\}$. The isometry g is *semisimple* if the infimum $\tau(g)$ is realized at some point x. If g is semisimple, define its *minset*, $\mathrm{Min}(g)$, to be the set of x which realize the translation length. If g is semisimple and $\tau(g) = 0$, then $\mathrm{Min}(g)$ is the fixed point set of g and g is said to be *elliptic*. If $\tau(g)$ is positive and is realized at some point x, then g is *hyperbolic* (also called *loxodromic*). If g is hyperbolic, then the union of geodesic segments of the form $[g^{n-1}x, g^n x]$ is isometric to $\mathbb{R}$; this line is called an *axis* for g.

Theorem 2.36 (cf. [35, II.6, Thm. 6.8]) *Let g be a hyperbolic isometry of a* CAT(0) *space.*

(i) *The axes of g are parallel to each other and their union is* $\mathrm{Min}(g)$.
(ii) $\mathrm{Min}(g)$ *is isometric to $Y \times \mathbb{R}$ for some convex subspace Y.*
(iii) *Every isometry h that commutes with g leaves $Y \times \mathbb{R}$ invariant and has the form (h', h'') where h' is an isometry of Y and h'' is a translation of $\mathbb{R}$.*

For H a subgroup of semisimple isometries of a CAT(0) space X put

$$\mathrm{Min}(H) = \bigcap_{g \in H} \mathrm{Min}(g).$$

Theorem 2.37 (The Flat Torus Theorem, cf. [35, II.7, Thm. 7.1]) *Let H be a free abelian group of rank n acting by semisimple isometries on a* CAT(0) *space X.*

(i) *The subspace* $\min(H)$ *is nonempty and splits as a product $Y \times \mathbb{E}^n$.*
(ii) *The quotient of the n-flat $\{y\} \times \mathbb{E}^n$ by H is an n-torus.*

Note that if a group of isometries G acts properly and cocompactly on a CAT(0) space, then each $g \in G$ is semisimple.

Using the Flat Torus Theorem it is fairly straightforward to show the following.

Theorem 2.38 (Solvable Subgroup Theorem, cf. [35, II.7,Thm. 7.8]) *Suppose G acts cocompactly and properly on a* CAT(0) *space X. Then any virtually solvable subgroup $H < G$ is finitely generated and contains an abelian subgroup of finite index.*

In other words, every solvable subgroup of a CAT(0) group is virtually free abelian.

2.4 Examples of NPC Square Complexes and Polygon Complexes

2.4.1 *Trees*

Any graph is a 1-dimensional NPC cube complex. Hence, any tree is a 1-dimensional CAT(0) cube complex (by Theorem 2.3 (i)).

Most of the examples in this book are higher dimensional analogs of group actions on trees. In the next four paragraphs, (T1)–(T4), we elucidate some of the aspects of group actions on trees that will be generalized in subsequent chapters.

(T1) (*Graphs of groups*). The theory of groups acting on trees is developed in the classic book of Serre [207]. The basic result is that if G acts on a tree T (without inversions), then we get a graph of groups decomposition of G over the orbit space T/G (which is a graph). For example, if T/G is an interval, then G splits as an amalgam $G \cong H_1 *_K H_2$, where K is an edge stabilizer and H_1, H_2 are the stabilizers of its end points. The theory of graphs of groups leads to the theory of complexes of groups, which was developed in Bridson–Haefliger [35] and which we will explain in Appendix A.

(T2) (*Buildings*). A tree is an example of a right-angled building (see Sects. 3.1.1 and 3.1.2). The associated Coxeter group is the infinite dihedral group; so, a tree is also a euclidean building of rank 2 (cf. Example 4.132 in Sect. 4.4.8). This aspect is developed in the second half of Serre's book [207]. For example, the regular tree T_{p+1} of degree $(p+1)$ is associated to the algebraic group $SL_2(K)$ where K is a field with discrete valuation and residue field $\mathbb{F}_p$ (for example, K could be $\mathbb{Q}_p$ or $\mathbb{F}_p((t))$, the formal Laurent series over $\mathbb{F}_p$).

(T3) (*Automorphism groups*). The full group of automorphisms, $\mathrm{Aut}(T)$, of a locally finite tree T is a locally compact topological group; hence, it has a Haar measure. It is totally disconnected, generally not discrete. For example, suppose that T is a regular tree of valence $n \geq 3$, that $v \in \mathrm{Vert}(T)$ and that $B(k)$ denotes the ball of radius of radius k centered at v. Let $G = \mathrm{Aut}(T)$ and let G_v be the stabilizer of v in G. We note that G_v is ∞-*transitive* in the sense of [44]. This means that for each $k \geq 0$, G_v acts transitively on the set of geodesics paths from v to a point in $\partial B(k)$. Indeed, G_v is the inverse limit, $\varprojlim \mathrm{Aut}_v(B(k))$, where $\mathrm{Aut}_v(B(k))$ is the group of automorphisms of the rooted tree $B(k)$ which fix v and where the bonding map $\mathrm{Aut}_v(B(k+1)) \to \mathrm{Aut}_v(B(k))$ is induced by geodesic contraction $B(k+1) \to B(k)$. The kernel of $\mathrm{Aut}_v(B(k+1)) \to \mathrm{Aut}_v(B(k))$ is a product of symmetric groups of order $n-1$. Hence, the inverse limit is infinite and so, is not discrete whenever $n \geq 3$. (This inverse limit is a good example of a totally disconnected group.) Similarly, automorphism groups of locally finite buildings are usually totally disconnected, locally compact groups.

(T4) (*Simplicity of automorphism groups*). An old theorem of Tits [217] asserts that in many cases the full automorphism group of a tree T is a simple group. This is a predecessor of many of the simplicity results in Sect. 4.4 concerning automorphism groups of buildings, as well as the Burger–Mozes results in Sect. 2.4.4 on the simplicity of certain discrete groups acting freely on a product of trees. The fact[1] that $\mathrm{Aut}(T) \curvearrowright T$ gives $\mathrm{Aut}(T)$ the structure of a graph of groups over the quotient graph Ω (except for the fact that $\mathrm{Aut}(T)$ might have elements which are inversions). Let $\mathrm{Aut}_+(T)$ denote the subgroup of $\mathrm{Aut}(T)$ generated by vertex stabilizers. In other words, $\mathrm{Aut}_+(T)$ is the direct limit of the vertex and edge stabilizers in the graph of groups. It follows that $\mathrm{Aut}(T)/\mathrm{Aut}_+(T)$ is the product of $\pi_1(\Omega)$ with cyclic groups of order 2. Tits' theorem is that the group $\mathrm{Aut}_+(T)$ is either the trivial group or a simple group. So, if Ω is a tree and there are no inversions, then $\mathrm{Aut}(T)$ is simple. That is to say, $\mathrm{Aut}(T)$ is a simple group unless it fails to be so for some obvious reason.

The tree T_{p+1} is the euclidean building associated to $SL_2(\mathbb{Q}_p)$, where $\mathbb{Q}_p$ denotes the p-adic completion of the rationals; so, the topological group $PSL_2(\mathbb{Q}_p)$ is a subgroup of $\mathrm{Aut}(T_{p+1})$, cf. Example 4.132 in Sect. 4.4.8. This gives another way to see that $\mathrm{Aut}(T_{p+1})$ is not discrete. Similarly, $\mathrm{Aut}(T_{p+1})$ contains $PSL_2(K)$ where K is the valuation field $\mathbb{F}_p((t))$, for $\mathbb{F}_p$ the finite field of order p.

Definition 2.39 A *lattice* in a locally compact topological group is a discrete subgroup of finite covolume (with respect to Haar measure). The lattice is *uniform* if it is cocompact.

For example, if a group G acts properly, cocompactly and faithfully on a tree T, then G is a uniform lattice in $\mathrm{Aut}(T)$.

2.4.2 *Products*

A product of graphs is naturally an NPC cube complex. To see this, suppose $X = \Omega_1 \times \cdots \times \Omega_n$ is a product of graphs. The cubes in X correspond to collections $\{c_1, \dots, c_n\}$, where c_i is either an edge or a vertex of Ω_i. The corresponding cube $\square(c_1, \dots, c_n)$ is isometric to $c_1 \times \cdots \times c_n$. Its dimension is the number of c_i which are edges. If $v = (v_1, \dots, v_n)$ is a vertex of X, then its link is the n-fold join:

$$\mathrm{Lk}(v, X) = \mathop{\ast}_{i=1}^{i=n} \mathrm{Lk}(v_i, \Omega_i) \tag{2.7}$$

Noting that the link of a vertex in a locally finite graph is a finite set, we see that the simplicial complex $\mathrm{Lk}(v, X)$, being a join of links of vertices in the graphs, is a join of finite sets of the form,

$$\mathrm{Lk}(v, X) = K_{k_1,\cdots,k_n} := (k_1 \text{ points}) * \cdots * (k_n \text{ points}), \tag{2.8}$$

[1] We write $G \curvearrowright X$ to mean that "G acts on X."

where $k_i = \text{Card}(\text{Lk}(v_i, \Omega_i))$ (where, $\text{Card}(A)$ denotes the cardinality of a set A). For example, in a product of two graphs each link is a complete bipartite graph. (By analogy, the join defined in (2.7) should be called a *complete n-partite complex*.) With the metric of a spherical join, each simplex of $\text{Lk}(v, X)$ is all right. By Lemma 2.25 (ii), $\text{Lk}(v, X)$ is a flag complex. So, by Gromov's Lemma 2.31, X is an NPC cube complex. A similar argument gives the following result.

Proposition 2.40 *The product of cube complexes is a cube complex. If $X = X_1 \times \cdots \times X_n$ is such a product and each X_i is NPC, then X is NPC.*

Remark 2.41 One can prove directly, without mentioning the link condition, that a product of NPC spaces is NPC. (See [35, Exercise 1.9 (c), p. 163 and Examples 1.15 (3), p. 167].)

Example 2.42 (The Cube Complex P_L) Given a finite simplicial complex L with vertex set I, we shall define in Sect. 3.1, a subcomplex of the cube $[-1, 1]^I$ such that the link at each vertex is L. It follows from Gromov's Lemma 2.28 that if L is a flag complex, then P_L is NPC.

Remark 2.43 The study of lattices in semisimple Lie groups is of central importance in number theory and differential geometry.

Suppose $\mathcal{G} = \mathcal{G}_1 \times \cdots \times \mathcal{G}_n$ is a product of simple Lie groups where $\mathcal{G}_k$ is the group of isometries of a symmetric space X_k. One way to get a lattice G in $\mathcal{G}$ is to take a product of lattices $G = G_1 \times \cdots \times G_n$. Such a lattice is called *reducible*. More generally, G is *reducible* if it has a subgroup of finite index that splits as a product. The lattice G is *irreducible* if it is not reducible. This terminology also makes sense for lattices in products of locally compact groups. For example one can find an irreducible lattice in $SL_2(\mathbb{Q}_p) \times SL_2(\mathbb{Q}_{p'})$ acting on $T_{p+1} \times T_{p'+1}$.

An important part of Margulis' proof of his Super Rigidity Theorem for lattices in semisimple Lie groups is his Normal Subgroup Theorem (cf. [175]). This asserts that if G is any irreducible lattice in a semisimple Lie group of noncompact type and rank ≥ 2, then any normal subgroup of G is either a finite group or a subgroup of finite index. In particular, when G has no nontrivial finite normal subgroup, the Normal Subgroup Theorem means that it is "just infinite," as defined below.

Definition 2.44 An infinite group G is *just infinite* if any nontrivial normal subgroup has finite index in G.

2.4.3 Higman Groups

Let H be the group defined by the presentation:

$$H = \langle x_1, x_2, x_3, x_4 \mid x_i x_{i+1} x_i^{-1} = x_{i+1}^2;\quad i \in \mathbb{Z}/4\rangle.$$

There are excellent expositions of the next result, e.g., [207, Prop. 6] and [50, Thm. 2.2]. More information also can be found in [35, II.12.17(6)] [176] and [149].

Proposition 2.45 (G. Higman [148])

(i) *Each subgroup of finite index in H is equal to H (and hence, H has no nontrivial finite quotient).*
(ii) *The group H is infinite.*

It follows from (i) that H is not residually finite (see Definition 2.51 in the next subsection).

The existence of a finitely generated infinite simple group was first proved as a corollary to the above proposition. In fact, by taking the quotient of H by a maximal normal subgroup we get the following.

Corollary 2.46 (G. Higman [148], also cf. [207, 1.4]) *There exists an infinite simple group generated by* 4 *elements.*

The proof of Proposition 2.45 is based on the following.

Lemma 2.47 (cf. [148])

(i) *The only positive integer n that divides $2^n - 1$ is $n = 1$.*
(ii) *In a finite group F the only element $g \in F$ that is conjugate to its square by an element of the same order as g is $g = 1$.*

Proof of Lemma 2.47 (following the argument in [50, Lemma 2.1])

(i) Suppose $n > 1$ divides $2^n - 1$ and let p be the smallest prime divisor of n. Then p divides $2^n - 1$. So, p is odd and the order of 2 in the multiplicative group $(\mathbb{Z}/p)^*$ is a divisor of n, as well as a divisor of $p - 1$. Since p is the smallest prime divisor of n, the integers n and $p - 1$ are relatively prime, a contradiction.
(ii) Suppose $x \in F$ is an element of the same order, n, as $g \in F$ and that x is such that $xgx^{-1} = g^2$. Then $g = x^n g x^{-n} = g^{2^n}$. So, $g^{2^n - 1} = 1$. The conclusion follows from (i).

Proof of Proposition 2.45 (i) Observe that H has an automorphism of order 4 that cyclically permutes the generators. Hence, if H has a nontrivial finite quotient, then it has a nontriivial finite quotient in which the image of each of the generators x_i has the same order n. Let $\overline{H}$ be such a quotient. Since x_i conjugates x_{i+1} to its square, Lemma 2.47 (ii) says that $n = 1$; hence, $\overline{H}$ is trivial.

(ii) Serre's proof that H is infinite in [207, pp. 9–10] uses the fact that H has the structure of a square of groups, that is, a polygon of groups where the underlying polygon is a square. (See Definition A.5 in Appendix A.1.) The edges are indexed by $\mathbb{Z}/4$. For $i \in \mathbb{Z}/4$, the edge stabilizer H_i is the infinite cyclic group generated by x_i. The vertex groups H_{12}, H_{23}, H_{34}, H_{41} are generated by x_i and x_{i+1} and each is isomorphic to Baumslag–Solitar group $BS(1, 2)$ (see the definition following this proof). For example, H_{12} is generated by x_1 and x_2 with the relation: $x_1 x_2 x_1^{-1} = x_2^2$. We consider the groups generated by the stars of opposite vertex groups H_{12} and H_{34}. By amalgamating H_{12} and H_{23} along H_2, we obtain $H_{123} = H_{12} *_{H_2} H_{23}$ and similarly, $H_{341} = H_{34} *_{H_4} H_{41}$. It can be shown that in H_{123} the subgroup generated by H_1 and H_3 is the free group on 2 generators, $G = H_1 * H_3$. Similarly, G is a subgroup of H_{341}. Since H is the direct limit of the square of groups, we see that H is the amalgam $H_{123} *_G H_{341}$. This is clearly infinite since it contains the free group on two generators $H_2 * H_4$.

The above proof shows that H has the structure of a square of groups. Each edge group is infinite cyclic, each vertex group is a copy of $BS(1, 2)$ and the group associated to the 2-cell is trivial. (Recall from the Introduction that the *Baumslag–Solitar group* $BS(n, m)$ is the group defined by the following presentation with 2 generators and 1 relation: $BS(n, m) = \langle x, y \mid x^n yx^{-n} y^{-m}\rangle$. When $n = 1$, the group $BS(1, m)$ is said to be a *solvable Baumslag–Solitar group*.)

Generalized Higman Groups Suppose $\alpha = (m_1, n_1), \dots (m_k, n_k)$ is a k-tuple of pairs of nonzero integers with $k > 3$. Following [149] define the *generalized Higman group* Hig_α by

$$\text{Hig}_\alpha := \langle a_1, \dots a_k \mid \{a_i a_{i+1}^{m_i} a_i^{-1} = a_{i+1}^{n_i}\}_{i \in \mathbb{Z}/k}\rangle.$$

As before, this gives a k-gon of groups for $k \geq 4$. (If $k = 3$, Hig_α might not be infinite.) We note that if each $(m_i, n_i) = (1, 1)$, then Hig_α is the RAAG, A_L, with L a k-cycle. It is proved in [35] that this polygon of groups for Hig_α is NPC provided $k \geq 4$ (see also Martin [176]). This amounts to showing that the link of each vertex is a flag complex, hence, with its all right metric is CAT(1). It follows that the complex of groups is "developable" with fundamental group Hig_α. Its universal cover is a CAT(0) square complex Y. The squares are obtained by subdividing the cone on the polygon into k squares each with a vertex at the cone point. (Note that the Hig_α-action on Y is not proper since the isotropy groups are not finite.) For further details and similar examples of such square complexes see Sect. 2.4.5 as well as Example 2.60 in Sect. 2.4.5.

Remark 2.48 Since the universal cover Y of this polygon of groups is contractible, it follows that the $K(\pi, 1)$-Question has a positive answer. (See Theorem A.24 in Appendix A.4.2.) Since the vertex groups and edge groups are type F, a corollary is that for $k \geq 4$, the group Hig_α is type F, i.e., $B\,\text{Hig}_\alpha$ has a model which is a finite 2-complex.

Remark 2.49 Horbez-Huang [149, Thm. 1.1] prove that the generalized Higman groups Hig_α are superrigid for measure equivalence. More precisely, if, for all i, $|m_i| \neq |n_i|$ and if $k \geq 5$, then any countable group G that is measure equivalent to Hig_α is virtually isomorphic to it.

R. Thompson gave the first examples of finitely presented, infinite simple groups. These were later extended to a family of such examples by G. Higman.

2.4.4 The Burger–Mozes Examples

There is an excellent exposition of the following material in Caprace's survey paper [50].

Theorem 2.50 (Burger–Mozes [43, 45]) *For every sufficiently large pair of even integers, (n, m), there is a finite NPC square complex $Y_{n,m}$ whose universal cover is a product of regular trees $T_n \times T_m$ and whose fundamental group $G_{n,m}$ is an infinite simple group. Moreover, $G_{n,m}$ is an irreducible lattice in* $\mathrm{Aut}(T_n) \times \mathrm{Aut}(T_m)$.

Definition 2.51 A group G is *residually finite* if for every $g \in G - \{1\}$, there is a finite index subgroup $H < G$ so that $g \in G - H$. (In other words, the elements 1 and g have distinct images in G/H, or more geometrically, the vertices 1 and g of the Cayley graph have distinct images in the quotient $\mathrm{Cay}(G, S)/H$.) Also, see [153].

The condition that G is residually finite is equivalent to the condition that the intersection of all normal finite index subgroups of G is trivial (or equivalently, that the natural map from G to its profinite completion is injective). In particular, an infinite simple group cannot be residually finite. Finitely generated linear groups are residually finite. There is the following well-known question of Gromov:

Question 2.52 (Gromov) Is every word hyperbolic group residually finite?

Precursors to Theorem 2.50 can be found in the work of Wise [233] and M. Bhattacherjee [24]. Paper [233] is essentially Wise's PhD thesis [232], in which he showed that there is a finite square complex X with no finite cover. In particular, for such an X, $\pi_1(X)$ is not residually finite. Somewhat earlier Bhattacherjee had proved an algebraic version: there is a free amalgam $L *_K H$ of finitely generated free groups with no finite index subgroups.

Theorem 2.50 is proved by first constructing a square complex $X = X_{n,m}$ with only one vertex so that the link of that vertex is the complete bipartite graph $K_{n,m}$, i.e., it is the join of n points with m points. it then follows that the universal cover of X is the product, $T_n \times T_m$. One shows that X can be chosen to have the following two properties:

(1) The group $G = \pi_1(X)$ is a cocompact lattice in $\mathrm{Aut}(T_n) \times \mathrm{Aut}(T_m)$ satisfying the following analog of Margulis' Normal Subgroup Theorem: any nontrivial normal subgroup of G has finite index.
(2) The group G is not residually finite.

Assuming (1) and (2), the proof of Theorem 2.50 goes as follows. Let $G_{n,m}$ denote the intersection of all finite index normal subgroups in G. Since G is not residually finite, $G_{n,m}$ is not trivial; so, by the Normal Subgroup Theorem any nontrivial normal subgroup of $G_{n,m}$ has finite index and hence, is equal to $G_{n,m}$. Thus, $G_{n,m}$ is simple and $Y_{n,m} = (T_n \times T_m)/G_{n,m}$ is the desired square complex. (Note that this shows that $G = \pi_1(X_{n,m})$ is virtually simple). The proof of the Normal Subgroup Theorem, (1), uses the following trick of Margulis: for a normal subgroup $N \lhd G$, a method for proving that G/N is finite is to show that it, simultaneously, is amenable and has Property (T).

The group $G_{n,m}$ is an irreducible lattice in $\text{Aut}(T_n) \times \text{Aut}(T_m)$. This means that the square complex $X_{n,m}$ is not finitely covered by a product of two graphs. There is the following natural question, the answer to which is not known.

Question 2.53 (cf. [50, Problem 4.31]) Suppose $T_1, \ldots, T_n$ are locally finite trees each with infinitely many ends and that $G < \text{Aut}(T_1) \times \cdots \times \text{Aut}(T_n)$ is a cocompact lattice. Can G be simple when $n \geq 3$?

Remark 2.54 There are arithmetic lattices in a product of groups of the form $SL_2(\mathbb{Q}_p)$ acting on a product of regular trees. By the Normal Subgroup Theorem such groups are just infinite. However, the condition of being not residually finite fails (since linear groups are residually finite).

By Caprace-Remy [52] there are simple groups acting as nonuniform lattices on the product of a Kac –Moody building with itself. (See the discussion at the end of Sect. 4.4.8.)

2.4.5 Nonpositively Curved Polygons of Groups

Suppose P is a polygon (i.e., a convex 2-cell) and $\mathcal{F}(P)$ is its poset of faces. A *polygon of groups* is a simple complex of groups over $\mathcal{F}(P)$. This terminology is explained in Appendix A.1. Roughly, it means that we are give a collection of groups $\{G_\sigma\}_{\sigma \in \mathcal{F}(P)}$ and monomorphisms $\phi_{\tau\sigma} : G_\sigma \to G_\tau$ defined whenever $\tau < \sigma$. We denote such a simple complex of groups by $G\mathcal{F}(P)$. The group G_P associated to the 2-dimensional face P is often required to be the trivial group. Let G denote the direct limit of this system of groups.

For each $x \in P$, let G_x denote the image of G_σ in G where σ is the face such that x lies in its relative interior. The *basic construction* on $G\mathcal{F}(P)$ is the polyhedron $D(G, P)$ given by $D(G, P) = (G \times P)/\sim$, where the equivalence relation $\sim$ is given by: $(g, x) \sim (h, y)$ if and only if $x = y$ and $gG_x = hG_x$. The group G acts on $D(G, P)$ with strict fundamental domain P. The isotropy subgroup at a point $x \in P$ is G_x. (For more details and the definition in greater generality, see Appendix A.2 or the paragraph surrounding (3.9) in Sect. 3.1.1.) By Theorem A.7 (vi)′ in Appendix A.2, $D(G, P)$ is simply connected.

Suppose e, f are the edges containing a vertex v of P. Then G_v acts on Lk_v (the link of $[1, v]$ in $D(G, P)$) with a 1-simplex as strict fundamental domain. (The endpoints of the 1-simplex correspond to e and f.) If G_v is generated by G_e and G_f, then the natural map $G_e * G_f \to G_v$ is surjective and Lk_v is a connected simplicial graph. We suppose this to be true. Then fundamental group of Lk_v is $\ker((G_e * G_f) \to G_v)$. The *girth* of Lk_v, denoted $\text{girth}(\text{Lk}_v)$, is the combinatorial length of the shortest cycle in Lk_v. (The girth of Lk_v is sometimes called its "systole.") If

$\text{girth}(\text{Lk}_v) \geq 2$, then Lk_v is a simplicial graph; if its girth is ≥ 4, then it is a flag complex.

There is an important dichotomy here. If each vertex group G_v is finite, then each vertex link Lk_v is a finite graph, $D(G, P)$ is a locally finite 2-complex and G acts geometrically on it. On the other hand, if some G_v is infinite, then the polyhedron $D(G, P)$ is not locally finite and the G-action is not proper. The reflection group actions in Examples 2.58 and 2.59 below are of the first type, while Examples 2.60 and 2.62 are not locally finite.

The polygon of groups $G\mathcal{F}(P)$ is *nonpositively curved* if the following conditions hold:

(a) P can be realized as a convex polygon in $\mathbb{E}^2$ or $\mathbb{H}^2$.
(b) If $\alpha_{v(i)}$ denotes the angle at $v(i)$, then $[\alpha_{v(i)}][\text{girth}(\text{Lk}_{v(i)})] \geq 2\pi$. (This is the Link Condition from Example 2.6 at the vertex $v(i)$.)

There are several important consequences of the nonpositive curvature of $G\mathcal{F}(P)$. First, $G\mathcal{F}(P)$ is *developable*. This means that the natural map from each local group G_σ to the direct limit G is injective (see the Appendix or [35, III.$\mathcal{C}$.2.11]). Second, $D(G, P)$ is CAT(0). Third, the $K(\pi, 1)$-Question for $G\mathcal{F}(P)$ has a positive answer. This means that if we use the natural maps to glue together copies of the BG_σ, the result is homotopy equivalent to BG, the Eilenberg –MacLane complex for the direct limit (see Theorem A.21 in Appendix A.4 or [138, Prop. 3.2.3]).

Lemma 2.55 *Suppose that $G\mathcal{F}(P)$ is an m-gon of groups, $m \geq 4$ and that for each vertex v of P,* $\text{girth}(\text{Lk}_v) \geq 4$. *When $m = 4$, $D(G, P)$ can be given the structure of a* CAT(0) *square complex. When $m \geq 5$, it can be given the structure of a piecewise hyperbolic* CAT(-1) *polygonal complex.*

Proof We can realize P as a convex right angled m-gon either in $\mathbb{E}^2$ when $m = 4$ or in $\mathbb{H}^2$ when $m \geq 5$.

Since an all right triangle is isometric to a subset of $\mathbb{S}^2$, Lemma 2.55 does not hold when $m = 3$. In this case we must choose the angles of P more carefully. So, suppose P is a triangle with vertices $v(i)$ for $i = 1, 2, 3$. Let $\alpha_{v(i)} \in (0, \pi/2]$ be a possible choice of angle at $v(i)$. The condition that the link at $v(i)$ be CAT(1) is that $[\alpha_{v(i)}][\text{girth}(\text{Lk}_{v(i)})] \geq 2\pi$, i.e., $\alpha_{v(i)} \geq 2\pi/\text{girth}(\text{Lk}_{v(i)})$.

Lemma 2.56 *Suppose P is a triangle with vertices $v(i)$, for $i = 1, 2, 3$ and that* $\text{girth}(\text{Lk}_{v(i)}) \geq 4$. *Then a necessary and sufficient condition to realize $G\mathcal{F}(P)$ as an NPC triangle of groups is that we can find numbers $\alpha_{v(i)} \in (0, \pi/2]$ so that*

$$\pi \geq \sum \alpha_{v(i)} \geq 2\pi \sum 1/\text{girth}(\text{Lk}_{v(i)}).$$

Proof The first inequality is the fact that the sum of the angles of a triangle in $\mathbb{E}^2$ or $\mathbb{H}^2$ is $\leq \pi$. The second inequality follows from the condition that $\text{Lk}_{v(i)}$ be large (see Definition 2.16).

Remark 2.57 The system of groups defined by an m-gon of groups is the same as the one defined by the graph of groups corresponding to the boundary m-cycle. However, the fundamental group of the graph of groups is not the direct limit. To see this, delete the interior of each copy of P from $D(G, P)$. The resulting graph is a cover of the m-cycle of groups; however, it is not simply connected.

Example 2.58 (Polygon Reflection Groups, cf. [82, §6.5]) Suppose P is a convex m-gon in $\mathbb{X}^2 = \mathbb{S}^2, \mathbb{E}^2$ or $\mathbb{H}^2$ so that the angle at $v(i)$ is π/m_i, where m_i is a natural number > 1. This is the data for a simple complex of groups over $\mathcal{F}(P)$: each edge group is $\mathbf{C}_2$ and the group associated to the vertex $v(i)$ is D_{m_i}, the dihedral group of order $2m_i$. (The group D_{m_i} is generated by reflections across the two edges meeting at $v(i)$.) The direct limit of this system of groups is the *Coxeter group* W_P. Since each local group is finite, the W_P action on $D(W_P, P)$ is proper. It turns out that W_P is isomorphic to the group of isometries of a space of constant curvature $\mathbb{X}^2$ generated by reflections across the sides of P. More precisely, the inclusion $P \hookrightarrow \mathbb{X}^2$ induces a W_P-equivariant homeomorphism $D(W_P, P) \to \mathbb{X}^2$ (see Lemma 3.7 (i) in the next section). This is a fundamental example to which we will return many times in this book.

Example 2.59 (Triangle Groups) If P is a triangle with angles π/m_i, then the Coxeter group W_P is called a *triangle group*. It will sometimes be denoted $W(m_1, m_2, m_3)$. (The definition of a "Coxeter group" is given by the presentation (4.5) in Sect. 4.1.2.) The triangle P can be realized as a triangle in $\mathbb{S}^2$, $\mathbb{E}^2$ or $\mathbb{H}^2$ as the sum of its angles is, respectively, $>$, $=$, or $< \pi$. In other words, $W(m_1, m_2, m_3)$ is spherical, euclidean or hyperbolic as the sum of reciprocals $1/m_1 + 1/m_2 + 1/m_3$ is $>$, $=$, or < 1, respectively.

Example 2.60 (The Higman Group and Its Generalizations) The proof of Proposition 2.45 (i) shows that Higman's group H is the direct limit of a square of groups where each edge group is infinite cyclic and each vertex group is a Baumslag–Solitar group, $BS(1, 2)$. If v is a vertex of the square P and e and f are the adjacent edges, then Lk_v is the quotient of the Bass-Serre tree for $G_e * G_f$ by $\ker((G_e * G_f) \to G_v)$. Martin proves in [176] that the girth of this simplicial graph is at least 4. Hence, this square of groups is NPC and its development $D(H, P)$ is a CAT(0) square complex.

Here are two more examples along the same line.

Example 2.61 (Graph Products over Cycles) We can define a complex of groups over $\mathcal{F}(P_m)$ as follows. For each edge e choose a group G_e. To a vertex $v = e \cap f$ associate the group $G_e \times G_f$. To the 2-cell associate the trivial group. The natural inclusions of the edge groups into vertex groups, then defines a polygon of groups. The resulting direct limit G is called the *graph product*. The graph in question is the m-cycle dual to ∂P_m. The link at the vertex $v = e \cap f$ is the join of G_e and G_f. (To avoid confusing this with the free product we do not denote the join by $G_r * G_f$.) Since this join is a complete bipartite graph, its girth is 4; so, the Link Condition holds. As before, if $m \geq 4$, this polygon of groups is nonpositively curved; it is called the *graph product complex*. (If $m = 3$, then G is the direct product of the 3 edge groups and the graph product complex should be thought of as lying over the 3-cube rather than a triangle.) For more general graphs, the graph product complex is explained in Sect. 3.1.2 of the next chapter. It is also explained there how, when $m \geq 4$, the development $D(G, P_m)$ is an example of a right-angled building. A chamber for $D(G, P_m)$ is the polygon P_m.

Example 2.62 (Two-dimensional Artin Groups over Polygons) A *dihedral Artin group* is an Artin group A_k whose associated Coxeter group is the dihedral group D_k. In other words, A_k has a presentation with two generators x and y and one relation of the form $xy \cdots = yx \cdots$ where on both sides of this equation there is an alternating word in x, y of length k. Now let P be an m-gon, $m \geq 4$ and let $\mathbf{k} : v(i) \mapsto k(i)$ be a function from Vert P to $\{2, 3, \dots\} \cup \{\infty\}$. Define a simple complex of groups $G\mathcal{F}(P)$ over P by putting an infinite cyclic group on each edge and a dihedral Artin group $A_{k(i)}$ on each vertex. The direct limit G is then an Artin group. By [10, Lemma 6], the link for the dihedral Artin group A_k has systole $\geq 2k$, i.e., girth$(\mathrm{Lk}_{v(i)}) \geq 2k_i$; so, $G\mathcal{F}(P)$ is a nonpositively curved polygon of groups. Hence, $D(G, P)$ is CAT(0). If W_P denotes the Coxeter group associated to G, then $D(G, P)/W_P$ is its associated "Deligne complex" (see Sect. 4.3.1). The theory of general Artin groups is the subject of Sect. 4.3. For general Artin groups, it is still not known if the natural metric on the Deligne complex is nonpositively curved. (See Conjecture 4.54 in Sect. 4.3.1.)

Chapter 3
Right-Angled Spaces and Groups

The notion of a "polyhedral product" of a collection of spaces (or pairs of spaces) indexed by the vertex set I of a simplicial complex L is used to define some of the main examples of cell complexes and groups that are discussed in this book. If L is a flag complex and I indexes a collection of copies of S^1, then the polyhedral product is the standard classifying space for the "right-angled Artin group" associated to L. More generally, if I indexes a collection of classifying spaces BG_i, then the polyhedral product is the classifying space for the "graph product" of the G_i. The universal cover of a polyhedral product is often a CAT(0) cube complex and isometry groups of these cube complexes are often graph products. For example, if the spaces that are indexed by I are cones over discrete sets, then the universal cover of the polyhedral product is a right-angled building (abbreviated as RAB). If each of the discrete sets is a group G_i, then the relevant isometry group of the RAB is the graph product of the G_i. When each of the factors is the cyclic group of order 2, the graph product is the "right-angled Coxeter group" (abbreviated as RACG) associated to L; when each factor is infinite cyclic, then the graph product is a right-angled Artin group (abbreviated as RAAG). In these cases the right-angled buildings are called, respectively, the "Davis–Moussong complex" of the RACG or the "Deligne complex" of the RAAG. In Sect. 3.2 we take a fairly deep dive into the theory of cocompact right-angled reflection groups on contractible manifolds. Here the flag complex L is a simplicial sphere (or more generally, a generalized homology sphere). The corresponding polyhedral product of intervals, P_L, is also known as the "generalized moment angle manifold." The contractible manifold is its universal cover. The fundamental group of P_L is the commutator subgroup of a RACG. Further quotients lead to the notion of a "small cover" of a right-angled Coxeter orbifold. Various related ideas, such as Haken manifolds and the Reflection Group Trick, are also explained in Sect. 3.2. In Sect. 3.3 we discuss a similar class of examples acting on CAT(0) cube complexes formed by "blowing up Coxeter zonotopes" and taking the universal covers. In general, the relevant groups are not Coxeter groups, rather they are "mock reflection groups."

M. W. Davis, *Infinite Group Actions on Polyhedra*, Ergebnisse der Mathematik
und ihrer Grenzgebiete. 3. Folge / A Series of Modern Surveys in Mathematics 77,
https://doi.org/10.1007/978-3-031-48443-8_3

Given a flag complex L, NPC cube complexes P_L and Z_L are defined as polyhedral products in Sect. 3.1.1. In both cases, there is a group G, which is a direct sum of discrete groups, and an action of G on the polyhedral product with strict fundamental domain the chamber K_L. The universal covers of P_L and Z_L are CAT(0) cube complexes. The group Γ of all lifts of the G-action to the universal cover is a graph product of discrete groups. In Sect. 3.1 we also define an NPC cube complex $\mathbb{T}_L$ which is a polyhedral product of copies of S^1. This cube complex is the classifying space for the right-angled Artin group associated to L. Similarly, a polyhedral product of classifying spaces BG_i, $i \in \mathrm{Vert}(L)$, is the classifying space $B\Gamma$ for the graph product of the G_i.

3.1 Polyhedral Products

Let L be a simplicial complex with vertex set I. The poset of simplices (including the empty simplex) is denoted by $\mathcal{S}(L)$. If σ is a simplex of L, then $I(\sigma)$ denotes its vertex set. Suppose $\mathbf{A} = \{(A_i, B_i)\}_{i \in I}$ is a collection of pairs of spaces with basepoints $*_i \in B_i$.

Definition 3.1 The *polyhedral product* of $\mathbf{A}$ with respect to L is the subset $\mathbf{A}^L$ of the product $\prod_{i\in I} A_i$ consisting of the points $(x_i)_{i\in I}$ satisfying the following two conditions:

(a) $x_i = *_i$ for all but finitely many i,
(b) $\{i \in I \mid x_i \notin B_i\}$ is a simplex of L.

(We could include condition (a) into the definition of $\prod_{i\in I} A_i$ and call it the *weak product.*) For pairs $M < M'$ of subcomplexes of L, the basepoints determine natural inclusions $\mathbf{A}^M \hookrightarrow \mathbf{A}^{M'}$. When L is finite, the polyhedral product inherits a topology from the (weak) product $\prod_{i\in I} A_i$. In general $\mathbf{A}^L$ acquires the topology of the direct limit, $\lim \mathbf{A}^M$, as M runs through the finite subcomplexes.

One can also view $\mathbf{A}^L$ as a union of products as follows. For each simplex σ in $\mathcal{S}(L)$ put

$$\mathbf{A}^\sigma = \prod_{i\in \mathrm{Vert}\,\sigma} A_i \times \prod_{i\notin \mathrm{Vert}\,\sigma} B_i$$

So, if σ is the empty simplex, then $A^\sigma = \prod B_i$. Also, note that if each $B_i = *_i$, then $\mathbf{A}^\sigma$ is just the product of the A_i, where i ranges over $\mathrm{Vert}\,\sigma$. As a union of products A^L is defined by

$$\mathbf{A}^L = \bigcup_{\sigma\in\mathcal{S}(L)} A^\sigma . \tag{3.1}$$

The topology on $\mathbf{A}^L$ also can be explained using (3.1). If Δ denotes the full simplex on I, then $\mathbf{A}^\Delta$ is the subset of the product of the A_i satisfying (a). So, $\mathbf{A}^\Delta$ is the union of all $\mathbf{A}^\sigma$ over all finite subsets of $\sigma \leq I$. Give $\mathbf{A}^\Delta$ the direct limit topology and give $A^L \leq A^\Delta$ the induced topology.

The main examples of this book are when:

(1) each A_i is the interval $[0, 1]$, $B_i = *_i = 1$,
(2) each A_i is the interval $[-1, 1]$, $B_i = \partial[-1, 1] = \{\pm 1\}$ and $*_i = 1$,
(3) each B_i is a discrete space E_i, $A_i = \operatorname{Cone} E_i$ and $*_i \in E_i$,
(4) each $A_i = S^1$ and $B_i = *_i = 1$.

In (3) $\operatorname{Cone} E_i$ means the cone on E_i, i.e., it is the quotient space $(E_i \times [0, 1])/\sim$. For example, if $E_i = \{\pm 1\}$, then $\operatorname{Cone} E_i$ can be identified with $[-1, 1]$. The *cone point* of E_i is the point 0_i corresponding to $0 \in [0, 1]$ Choose an element $*_i \in E_i$ to be the basepoint. Let $\mathbf{Cone\,E}$ denote the I-tuple $(\operatorname{Cone} E_i, E_i, *_i)_{i \in I}$.

Denote the polyhedral products in (1), (2), (3), (4) by K_L, P_L, Z_L, and $\mathbb{T}^L$, respectively, i.e.,

$$K_L = ([0, 1], 1)^L, \tag{3.2}$$

$$P_L = ([-1, 1], \{\pm 1\})^L, \tag{3.3}$$

$$Z_L = (\mathbf{Cone\,E}, \mathbf{E})^L, \tag{3.4}$$

$$\mathbb{T}_L = (S^1, 1)^L. \tag{3.5}$$

When the pairs (A_i, B_i) are independent of i so that $(A_i, B_i) = (A, B)$, we write simply $(A, B)^L$ instead of using boldface (for example, this is the case in (3.2), (3.3) and (3.5)).

3.1.1 The Cube Complexes K_L, P_L, Z_L, and $\mathbb{T}_L$

Since polyhedral products are subspaces of products, P_L is a cubical subcomplex of the cube $[-1, 1]^I$, while $\mathbb{T}_L$ is a subcomplex of the torus, $(S^1)^I$. Regarding S^1 as a 1-cube complex with one edge and one vertex, the product $(S^1)^I$ becomes a cube complex and $\mathbb{T}_L$ is a cubical subcomplex. The cone, $\operatorname{Cone} E_i$, is the union of intervals, one for each point of E_i, glued together at the cone point 0_i. So, $\operatorname{Cone} E_i$ is a 1-cube complex. Hence, the (weak) product $\prod_{i \in I} \operatorname{Cone} E_i$ is a cube complex. Since the polyhedral product Z_L, defined by (3.4), is a subcomplex, it is also a cube complex. Each of these cube complexes, K_L, P_L, Z_L and $\mathbb{T}_L$ has the same dimension, namely, $1 + \dim(L)$.

The Complex P_L Let $\mathbf{C}_2 = \{\pm 1\}$ be the cyclic group of order 2 and denote by $(\mathbf{C}_2)^I$ (or by $\sum_{i \in I} \mathbf{C}_2$) the direct sum of I copies of $\mathbf{C}_2$. Similarly, $[-1, 1]^I$ is the set of all (t_i) in the product satisfying condition (a) in Definition 3.1. We have

$(\mathbf{C}_2)^I \curvearrowright [-1,1]^I$ as a reflection group. (As in the footnote in Sect. 2.4.2, $G \curvearrowright X$ means "the group G acts on a space X.") Each element of I can be regarded as a standard basis vector for the euclidean space $\mathbb{R}^I$. So, each $(k-1)$-face σ of the full simplex on I determines a k-dimensional subspace $\mathbb{R}^\sigma$ spanned by the vertex set $I(\sigma)$ of σ. The cube complex P_L is stable under the $(\mathbf{C}_2)^I$-action on $[-1,1]^I$. Any k-dimensional face of $[-1,1]^I$ is parallel to some $\mathbb{R}^\sigma$. We denote such a face by $\square^\sigma$. Given σ, the corresponding set of k-cubes parallel to $\square^\sigma$ forms a single $(\mathbf{C}_2)^I$-orbit which we denote by $\mathrm{orb}(\square^\sigma)$. In other words, $\mathrm{orb}(\square^\sigma) \cong (\mathbf{C}_2)^{I-I(\sigma)} \times [-1,1]^{I(\sigma)}$. Thus,

$$P_L = \bigcup_{\sigma \in \mathcal{S}(L)} \mathrm{orb}(\square^\sigma). \tag{3.6}$$

In particular, $\mathrm{orb}(\square^\emptyset)$ means the $(\mathbf{C}_2)^I$-orbit of a vertex, i.e., $\mathrm{orb}(\square^\emptyset) = \{\pm 1\}^I$.

Example 3.2 (Examples of P_L) See also [81], [84].

(i) If L consists of n vertices and no higher dimensional cells, then P_L is the 1-skeleton of an n-cube.
(ii) If $L = \partial \Delta^n$ is the boundary complex of an n-simplex, then P_L is the boundary complex of an $(n+1)$-cube, i.e., P_L is homeomorphic to S^n
(iii) Suppose L is an m-circuit (that is, a circle triangulated as the boundary of an m-gon). The square complex P_L is a closed 2-manifold (since the link of each vertex is S^1 and the link of each edge is S^0). It is easily seen to be connected and orientable. Its Euler characteristic is $2^m - m2^{m-1} + m2^{m-2} = 2^m(1-m/4)$. This example appeared already in Coxeter's 1938 paper [68, p. 57]. (My thanks to Alex Suciu for pointing out this reference to me.) If $m = 4$, P_L is a flat 2-torus.

If the 1-skeleton L^1 of L is not a complete graph, then P_L is not simply connected. In general, $\pi_1(P_L)$ depends only on L^1 (cf. [85, Lemma 2.9]).

The Complex Z_L The polyhedral product, $Z_L := (\mathbf{Cone\,E})^L$, defined by (3.4), is a generalization of the previous example P_L of a polyhedral product of intervals.

Example 3.3 Suppose $I = \{1, 2\}$ and that for $i \in I$, E_i is a finite set of m_i points. If L is a 1-simplex σ^1, then the polyhedral product $Z_{\sigma^1} = \mathrm{Cone}\, E_1 \times \mathrm{Cone}\, E_2$ is a square complex. Each square has the form $\mathrm{Cone}\, e_1 \times \mathrm{Cone}\, e_2$, where $e_i \in E_i$, which is a cone on $(\mathrm{Cone}\, e_1 \times e_2) \cup (e_1 \times \mathrm{Cone}\, e_2)$. The square complex Z_{σ^1} also has the structure of cone, namely, it is the cone on

$$Z_{\partial\sigma^1} = (\mathrm{Cone}\, E_1 \times E_2) \cup (E_1 \times \mathrm{Cone}\, E_2).$$

The space $Z_{\partial\sigma^1}$ can be identified with the join $E_1 * E_2$ (this is the complete bipartite graph K_{m_1,m_2}). More generally, suppose σ is an $(n-1)$-simplex with vertex set $I = \{1, \dots, n\}$ and that E_i is a finite set of cardinality m_i. Then Z_σ is the product,

Cone $E_1 \times \cdots \times$Cone E_n, which is a cube complex. Moreover, $Z_{\partial\sigma}$ is homeomorphic to the join, $E_1 * \cdots * E_n$.

The Fundamental Chamber K_L Suppose each E_i is a singleton, necessarily the basepoint. The polyhedral product becomes

$$K_L := (\text{Cone} *_i, *_i)^L. \tag{3.7}$$

Since $(\text{Cone} *_i, *_i) \cong ([0, 1], 1)$, K_L is naturally a cube complex. Since each cube contains the basepoint $* = (*_i)_{i\in I}$, K_L is a cone—in fact it can be identified with the cone on the barycentric subdivision of L. Note that K_L is a subcomplex of Z_L. In particular, it is a subcomplex of P_L. We have that P_L is a subcomplex of $[-1, 1]^I$ and K_L can be identified with the intersection $P_L \cap [0, 1]^I$. So, it is a strict fundamental domain, in the sense defined below, for the $(\mathbf{C}_2)^I$-action on P_L. For this reason, K_L will be called the *fundamental chamber*.

Definition 3.4 As before, given a simplicial complex L, its poset of simplices (including the empty simplex) is denoted $\mathcal{S}(L)$. For each integer $k \geq -1$, let $L^{(k)}$ denote the set of k-simplices in L.

To simplify notation, put $K = K_L$ and for each vertex $i \in I$, define the *mirror* K_i to be the intersection of K with the hyperplane $x_i = 0$. Similarly, for each simplex $\sigma \in \mathcal{S}(L)$, put

$$K_\sigma = \bigcap_{i\in I(\sigma)} K_i = \{x \in K \mid x_i = 0 \text{ for all } i \in I(\sigma)\}. \tag{3.8}$$

The Standard Classifying Space $\mathbb{T}_L$ for a RAAG The cube complex $\mathbb{T}_L$, defined by (3.5), is a polyhedral product of circles. If $L = \Delta$ is an $(n-1)$-simplex, then $\mathbb{T}_\Delta$ is the n-torus, while $\mathbb{T}_{\partial\Delta}$ is the $(n-1)$-skeleton of the n-torus. As we shall see in Proposition 3.27, $\pi_1(\mathbb{T}_L)$ is the right-angled Artin group A_L associated to the 1-skeleton L^1 of L. If L is a flag complex, then $\mathbb{T}_L$ is the *standard classifying space* of A_L. It is also called the *Salvetti complex* of A_L. Thus, $\mathbb{T}_L$ is a subcomplex of $\mathbb{T}^\Delta$ where Δ denotes the full simplex on I. In other words, $\mathbb{T}_L$ is the union of subtori $\mathbb{T}_\sigma$, where $\sigma \in \mathcal{S}(L)$, cf. (3.1). (There is a canonical homomorphism from A_L to the right-angled Coxeter group W_L associated to L^1. We will sometimes use the term *Salvetti complex* of A_L to mean the covering space of $\mathbb{T}_L$ corresponding to $\ker(A_L \to W_L)$. The definition for general Artin groups is given in Sect. 4.3.3.)

The Basic Construction

Definition 3.5 Suppose G is a discrete group acting on a space X. A closed subspace $Y < X$ is a *strict fundamental domain* if it intersects each G-orbit in exactly one point.

The proof of the next lemma is left as an exercise for the reader.

Lemma 3.6 *Suppose* $G \curvearrowright X$ *with strict fundamental domain* Y. *Let* $\pi : X \to X/G$ *be projection to the orbit space and let* $\overline{\pi} = \pi|_Y$. *Then* $\overline{\pi} : Y \to X/G$ *is a homeomorphism.*

One can reconstruct the G-action on X from the group G, the strict fundamental domain Y, and the knowledge of the isotropy subgroups G_y at each point $y \in Y$ (where $G_y = \{g \in G \mid gy = y\}$). Define an equivalence relation $\sim$ on $G \times Y$ by

$$(g, y) \sim (g', y') \iff y = y' \text{ and } g^{-1}g' \in G_y . \tag{3.9}$$

The *basic construction* is the G-space $D(G, Y)$ defined by $D(G, Y) = (G \times Y)/\sim$, with the quotient topology. Let $[g, y]$ denote the equivalence class of (g, y).

Lemma 3.7 (Properties of the Basic Construction)

(i) *The natural map* $y \mapsto [1, y]$ *from* Y *into* $D(G, Y)$ *is an embedding (which we regard as an inclusion).*
(ii) $G \curvearrowright D(G, Y)$ *and* Y *is a strict fundamental domain.*
(iii) *The map* $D(G, Y) \to X$ *defined by* $[g, y] \mapsto gy$ *is a* G*-equivariant homeomorphism.*
(iv) *The map* $r : D(G, Y) \to Y$ *defined by* $[g, y] \mapsto Y$ *is a retraction. (The map* r *is essentially the orbit projection.)*
(v) *The* G*-action on* $D(G, Y)$ *is proper if and only if* Y *is Hausdorff and for each* $y \in Y$, G_y *is a finite subgroup of* G.

Links of Vertices in These Cube Complexes The most salient feature of P_L is that the link of each vertex can be identified with L. We state this as the following lemma.

Lemma 3.8 *For each vertex* v *of* P_L, $\mathrm{Lk}(v, P_L) = L$.

Consider the 1-cube complex, Cone E_i. Let 0_i denote the cone point. There are two types of vertices in Cone E_i – the cone point 0_i is a vertex, as is each element $e \in E_i$. As for links, $\mathrm{Lk}(0_i, \text{Cone } E_i) = E_i$, while for each $e \in E_i$, $\mathrm{Lk}(e, \text{Cone } E_i)$ is a point. Next consider the general case, $Z_L := (\textbf{Cone E})^L$. Any vertex v of Z_L has the form $v = (v_i)_{i \in I}$, where v_i is a vertex of Cone E_i, i.e., either $v_i \in E_i$ or $v_i = 0_i$. Put

$$\tau(v) = \{i \in| I | v_i = 0_i\}. \tag{3.10}$$

So, $\tau(v)$ is a simplex of L (possibly the empty simplex). This proves the following lemma which describes links in Z_L.

Lemma 3.9 *For a vertex v of Z_L let $\tau(v)$ be defined by* (3.10). *The link of v in Z_L is the join*

$$\mathrm{Lk}(v, Z_L) = \mathrm{Lk}(\tau(v), L) * \circledast_{i \in I(\tau(v))} E_i. \tag{3.11}$$

In particular, if each v_i lies in some E_i, then $\mathrm{Lk}(v, Z_L) = L$.

Remark 3.10 If each E_i consists of two points, then $Z_L = P_L$. However, the cubical structure on P_L is coarser than the one we get by identifying it with Z_L. The reason is that if we identify $[-1, 1]$ with $\mathrm{Cone}\{\pm 1\}$, then we should subdivide $[-1, 1]$ into two intervals, $[-1, 0]$ and $[0, 1]$. Thus, when thought of as Z_L, each n-cube in P_L is subdivided into 2^n cubes.

Definition 3.11 (Compare Definition 3.1 and Formula (3.1)**)** Suppose $\{X_i\}_{i \in I}$ is a collection of spaces indexed by the set I of vertices of a simplicial complex L. For any simplex σ of L, let $X(\sigma)$ denote the join $\circledast_{i \in I(\sigma)} X_i$. Analogously to the definition of polyhedral product in (3.1), the *polyhedral join over L* of $\{X_i\}_{i \in I}$ is defined by

$$\circledast_L X_i := \bigcup_{\sigma \in \mathcal{S}(L)} X(\sigma). \tag{3.12}$$

Lemma 3.12 *The link of each vertex v in $\mathbb{T}_L$ is a polyhedral join of* 0*-spheres,* $\mathrm{Lk}(v, \mathbb{T}_L) = \circledast_L S^0$.

In the sequel this polyhedral join of 0-spheres will be called the *octahedralization* of L. (cf. [12] or [105, §8]).

Lemma 3.13 *Suppose L is a flag complex. Then the polyhedral join defined by* (3.12) *of the $\{X_i\}_{i \in I}$ over L is a flag complex if*

- *each X_i is discrete or if*
- *each X_i is a flag complex.*

In particular, the octahedralization of a flag complex is a flag complex.

So, the next result follows from Gromov's Lemma 2.31.

Theorem 3.14 *The cube complexes K_L, P_L, Z_L and $\mathbb{T}_L$ are NPC if and only if L is a flag complex.*

Proof The issue is to show that links of vertices in P_L, Z_L and $\mathbb{T}_L$ are flag complexes if and only if L is a flag complex. So, suppose L is a flag complex. Since the link of each vertex in P_L is L (by Lemma 3.8), this is immediate for P_L. By Lemma 3.9, the link of a vertex v of Z_L has the form $\mathrm{Lk}(\tau(v), L) * \circledast_{i \in I(\tau(v))} E_i$. By part (iii) of Lemma 2.25, $\mathrm{Lk}(\tau(v), L)$ is a flag complex and by part (ii), so is $\circledast_{i \in I(\tau(v))} E_i$. Another application of part (ii) gives that the join, $\mathrm{Lk}(\tau(v), L) * \circledast_{i \in I(\tau(v))} E_i$, is a flag. complex. The complex K_L is the special case of Z_L where each E_i is a singleton. As for $\mathbb{T}_L$, if L is a flag complex, then, by Lemma 3.13, so is $*_L S^0$; hence, by Lemma 3.12, the link of each vertex in $\mathbb{T}_L$ is a flag complex.

Here is another straightforward application of Gromov's Lemma.

Lemma 3.15 *For each $i \in I$ suppose A_i is an NPC cube complex and its basepoint $*_i \in A_i$ is a vertex of A_i. Put $\mathbf{A} = (A_i, *_i)$. Then $\mathbf{A}^L$ is naturally a cube complex. Moreover, it is NPC if and only if L is a flag complex.*

Universal Covers More important for us than the cube complexes P_L, Z_L and $\mathbb{T}_L$ will be their universal covers, denoted by $\tilde{P}_L$, $\tilde{Z}_L$ and $\tilde{\mathbb{T}}_L$, respectively. If L is a flag complex, then by the Cartan–Hadamard Theorem (Theorem 2.3 (i)), these universal covers are CAT(0) cube complexes. The complex $\tilde{P}_L$ is the *Davis–Moussong complex* for the right-angled Coxeter group associated to L, when each E_i has a least two points the complex $\tilde{Z}_L$ is a *right-angled building*, and $\tilde{\mathbb{T}}_L$ is the universal cover of the Salvetti complex for the right-angled Artin group associated to L. In order to describe the groups which act on these complexes we need the notion of a "graph product" of groups, developed in the next subsection.

Right-Angled Coxeter Groups The group $(\mathbf{C}_2)^I$ acts on $[-1, 1]^I$ as a group generated by $\{r_i\}_{i \in I}$, where r_i is the reflection on $[-1, 1]^I$ across the hyperplane $\{x_i = 0\}$. Since the subspace P_L of $[-1, 1]^I$ is stable under the $(\mathbf{C}_2)^I$-action, $(\mathbf{C}_2)^I$ is also a reflection group on P_L. Let W_L denote the group of all lifts of the $(\mathbf{C}_2)^I$-action to the universal cover $\tilde{P}_L$. (In other words, an element of W_L is a lift of an element $g \in (\mathbf{C}_2)^I$ to $\tilde{P}_L$.) This gives an exact sequence

$$1 \to \pi_1(P_L) \to W_L \to (\mathbf{C}_2)^I \to 1. \tag{3.13}$$

We claim that W_L is a Coxeter group. As in (3.7), let $K = K_L$ be the fundamental chamber for $(\mathbf{C}_2)^I$-action on P_L and let $K_i = K_L \cap \{x_i = 0\}$. Choose a component of the inverse image of K in $\tilde{P}_L$ and identify it with K. (We can do this since K is simply connected). Let s_i be the lift of r_i to $\tilde{P}_L$ which fixes K_i. Then s_i is an involution. Moreover, $(s_i s_j)^2 = 1$, whenever $\{i, j\} \in \operatorname{Edge} L^1$. Consider the W_L-action on the cube complex for $\tilde{P}_L$ Since $(\mathbf{C}_2)^I$ acts simply transitively on $\operatorname{Vert}([-1, 1]^I)$, the group W_L is simply transitive on $\operatorname{Vert}(\tilde{P}_L)$. Hence, the 1-skeleton of $\tilde{P}_L$ is a Cayley graph for W_L and its 2-skeleton is a Cayley 2-complex. To see that this gives a presentation of the desired form, consider the vertex $v = (1, 1, \dots, 1) \in P_L$ and let $\tilde{v}$ be a lift to $\tilde{P}_L$. The edges emanating from $\tilde{v}$ have the form $[\tilde{v}, s_i \tilde{v}]$. This shows that $\{s_i\}_{i \in I}$ is a set of generators for W_L. Moreover, for each $\{i, j\} \in \operatorname{Edge} L^1$, there is a square at $\tilde{v}$ with boundary edges labeled $s_i s_j s_i s_j$. It follows that W_L is a right-angled Coxeter group with $\{s_i\}_{i \in I}$ as a fundamental set of generators and with relations as defined in the previous sentence. Such a Coxeter group is said to be "right-angled." In the next subsection we generalize this construction to graph products of groups and to RABs.

3.1.2 Graph Products of Groups

Next we want to explain the close connection between five topics:

- graph products of groups,
- the fundamental group of a polyhedral product,
- polyhedral products of pairs of spaces of the form (Cone G_i, G_i) where G_i is a discrete group,
- polyhedral products of classifying spaces $(BG_i, *_i)$ and the classifying space of the graph product,
- right-angled buildings (or RABs) with chamber-transitive automorphism groups.

Suppose $\{G_i\}_{i\in I}$ is a collection of groups indexed by the vertex set I of a simplicial graph L^1.

Definition 3.16 The *graph product*, denoted $\prod_{L^1} G_i$, is the quotient of the free product of the G_i by the normal subgroup generated by all commutators of the form $[g_i, g_j]$, with $\{i, j\} \in L^{(1)}$, where $L^{(1)}$ (= Edge L^1) is the set edges of L^1, $g_i \in G_i$ and $g_j \in G_j$. (Such groups were first studied by Droms [109].)

As examples, if $L^{(1)} = \emptyset$, the graph product is the free product, while if L^1 is a complete graph, then the graph product is the direct sum: $\prod_{L^1} G_i = \sum_{i\in I} G_i$. (Here, the *direct sum*, $\sum_{i\in I} G_i$, means the subset of the direct product where only finitely many coordinates are not equal to the identity element of G_i. Often we omit the subscript on $\sum$ and write simply $\sum G_i$.) There is a canonical epimorphism $p : \prod_{L^1} G_i \to \sum G_i$.

As in [207, p. 1], a *system of groups* simply means a family of groups $\{G_a\}_{a\in A}$ indexed by some set A together with a subset F_{ab} of $\mathrm{Hom}(G_a, G_b)$ for each pair $(a, b) \in A \times A$. Given a system of groups, one can then define the notion of its *direct limit* G as in [207, p. 1]: it has the universal property that given any collection of homomorphisms $\{\theta_a : G_a \to H\}$ to a group H that are compatible with the bonding homomorphisms F_{ab}, there is unique homomorphism $G \to H$ making the obvious diagrams commute. (See Appendix A.1.)

For example, if $\{G_i\}_{i\in I}$ is a collection of groups indexed by the vertex set I of a simplicial graph L^1, then we get a system of groups

$$G_i,\ i \in I, \quad G_i \times G_j,\ \{i, j\} \in L^{(1)},$$

indexed by the cells of L^1, where the bonding homomorphisms are the natural inclusions $G_i \hookrightarrow G_i \times G_j$ defined whenever $\{i, j\} \in L^{(1)}$. The graph product $\prod_{L^1} G_i$ then can be described as the direct limit of this system of groups. (N.B.: This is *not* a graph of groups.)

Let L be the flag complex determined by L^1. There is a larger system of groups indexed by the set $\mathcal{S}(L)$ of simplices in L including the empty simplex. For any k, let $L^{(k)}$ denote the set of k-simplices in L. For $\sigma \in \mathcal{S}(L)$, the group G_σ is defined to be the direct sum $\Sigma_{i\in I(\sigma)} G_i$; if τ is a face of σ, we have a natural inclusion

$G_\tau \hookrightarrow G_\sigma$. For $\mathbf{G} = (G_i)_{i\in I}$, this defines a system of groups,

$$\mathcal{G}_L(\mathbf{G}) := \{G_\sigma\}_{\sigma\in\mathcal{S}(L)}, \qquad \{G_\tau \hookrightarrow G_\sigma\}_{\tau<\sigma} \tag{3.14}$$

called the *graph product complex* in Example 3.19 below. The direct limit of this system is the same graph product as before. As explained in Appendix A, $\mathcal{G}_L(\mathbf{G})$ is a nice example of a "simple complex of groups," as defined in Appendix A.1. (Later when using this simple complex of groups, we shall index the groups by the opposite poset $\mathcal{S}(L)^{\mathrm{op}}$.) There are two cube complexes associated to this simple complex of groups. First we have the polyhedral product $Z_L = (\mathbf{Cone\,G}, \mathbf{G})^L$ defined as a subset of the product $(\mathbf{Cone\,G}, \mathbf{G})^\Delta$ as in (3.4) (with $\mathbf{E} = \mathbf{G}$). The space Z_L is a union of pieces of the form $(\mathrm{Cone}\,G_i, G_i)^\sigma := \prod_{i\in I(\sigma)} \mathrm{Cone}\,G_i$. Since it is a product of cones over discrete sets, $(\mathrm{Cone}\,G_i, G_i)^\sigma$ is a cubical complex of dimension equal to $\mathrm{Card}(I(\sigma))$. There is one copy of $(\mathrm{Cone}\,G_i, G_i)^\sigma$ for each coset of G_σ in $\sum G_i$. Each copy of $(\mathrm{Cone}\,G_i, G_i)^\sigma$ contains one top-dimensional cube for each element of G_σ.

Thus, Z_L is a union of cubes, one for each element of $(\Sigma G_i \times \mathcal{S}(L)$; the dimension of the cube corresponding to (g, σ) is $\dim\sigma + 1$. The group $\sum G_i$ acts on Z_L having as strict fundamental domain the chamber K_L that was defined by (3.2). Hence,

$$Z_L = D(\Sigma G_i, K_L) = (\Sigma G_i, K_L)/\sim. \tag{3.15}$$

Here, ΣG_i means the direct sum of the G_i and the equivalence relation $\sim$ is defined by (3.9). Consider the universal cover $\tilde{Z}_L$ of Z_L. Let $\Gamma = \prod_{L^1} G_i$ denote the graph product.

Example 3.17 (Right-Angled Coxeter Groups) If each G_i is $\mathbf{C}_2$, the cyclic group of order two, then the graph product is the *right-angled Coxeter group* associated to L^1.

Example 3.18 (Right-Angled Artin Groups) If each G_i is the infinite cyclic group, then $\prod_{L^1}\mathbb{Z}$ is the *right-angled Artin group* associated to L^1. (For an introduction to RAAGs, see [54].)

The terms "right-angled Coxeter group" and "right-angled Artin group" will often be abbreviated by RACG and RAAG, respectively.

Relative Graph Products Let $G = \sum_{i\in I} G_i$ denote the direct sum. Suppose each G_i acts on a set E_i. Then $G \curvearrowright (\mathbf{Cone}(\mathbf{E}))^\Delta$, where Δ means the full simplex on I so that $(\mathbf{Cone}(\mathbf{E}))^\Delta$ is the product of the E_i. (A "simplex" of Δ means the span of a finite subset of I even when I is not assumed to be finite.) The polyhedral product $Z_L = (\mathbf{Cone\,E})^L$ is a subspace of $(\mathbf{Cone}(\mathbf{E}))^\Delta$ that is stable under the G-action. Choose basepoints $*_i \in E_i$ so that the I-tuple $* = (*_i)_{i\in I}$ is a basepoint for Z_L. Suppose we are given a G_i-set E_i for each $i \in I$. Denote the I-tuple $(G_i, E_i)_{i\in I}$ by $(\mathbf{G}, \mathbf{E})$ and put $(\mathbf{Cone}(\mathbf{E}), \mathbf{E}) = (\mathrm{Cone}(E_i), E_i, *_i)_{i\in I}$, where $*_i \in E_i$ is a basepoint. Let L be a simplicial complex with L^1 as its 1-skeleton. Then

$G \curvearrowright (\mathbf{Cone\,E})^{\Delta}$, where Δ means the full simplex on I. The polyhedral product $Z_L = (\mathbf{Cone\,E})^L$ is a subspace of $(\mathbf{Cone\,E})^{\Delta}$ that is stable under the G-action. To simplify the discussion, suppose G_i acts effectively on each orbit in E_i with B_i the isotropy subgroup at $*_i$. (This means that if G_i/B_i is such an orbit, then the subgroup B_i is malnormal.) Then G acts effectively on the polyhedral product $Z_L := (\mathbf{Cone(E)}, \mathbf{E}))^L$. Let $\tilde{Z}_L$ denote the universal cover and let Γ_L denote the group of all lifts of the G-action to $\tilde{Z}_L$. The group Γ_L is called the *relative graph product* or "generalized graph product" of the E_i. (N.B. Γ_L depends only on the 1-skeleton L^1.) We have a short exact sequence (cf. (3.13)):

$$1 \to \pi_1(Z_L) \to \Gamma_L \to G \to 1\,. \tag{3.16}$$

(If G_i does not act effectively on E_i but has kernel N_i, then in (3.16) we should replace G by G/N where N is the product of the N_i. The definition of Γ_L should then be modified by an appropriate extension by N. See [85, §2.3] for details.) We can also use the notation $\prod_{L^1}(G_i, E_i)$ for Γ_L. The notion of a relative graph product goes back to Januszkiewicz–Świątkowski [158] and Haglund [140]. Further information can be found in [85, 98].

Example 3.19 (The Graph Product Complex) Suppose G_i is transitive on E_i so that $E_i = G_i/B_i$, for $i \in I$. Also suppose B_i is malnormal in G_i. For each $\sigma \in L$, let $I(\sigma)$ denote the vertex set of σ. Put $\mathbf{G} := (G_i, G_i/B_i)_{i\in I}$. As in (3.14) we get a simple complex of groups $\mathcal{G}_L(\mathbf{G}) = \{G_\sigma\}_{\sigma\in\mathcal{S}(L)}\}$ over the poset $\mathcal{S}(L)^{\mathrm{op}}$, where $G_\sigma = \sum_{i\in I} J_i(\sigma)$ is the direct sum of groups $J_i(\sigma)$ with

$$J_i(\sigma) = \begin{cases} G_i, & \text{if } i \in I(\sigma), \\ B_i, & \text{if } i \notin I(\sigma). \end{cases}$$

For $\sigma = \emptyset$, we have $G_\emptyset = \sum B_i$. Whenever $\tau < \sigma$, there is a natural inclusion $G_\tau \to G_\sigma$. the direct limit of system of groups $\{G_\sigma\}_{\sigma(L)}$ coincides with the relative graph product Γ_L, defined in Lemma 3.21 below. In fact, it is only necessary to use the simplices σ with $\mathrm{Card}(I(\sigma)) \le 2$ to get this direct limit. As we shall see below in Sect. 4.4.3, if each B_i is a proper subgroup of G_i, then the universal cover $\tilde{Z}_L$ is a regular "right-angled building" (abbreviated RAB).

Example 3.20 Here is a particular case of the previous example. Suppose L^1 is a graph with vertex set $\{1, 2\}$ and with no edges. The simple complex of groups $\mathcal{G}_L$ gives the data for an interval of groups. The vertex groups are $G_1 \times B_2$ and $B_1 \times G_2$ while the edge group is $B_1 \times B_2$. The polyhedral product Z_L is the join $G_1/B_1 * G_2/B_2$, which is a complete bipartite graph. (Note that this is a subset of $\mathrm{Cone}(G_1/B_1) \times \mathrm{Cone}(G_2/B_2)$.) It follows that the relative graph product Γ_L is an extension of $G_1 \times G_2$ by a free group (the fundamental group of the join).

Lemma 3.21 *With notation as above, let $G = \sum G_i$ be the direct sum and Γ_L be relative the graph product defined above. Let $Z_L = (\mathbf{Cone\,E}, \mathbf{E})^L$ be the polyhedral product and let $\tilde{Z}_L$ be its universal cover. For each $i \in I$, suppose the G_i-action on E_i is transitive. Then the relative graph product Γ_L is the group of all lifts of the G-action to $\tilde{Z}_L$. Hence, there is a short exact sequence,*

$$1 \to \pi_1(Z_L) \to \Gamma_L \to G \to 1 \,.$$

Remark 3.22 The ordinary graph product $\prod_{L^1} G_i$ fits within the context of this lemma. Indeed, form the polyhedral product Z_L, where $(A_i, B_i, *_i) = (\mathrm{Cone}(G_i), G_i, *_i)$. The lemma asserts that the ordinary graph product is the group of lifts of the G-action on Z_L to the universal cover $\tilde{Z}_L$.

Remark 3.23 It can be shown as in Example 3.20 that $\pi_1(Z_{L^1})$ is a free group generated by missing edges $\{i, j\}$. Moreover, since $\pi_1(Z_L) \cong \pi_1(Z_{L^1})$, we see that $\pi_1(Z_L)$ is a free group. So, Lemma 3.21means that the relative graph product is an extension of G by a free group.

Proof of Lemma 3.21 Let H be the group of all lifts of the G-action to $\tilde{Z}_L$. We will show that H is equal to the graph product, Γ_L. Let $\varphi : H \to G$ be the canonical projection. Let c_i be the cone point of Cone E_i and let $\tilde{c}_i$ be a lift of it to $\tilde{Z}_L$. The subgroup H_i consisting of all lifts which fix the point $\tilde{c}_i$ projects isomorphically onto G_i. If $\{i, j\} \in L^{(1)}$, then Cone $E_i \times$ Cone E_j is a subcomplex of Z_L and each component of the inverse image of this subcomplex in $\tilde{Z}_L$ projects isomorphically to it. One such component contains the basepoint $\tilde{*}$. This component contains the square $[\tilde{*}_i, \tilde{c}_i] \times [\tilde{*}_j, \tilde{c}_j]$ as a subcomplex. The subgroup $H_i \times H_j$ fixes the point $\tilde{c}_i \times \tilde{c}_j$ and projects isomorphically to $G_i \times G_j$. So, $H_i \times H_j$ is the group of all lifts of elements of $G_i \times G_j$ that fix $\tilde{c}_i \times \tilde{c}_j$. So, whenever $\{i, j\}$ is an edge of L^1, the subgroups H_i and H_j commute. If $\{i, j\}$ is not an edge of L^1, then $\partial(\mathrm{Cone}\, E_i \times \mathrm{Cone}\, E_j) = E_i * E_j$ is a subcomplex of Z_L and by Example 3.20, a component of its inverse image in $\tilde{Z}_L$ has fundamental group

$$(G_i \times B_j) *_{B_i \times B_j} (B_i \times G_j)$$

So, the isotropy subgroups corresponding to the edges $\{i, j\}$ of L^1 are local groups of the simple complex of groups in Example 3.19. The relative graph product is the direct limit of this simple complex of groups over $\mathcal{S}(L^1)^{\mathrm{op}}$. It follows from this and the universal property of direct limits (cf. Appendix A.1), that the natural inclusions, $G_i \cong H_i \hookrightarrow H$ and extend to a homomorphism $\phi : \Gamma_L \to H$ from the relative graph product to the group of lifts. Since G_i is transitive on E_i, it follows that H is transitive on the inverse image of E_i in Z_L. A straightforward argument then implies that $\phi : \Gamma_L \to H$ is an isomorphism.

Assume for the remainder of this section that, as in the lemma, $G_i = E_i$. Then Γ acts on $\tilde{Z}_L$ with strict fundamental domain K $(= K_L)$. Hence, $\tilde{Z}_L$ can be defined by the basic construction. We state this as the following theorem.

Theorem 3.24 *The space $\tilde{Z}_L$ is given by the basic construction:*

$$\tilde{Z}_L = D(\Gamma_L, K) = (\Gamma_L \times K)/\sim,$$

where the equivalence relation $\sim$ is defined by (3.9).

Right-Angled Buildings As is explained in Sects. 4.4.1 and 4.4.2, there are certain combinatorial objects called "buildings." Such buildings have geometric realizations as polyhedra. The polyhedron $\tilde{Z}_L$ is the "standard realization" of a right-angled building (or "RAB"). (By definition, in order for $\tilde{Z}_L$ to be a building, each E_i must have at least two elements.) There is projection $p : \tilde{Z}_L \to K_L$. The space $\tilde{Z}_L$ decomposes into chambers each of which projects homeomorphically onto K_L. The inverse image of the cone point $\mathbf{0} := (\mathbf{0_i})_{\mathbf{i}\in\mathbf{I}}$ is the set of "center points" of the chambers. The relative graph product Γ_L acts simply transitively on the set of centers; hence, Γ_L acts transitively on the set of chambers. A collection of center points of chambers forms an "$\{i\}$-residue" of $\tilde{Z}_L$ if they are the set of extremal vertices of a copy of Cone E_i. (In other words, two chambers belong to the same $\{i\}$-residue, if they belong to the same coset in Γ_L/G_i.)

If for each i, $\mathrm{Card}(E_i) = 2$, then $\tilde{Z}_L$ is the Davis–Moussong complex and the associated graph product is the RACG, W_L. If each G_i is infinite cyclic, then the graph product is the RAAG, A_L, and the polyhedron $\tilde{Z}_L$ is the standard realization of the RAB associated to this RAAG.

Example 3.25 Let L be a finite dimensional flag complex with vertex set I. For each $i \in I$, suppose G_i is a nontrivial finite group. Let $Z_L = \{(\mathrm{Cone}\, G_i, *_i)\}^L$ be the polyhedral product and let $\tilde{Z}_L$ be the associated RAB. Let Γ be the graph product. Then Γ acts properly and each finite subgroup of Γ projects isomorphically onto a finite subgroup of $G = \sum_{i\in I} G_i$. It follows that $\tilde{Z}_L$ is the universal proper Γ-space $\underline{E}\Gamma$. (See Definition 7.8.)

The Fundamental Group of a Polyhedral Product As in Definition 3.1, $\mathbf{A} = \{(A_i, B_i, *_i)\}_{i\in I}$ and $\mathbf{A}^L$ denotes the polyhedral product. Each A_i is assumed path connected. Put $G_i := \pi_1(A_i, *_i)$ so that $E_i := \pi_1(A_i, B_i, *_i)$ is a G_i-set. Call $(\mathbf{G}, \mathbf{E}) := (G_i, E_i)_{i\in I}$ the *fundamental group data* for $\mathbf{A}$. The next result shows that the fundamental group of a polyhedral product is equal to the relative graph product of its fundamental group data.

Proposition 3.26 (cf. [85, Theorem 2.18, p. 248]) *Let $\Gamma_L = \prod_{L^1}(G_i, E_i)$ denote the relative graph product defined by* (3.16). *Then*

$$\pi_1(\mathbf{A}^L) = \Gamma_L\,.$$

Sketch of Proof Let $\tilde{A}_i$ be the universal cover of A_i, let $\tilde{B}_i$ be the inverse image of B_i in $\tilde{A}_i$ and $(\tilde{\mathbf{A}}, \tilde{\mathbf{B}}) = (\tilde{A}, \tilde{B})_{i \in I}$. The polyhedral product $Z_L(\tilde{\mathbf{A}}, \tilde{\mathbf{B}})$ is a covering space of $Z_L(\mathbf{A}, \mathbf{B})$. Since $\tilde{A}_i$ is simply connected and the set of components of $\tilde{B}_i$ is E_i, the natural map $Z_L(\tilde{\mathbf{A}}, \tilde{\mathbf{B}}) \to Z_L(\mathbf{Cone}\, E, E)$ induces an isomorphism on fundamental groups. Hence, the universal cover of $Z_L(\tilde{\mathbf{A}}, \tilde{\mathbf{B}})$ is induced from the universal cover of $Z_L(\mathbf{Cone}\, E, \mathbf{E})$. Since $\pi_1(\mathbf{A}^L)$ and Γ_L are both lifts of appropriate G-actions to these universal covers, the proposition follows.

Polyhedral Products of Classifying Spaces Next consider a polyhedral product with respect to L, where $(A_i, B_i) = (BG_i, BH_i)$, where BG_i denotes the classifying space of a discrete group G_i and H_i is a proper subgroup of G_i. In Appendix A.4.2 we will prove the following.

Proposition 3.27 *Suppose L is a flag complex and that $(\mathbf{BG}, \mathbf{BH}) = \{(BG_i, BH_i, *_i)\}_{i \in I}$. Then the polyhedral product $(\mathbf{BG}, \mathbf{BH})^L$ is the classifying space for the relative graph product $\Gamma_L = \prod_{L^1}(G_i, H_i)$, i.e.,*

$$B\Gamma_L = (\mathbf{BG}, \mathbf{BH})^L .$$

In particular, if each H_i is the trivial subgroup, then the classifying space of the ordinary graph product is the polyhedral product of the BG_i.

Example 3.28 (Standard Classifying Spaces for RAAGs**, cf. Example 3.18)** A basic example to keep in mind for the above proposition is the case where each $G_i = \mathbb{Z}$ is infinite cyclic. The graph product Γ is the RAAG, A_L. Each $BG_i = S^1$. For each simplex $\sigma \in \mathcal{S}(L)$, the polyhedral product $(S^1, *_i)^\sigma$ is a torus with basis $\operatorname{Vert}\sigma$. The proposition identifies the classifying space $B\Gamma$ with the polyhedral product $\mathbb{T}_L := (S^1)^L$ which is a subcomplex of the full torus on I $(= \operatorname{Vert} L)$. An alternative proof that $BA_L = \mathbb{T}_L$ is provided by Theorem 3.14, which states that $\mathbb{T}_L$ is NPC. So, $\mathbb{T}_L$ is aspherical and therefore, is equal to the classifying space of its fundamental group.

Proof of Proposition 3.27 For simplicity of notation, we prove only the absolute case where each H_i is trivial. Let $p_i : EG_i \to BG_i$ be the universal cover. The induced map of polyhedral products, $(EG_i, p^{-1}(*_i))^L \to (BG_i, *_i)^L$, is then a covering projection. Since EG_i is contractible, $(EG_i, p_i^{-1}(*_i))$ and $(\operatorname{Cone} G_i, G_i)$ are homotopy equivalent pairs; so, the polyhedral products $Z_L = (\operatorname{Cone} G_i, G_i)^L$ and $Y_L = (EG_i, p_i^{-1}(*_i))^L$ also are homotopy equivalent. Hence, their universal covers $\tilde{Z}_L$ and $\tilde{Y}_L$ are homotopy equivalent. Moreover, by Lemma 3.21, the group of lifts of the $(\sum_{i \in I} G_i)$-action on Y_L to $\tilde{Y}_L$ is identified with the graph product Γ. So, the composition $\tilde{Y}_L \to Y_L \to (BG_i, *_i)^L$ is the universal cover and Γ is its group of deck transformations. Since L is a flag complex, it follows from Theorem 3.14 that Z_L is NPC. By the Cartan–Hadamard Theorem, $\tilde{Z}_L$ is contractible. Alternatively, it is proved in [83] that $\tilde{Z}_L$ is the standard realization of a right-angled building and hence, is contractible (since the standard realization of any building is CAT(0), cf. Theorem 4.101 or [77]). Since $\tilde{Y}_L$ is homotopy equivalent to $\tilde{Z}_L$, it also is contractible. Hence, $B\Gamma \sim \{(BG_i, *_i)\}^L$.

3.1.3 Wreath-Graph Products

Here we restrict to the case where each factor in the direct sum $\sum_{i\in I} G_i$ is the same group G. Let J be a group of permutations of the index set I. Then J acts on the direct sum $\sum_{i\in I} G$ by permuting the factors. The (unrestricted) *wreath product* $G \int J$ means the semidirect product $\sum G \rtimes J$. The notion of "restricted" wreath product (which we shall never use in this book) means that the direct sum is replaced by the direct product. Suppose that L^1 is a simplicial graph on vertex set I, that J is a group of automorphisms of L^1 and that $\Gamma = \prod_{L^1} G$ is the graph product. The J-action on L^1 naturally induces a J-action on Γ via group automorphisms. The *wreath-graph product* of G and J is the semidirect product $\Gamma \rtimes J$. (This terminology was introduced in [167].) The group $\Gamma \rtimes J$ naturally acts on

(1) the universal cover $\tilde{Z}_L$ of the polyhedral product $Z_L := (\mathbf{Cone\,G}, \mathbf{G})^L$, and on
(2) the universal cover $E\Gamma$ of $B\Gamma$.

To understand the $(\Gamma \rtimes J)$-action in case (1), first note that the ordinary wreath product $\sum G \rtimes J$ acts on $(\mathbf{Cone\,G}, \mathbf{G})^{\Delta}$, where Δ means the simplex on I. The subspace $(\mathbf{Cone\,G}, \mathbf{G})^L$ is stable under $\sum G \rtimes J$. As in Lemma 3.21, the group of lifts of this action to $\tilde{Z}^L$ is the wreath-graph product $\Gamma \rtimes J$. (To see this, it is sufficient to note that since J fixes the basepoint of $(\text{Cone}\,\mathbf{G}, \mathbf{G})^L$, the group of lifts of J at a lifted basepoint in $\tilde{Z}_L$ projects isomorphically to J.) A case of primary interest in (1) is when L is a finite flag complex and J is a finite group of automorphisms of L. For example, when $G = \mathbb{Z}$, then Γ is the RAAG, A_L, and we are dealing with the wreath-graph product $A_L \rtimes J$. Such wreath-graph products are used by Leary–Nucinkis in their examples of finite extensions of Bestvina–Brady groups with exotic finiteness properties. (See Theorem 7.19 and Corollary 7.20 in Sect. 7.4.2.)

When both G and J are finite groups, (1) yields the following.

Proposition 3.29 *Suppose G and J are both finite groups. Then $\tilde{Z}_L$ is a model for $\underline{E}(\Gamma \rtimes J)$, where $\underline{E}H$ denotes the classifying space for proper actions of the group H. If L is a finite flag complex, then $\Gamma \rtimes J$ acts cocompactly on $\underline{E}(\Gamma \rtimes J)$.*

Proof Since G is finite, $\sum G \rtimes J$ acts properly on Z_L; hence, $\Gamma \rtimes J$ acts properly on $\tilde{Z}_L$. The point is that $\Gamma \rtimes J$ acts by isometries on the CAT(0) cube complex $\tilde{Z}_L$. So, the fixed point set of any finite subgroup of $\Gamma \rtimes J$ is CAT(0) and hence, is contractible.

In case (2), by using the construction in the proof of Proposition 3.27, we see that $\Gamma \rtimes J$ acts on $E\Gamma$ as follows: the covering space $(EG, p^{-1}(*))^L \to (BG, *)^L$ corresponds to the homomorphism $\pi_1((BG, *)^L) = \Gamma \to \sum G$; the ordinary wreath product $\sum G \rtimes J$ acts on $(EG, p^{-1}(*))^L$; and $\Gamma \rtimes J$ is the group of lifts of $\sum G \rtimes J$ to the universal cover $E\Gamma$. We are primarily interested in case (2) when J is finite and BG is finite-dimensional (so that G is torsion-free).

Proposition 3.30 *Suppose J is a finite group of automorphisms of L and that BG is finite dimensional. Then the action of $\Gamma \rtimes J$ on the universal cover of $(BG, *)^L$ is a model for $\underline{E}(\Gamma \rtimes J)$. If BG is a finite complex, then the $(\Gamma \rtimes J)$-action is cocompact on $E\Gamma$.*

When $\Gamma = A_L$, Proposition 3.30 is a result of Crisp [69] (cf. Theorem 7.18 in Sect. 7.4.2).

Proof By Proposition 3.27, $E\Gamma$ is the universal cover of $(BG, *)^L$. Since J is finite and G is torsion-free, we have that $\Gamma \rtimes J$ is virtually torsion-free and its action on $E\Gamma$ is proper. By Proposition 3.29, $\tilde{Z}_L = \underline{E}(\Gamma \rtimes J)$ and there is a $(\Gamma \rtimes J)$-equivariant map $\varphi : E\Gamma \to \tilde{Z}_L$ that classifies the proper $(\Gamma \rtimes J)$-action on $E\Gamma$. Since $E\Gamma$ and $\tilde{Z}_L$ are both contractible, φ is a homotopy equivalence. Hence, for any finite subgroup $J' \leq J$, φ restricts to a homotopy equivalence $(E\Gamma)^{J'} \to (\tilde{Z}_L)^{J'}$. So, $(E\Gamma)^{J'}$ is also contractible and therefore, $E\Gamma$ with its $(\Gamma \rtimes J)$-action is a model for $\underline{E}(\Gamma \rtimes J)$.

Remark 3.31 We are also interested in wreath-graph products when the flag complex $\tilde{L}$ is a covering space of a finite flag complex L with J a (possibly infinite) group of covering transformations. When $\Gamma = W_{\tilde{L}}$ the wreath-graph product $\Gamma \rtimes J$ is relevant to the "Reflection Group Trick" (discussed in Sect. 3.2.6 below).

Replacing the Fundamental Chamber by an Aspherical Complex Here we develop an idea which will be used in connection with the Reflection Group Trick. According to Theorem 3.24, $Z_L = D(\sum G, K)$, where $\sum G$ denotes the weak direct product of copies of G. Recall that $K = |\mathcal{S}(L)|$ is the cone on the barycentric subdivision of L and that K is stratified by the dual cones, K_σ $(= |\mathcal{S}(L)_{>\sigma}|)$. Let ∂K denote the union of the K_σ, for $\sigma \in \mathcal{S}(L)_{>\emptyset}$, so that ∂K is homeomorphic to the barycentric subdivision of L. The idea in this paragraph is to replace the fundamental chamber K by an aspherical space X which contains ∂K as a subspace. Instead of taking the basic construction $D(\sum G, K)$ with fundamental chamber K, we replace each copy of K by a copy of X. In Theorem 3.32 below we will show that if $D(\sum G, K)$ is aspherical, then so is $D(\sum G, X)$. (In our actual use of this construction in Sect. 3.2.6, X will be a compact aspherical manifold with $\partial X = \partial K$ and the graph product Γ will be a RACG.)

Suppose we are given a map $f : \partial K \to B\pi$, where $B\pi$ is an aspherical CW complex with fundamental group π. Let X denote the mapping cylinder of f, i.e., X is $\partial K \times [0, 1]$ with $\partial K \times 1$ glued to $B\pi$ via f. Then X is a stratified space; its strata are indexed by $\mathcal{S}(L)^{\mathrm{op}}$ as follows: a stratum that is contained in $\partial K \times 0$ coincides with a stratum of X indexed by the same element $\sigma \in \mathcal{S}(L)_{>\emptyset}$, while the pure stratum corresponding to $\emptyset$ is the subset $((0,1] \times \partial K) \cup_f B\pi$ of X. If we replace the fundamental chamber K by X, then the space $\tilde{Z}_L$ is replaced by $D(\Gamma, X)$. By Lemma 3.7 (iv), the orbit projection $D(\Gamma, X) \to X$ is a retraction. Let Γ' be the kernel of $\Gamma \to \sum G$. If G is finite, then Γ' is a torsion-free subgroup of finite index in Γ. In Theorem 3.32 below we use a variation of the wreath-graph product construction to show that $D(\sum G, X)$ $(= D(\Gamma, X)/\Gamma')$ is aspherical. Let $p : \tilde{X} \to X$ be the projection map of the universal cover. Since $L = \partial K \times 0$, we can regard it as a subspace of X. Put $\tilde{L} = p^{-1}(L)$. Then $p : \tilde{L} \to L$ is a covering space (not necessarily connected). The simplicial structure on L lifts to a simplicial structure on $\tilde{L}$ and makes $\tilde{L}$ into a flag complex. Let $\tilde{\Gamma} = \prod_{\tilde{L}^1} G$ be the corresponding graph product. The group π acts on $\tilde{X}$ via deck transformations; hence, $\pi \curvearrowright \tilde{L}$. So, π acts on $\tilde{\Gamma}$ by automorphisms and we can form the wreath-graph product $\tilde{\Gamma} \rtimes \pi$. Let $\tilde{\Gamma}'$ be the kernel of the natural epimorphism $\tilde{\Gamma} \to \sum_I G$. Still supposing G is finite, we see that $\tilde{\Gamma}'$ is torsion-free subgroup of finite index in $\tilde{\Gamma}$ and that the same is true for the subgroup $\tilde{\Gamma}' \rtimes \pi$ of the wreath-graph product $\tilde{\Gamma} \rtimes \pi$. (In contrast to Proposition 3.30, note that the graph product $\tilde{\Gamma}$ has infinitely many factors and that the group π of deck transformations of $\tilde{L}$ is an infinite group of permutations of the index set.)

Theorem 3.32 (cf. [82, p. 169]) *Suppose G is finite. Then*

(i) *The space $D(\sum G, X)$ is aspherical and it retracts onto X $(= B\pi)$.*
(ii) *The wreath-graph product $\tilde{\Gamma} \rtimes \pi$ $(= \prod_{\tilde{L}^1} G \rtimes \pi)$ acts properly on the universal cover $\tilde{D}$ of $D(\sum G, X)$. The fundamental group of $D(\sum G, X)$ is $\tilde{\Gamma}' \rtimes \pi$, which is a subgroup of finite index in the wreath-graph product. If π is type F, then so is $\tilde{\Gamma}' \rtimes \pi$. (Recall that π is* type F *if $B\pi$ has the homotopy type of a finite complex.)*

Our principal interest in this theorem lies with the Reflection Group Trick which will be discussed in Sect. 3.2.6 below. In this special case, G is $\mathbf{C}_2$, X is a compact manifold with boundary, and $D(\sum G, X)$ can then be seen to be a closed aspherical manifold.

Proof of Theorem 3.32 Since $\tilde{X}$ is contractible, $D(\tilde{\Gamma}, \tilde{X})$ is homotopy equivalent to the RAB with fundamental chamber Cone$(\tilde{L})$; hence, $D(\tilde{\Gamma}, \tilde{X})$ is contractible. It is straightforward to check that there is a natural action of $\tilde{\Gamma} \rtimes \pi$ on $D(\tilde{\Gamma}, \tilde{X}) = (\tilde{\Gamma} \times \tilde{X})/\sim$. (See [82, p. 169].) If $g = (\gamma, \alpha) \in \tilde{\Gamma} \rtimes \pi$ and $[\gamma', x] \in (\tilde{\Gamma} \times X)/\sim$, then the action is defined by $g \cdot [\gamma', x] = [\gamma\alpha(\gamma'), \gamma' x]$. Similarly, it is straightforward that the natural projection $(\tilde{\Gamma} \times \tilde{X})/\sim \longrightarrow (\sum G \times X)/\sim$ is a covering projection. Since $D(\tilde{\Gamma}, \tilde{X})$ is simply connected, it is the universal cover of $D(\sum G, X)$ and since it is contractible, $D(\sum G, X)$ is aspherical.

3.2 Right-Angled Reflection Groups on Manifolds

3.2.1 Some Basic Notions Concerning Manifolds

Spheres, PL-spheres, and Generalized Homology Spheres Let L be an $(n-1)$-dimensional simplicial complex. Then L is a PL-sphere if it is PL-homeomorphic to S^{n-1}. Equivalently, L is a PL sphere if it has a subdivision isomorphic to some subdivision of the boundary complex of an n-simplex. If L is a PL-sphere, then the link of any simplex in L is also a PL-sphere. To say that L is a PL-sphere is nearly the same as saying that for any k-simplex $\sigma \in \mathcal{S}(L)$, $\mathrm{Lk}(\sigma, L)$ is homeomorphic to S^{n-k-2} (or even homotopy equivalent to S^{n-k-2}). The only discrepancy is that the PL version of the 4-dimensional Poincaré Conjecture is not known to be true; it is possible that there exists a triangulation of S^4 which is not PL homeomorphic to S^4. So, let us say that L is *nearly a PL-sphere* if the weaker condition holds that $\mathrm{Lk}(\sigma, L)$ be homeomorphic to S^{n-k-2} for each $\sigma \in \mathcal{S}(L)$ (including $\sigma = \emptyset$). A simplicial complex M is a *homology* $(n-1)$*-manifold*, if the link of each k-simplex σ, with $k \geq 0$, has the same homology as does S^{n-k-2}. The simplicial complex L^{n-1} is a *generalized homology sphere*, or a GHS^{n-1} for short, if it is a closed homology $(n-1)$-manifold that has the same homology as S^{n-1}.

Suppose a simplicial complex M is a homology manifold. Each simplex σ in M has a *dual cone*, D_σ, defined as the cone on the barycentric subdivision of $\mathrm{Lk}(\sigma, M)$. Thus, each dual cone (to a nonempty simplex) is the cone on a generalized homology sphere. If each such link is a nearly a PL-sphere (resp., a PL-sphere), then each dual cone is a cell (resp., a PL-cell), called the *dual cell*.

A polyhedral homology manifold that is not a PL manifold can still be a topological manifold. Examples are provided by the Double Suspension Theorem of Cannon and Edwards. This states that the double suspension of a homology sphere is a topological manifold. The definitive result is the following.

Theorem 3.33 (Edwards' Polyhedral Homology Manifold Characterization Theorem) *A polyhedral homology manifold of dimension* ≥ 3 *is a topological manifold if and only if the link of each vertex is simply connected.*

This is due to Edwards [111] in dimensions ≥ 5. When the dimension is ≤ 3, every polyhedral homology manifold is a PL-manifold. The link of any vertex in a 4-dimensional polyhedral homology manifold is a homology 3-sphere and if such a link is required to be simply connected, then it must be PL-homeomorphic to S^3 (by the 3-dimensional Poincaré Conjecture). Hence, the theorem is also true in dimension 4.

Resolutions In the following subsections in this chapter we shall have occasion to regard the dual of a cone on a GHS^{n-1} as being analogous to a simple n-polytope. First suppose that L is a triangulation of an $(n-1)$-sphere. If L is a PL-sphere, then for each simplex $\sigma \in \mathcal{S}(L)$, the dual cone D_σ is a PL-disk. (This is also true when σ is the empty simplex.) Since the decomposition of D $(= D_\emptyset)$ into dual cells resembles the stratification of a simple convex polytope by its faces, we shall call D a *simple PL-cell.* Similarly, if L is nearly a PL-sphere, then D will be called a *simple n-cell* (cf. [89, Prop. 2.2]). What happens if the simplicial complex L is only required to be a GHS^{n-1}? As we explain in Proposition 3.35 below, the answer is that $\mathrm{Cone}(L)$ can be "resolved" to a *simple homotopy n-cell* C. This means that C is a manifold with faces, where the face C_σ corresponding to a k-simplex σ is a contractible $(n-k-1)$-manifold with corners and that the poset of faces of C is dual to the simplicial complex L (cf. [89, Prop. 2.3]). The manifold with faces C is a "resolution" of the polyhedral homology manifold $\mathrm{Cone}(L)$. The proof of this depends on the following.

Theorem 3.34 *Let Σ^{n-1} be a homology sphere (i.e., a closed topological manifold with the same homology as S^{n-1}). Then Σ^{n-1} bounds a compact contractible n-manifold C^n. If $n \neq 4$ and Σ^{n-1} is a PL homology sphere, then we can take C^n to be a contractible PL manifold.*

Remarks on the Proof The fact that every homology 3-sphere bounds a topological contractible 4-manifold is proved in [121, Thm. 1.4, p. 367]. When the dimension of the homology sphere is ≥ 4, this is a consequence of surgery theory.

Proposition 3.35 ([73, Thm. 12.2] or [82, Thm. 10.8.3]) *If L is a simplicial complex and a* GHS^{n-1}, *then* $\mathrm{Cone}(L)$ *can be resolved to a simple homotopy cell* C^n.

Sketch of Proof The complex $\mathrm{Cone}(L)$ can decomposed into dual cones D_σ, $\sigma \in \mathcal{S}(L)$, where if σ has codimension k in L, then ∂D_σ $(= \mathrm{Lk}(\sigma, L))$ is a GHS^{k-1}. For $k \leq 3$ every GHS^{k-1} is a standard sphere and consequently, D_σ is a disk whenever $\mathrm{codim}(\sigma) \leq 3$. For such a σ, put $C_\sigma = D_\sigma$. If $\mathrm{codim}(\sigma) = 4$, then ∂D_σ is a PL homology 3-sphere and by Theorem 3.34, it bounds a contractible topological 4-manifold C_σ. Replace each D_σ in the dual 4-skeleton by C_σ and continue. For $\mathrm{codim}\,\sigma = 5$ we will already have constructed ∂C_σ. It is a topological 4-manifold with the same homology as S^4 (moreover, this 4-manifold can be given a PL-structure). It bounds a contractible 5-manifold C_σ (again by Theorem 3.34). Continuing in this fashion, there results a simple homotopy n-cell C, whose poset of faces is $\mathcal{S}(L)^{\mathrm{op}}$, the dual poset to $\mathcal{S}(L)$.

Open Contractible Manifolds In dimension $n \geq 3$, with $n \neq 4$, an open contractible manifold U^n is PL homeomorphic to $\mathbb{R}^n$ if and only if it is simply connected at infinity. (For $n \geq 5$, this a theorem of Stallings, and for $n = 3$, it follows from the Poincaré Conjecture.) In particular, for $n \neq 4$, such a U^n has a unique smooth or PL structure. For $n = 4$, U^4 is still homeomorphic to $\mathbb{R}^4$ by work of Freedman [121]. (On the other hand, Gompf showed that $\mathbb{R}^4$ has uncountably many smooth or PL structures.)

If a contractible space X is simply connected at infinity, then its *fundamental group at infinity*, $\pi_1^\infty(Y)$, must be trivial. This group is defined as the inverse limit of $\pi_1(Y - K)$ where K ranges over the compact subsets of Y.

When is a CAT(0) polyhedral homology n-manifold Y simply connected at infinity? It is instructive to consider the case when Y only has PL singularities at its vertices. In other words, when the link of a vertex is allowed to be a nonsingular homology $(n-1)$-sphere, but the links of all positive dimensional simplices are required to be standard PL spheres. Write Y as an increasing union of metric balls, $B_1 < \cdots < B_k < \cdots$, where ∂B_k contains no vertex of Y. Geodesic retraction provides a deformation retraction $(Y - B_k) \to \partial B_k$. Hence,

$$\pi_1^\infty(Y) = \varprojlim \pi_1(\partial B_k). \tag{3.17}$$

If there are m vertices in B_k with nonsimply connected links, $L_1, \dots, L_m$, then ∂B_k is homeomorphic the m-fold connected sum, $\partial B_k = L_1 \# \cdots \# L_m$. Hence, $\pi_1(\partial B_k)$ is the m-fold free product, $\pi_1(L_1) * \cdots * \pi_1(L_m)$. In particular, the inverse limit in (3.17) is nontrivial. (For further details of this argument see [91, Prop. (3d.3)].)

For example, in [91, Theorem 5a.1] we hyperbolized the E_8 homology 4-manifold, $X^4(E_8)$, to obtain an NPC polyhedral homology 4-manifold $\mathcal{H}(X^4(E_8))$ with only one singular vertex whose link is Poincaré's homology 3-sphere Σ^3. (This is explained in Theorem 6.7 of Sect. 6.2.2.) In the universal cover of $\mathcal{H}(X^4(E_8))$ each metric sphere is a connected sum of copies of Σ^3. So, the universal cover is not simply connected at infinity.

3.2.2 *Aspherical Manifolds Not Covered by Euclidean Space*

In the next two subsections we will use RACGs to construct examples of aspherical manifolds that are not covered by euclidean space. In both cases we begin with a GHS^{n-1}, L. In the first subsection, L is arbitrary – we resolve Cone L to a contractible manifold with boundary, C, and then apply the basic construction to get $D((\mathbf{C}_2)^I, C)$. It is homotopy equivalent to P_L and its fundamental group at infinity is isomorphic to that of $\tilde{P}$ and hence, is often nontrivial. In Sect. 3.2.3 we do not change P_L but we require that the generalized homology sphere L be simply connected. It turns out that when this is the case the NPC homology manifold P_L, although not a PL manifold, will still be a topological manifold by Edwards' Theorem 3.33.

In formula (3.3) of Sect. 3.1, the cube complex P_L was defined as the polyhedral product: $P_L = ([-1,1], \{\pm 1\})^L$. The group $(\mathbf{C}_2)^I$ acts on P_L with fundamental domain K_L; so, $P_L = D((\mathbf{C}_2)^I, K_L)$. (Earlier we used the notation $\sum \mathbf{C}_2$ for $(\mathbf{C}_2)^I$.)

When is P_L a manifold? There is an obvious sufficient condition: if L is homeomorphic to S^{n-1}, then P_L is a topological n-manifold. The reason is that if $L = S^{n-1}$, then a neighborhood of any vertex of P_L is homeomorphic to an n-disk (the cone on S^{n-1}). In fact, it follows from Edwards' Theorem 3.33 that a necessary and sufficient condition for P_L to be a topological manifold is that L is a GHS^{n-1} and that L is simply connected whenever $n \geq 3$. This gives the following theorem.

Theorem 3.36 (cf. [82, Thm. 10.6.1 (ii)]) *The cube complex P_L is a topological n-manifold if and only if L is a* GHS^{n-1} *that is simply connected whenever $n \geq 3$.*

The dual cone to a cubical cell $\square^\sigma$ of P_L is a translate of a stratum of K_L, i.e., it has the form $K_{I(\sigma)}$, where $I(\sigma) = \mathrm{Vert}(\sigma)$. For a simplex σ of L the fixed set of the subgroup $(\mathbf{C}_2)^{I(\sigma)}$ is the result of applying the basic construction to the subgroup $(\mathbf{C}_2)^{I-I(\sigma)}$ with $K_{I(\sigma)})$ as fundamental chamber. In other words, it is $D((\mathbf{C}_2)^{I-I(\sigma)}, K_{I(\sigma)})$.

The next two propositions address what happens when L is a PL sphere or even nearly a PL sphere. (I would like to thank Frédéric Haglund and Daniel Wise for suggesting that I include the second proposition which explains when the Davis–Moussong complex is homeomorphic to euclidean space.)

Proposition 3.37 *Suppose L is nearly a PL $(n-1)$-sphere. Then P_L is a manifold; each dual cone is a cell; and for each simplex σ of codimension k in* $\mathrm{Cone}(L)$, *the fixed set of $(\mathbf{C}_2)^{I(\sigma)}$ is a locally flat, k-dimensional submanifold of P_L. If L is a PL-sphere, then P_L as well as each of the fixed point sets is a PL manifold.*

Proposition 3.38 (cf. Davis–Januszkiewicz [91] and Stone [211]) *Suppose L is nearly a PL $(n-1)$-sphere. Let $\tilde{P}_L$ be the universal cover of P_L.(Note that $\tilde{P}_L$ is the Davis–Moussong complex for the* RACG, *W_L.) Then $\tilde{P}_L$ is homeomorphic to $\mathbb{R}^n$.*

Comments on Proof of Proposition 3.38 When L is actually a PL sphere, the result is due to Stone [211] and in this case, it is stated and proved in [91, Thm. 3b.2]. The proof in [91] uses the theory of cell-like maps and it only depends on the fact that L is nearly a PL sphere, i.e., for each simplex $\sigma \in \mathcal{S}(L)$, the link $\mathrm{Lk}(\sigma, L)$ is required to be homeomorphic to $S^{n-2-\dim\sigma}$, but not necessarily PL homeomorphic to a sphere. A major point in this argument is that for any $v \in \mathrm{Lk}(\sigma, L)$ and $r \in (0, \pi)$, the open ball $B_v(r)$ in $\mathrm{Lk}(\sigma, L)$ of radius r is homeomorphic to a euclidean ball of the appropriate dimension. From this one gets that for any $x \in \tilde{P}_L$ and $r \in (0, \infty)$, $B_x(r)$ is homeomorphic to a euclidean n-ball. Alternatively, a similar argument shows that for each nonempty simplex σ in L, the complement $L-\sigma$ is contractible. By [101, Thm. 4.3] this implies that, for $n \neq 2$, the Davis–Moussong complex is simply connected at infinity. As was pointed out in the previous subsection, this implies that it is homeomorphic to $\mathbb{R}^n$.

If L is only required to be a GHS^{n-1} (not necessarily simply connected), then the cube complex P_L need not be a manifold. However, by replacing the fundamental chamber K_L (the dual cone to the empty simplex) by its resolution by a simple homotopy cell C, we do get a manifold:

$$M_L = D((\mathbf{C}_2)^I, C). \tag{3.18}$$

The group $(\mathbf{C}_2)^I$ acts on M_L with fundamental chamber C. A strata-preserving homotopy equivalence $C \to K_L$ induces a $(\mathbf{C}_2)^I$-equivariant homotopy equivalence $M_L \to P_L$. Since C is a manifold with corners, it can be regarded as an orbifold whose local groups are subgroups of $(\mathbf{C}_2)^I$ (see [86]). It follows that $(\mathbf{C}_2)^I$ is a locally linear reflection group on M_L. This gives the following non-cubical version of Proposition 3.37.

Proposition 3.39 *Suppose L is a* GHS^{n-1}. *Then P_L is $(\mathbf{C}_2)^I$-equivariantly homotopy equivalent to the manifold, M_L, defined by* (3.18). *Moreover, for each simplex σ of codimension k in* Cone(L), *the fixed set of* $(\mathbf{C}_2)^{I(\sigma)}$ *is a locally flat, k-dimensional submanifold of M_L.*

Next suppose L is a flag complex so that P_L is an NPC cube complex and consequently, is aspherical. The homotopy equivalent manifold M_L is also aspherical. Let W_L be the RACG associated to L and let Γ_L be the kernel of the natural epimorphism $W_L \to (\mathbf{C}_2)^I$ (so, Γ_L is the commutator subgroup of W_L). Then $\Gamma_L = \pi_1(P_L) = \pi_1(M_L)$. (In Sect. 3.2.5, we will see that M_L is a "generalized Haken manifold" in the sense of [89] or [119].) Let $\tilde{P}_L$ and $\tilde{M}_L$ denote the universal covers of P_L and M_L, respectively. The next theorem is the main result of [73].

Theorem 3.40 ([73] or [82, Thm. 10.5.1]) *If L^{n-1} is a PL homology sphere which is not simply connected and $n \geq 4$, then the open contractible n-manifold $\tilde{M}_L$ is not simply connected at infinity and so, is not homeomorphic to euclidean n-space. Hence, in each dimension $n \geq 4$, there are closed aspherical n-manifolds that are not covered by euclidean space.*

Proof The strata-preserving homotopy equivalence $C \to K_L$ leads to a proper homotopy equivalence $\tilde{P}_L \to \tilde{M}_L$. Since $\tilde{M}_L$ and $\tilde{P}_L$ are proper homotopy equivalent, to show that $\pi_1^\infty(\tilde{M}_L)$ is not trivial, it suffices to show that $\pi_1^\infty(\tilde{P}_L)$ is not trivial. Let B be a metric ball in $\tilde{P}_L$ containing m copies of the cone point of K_L in its interior and with no cone points on ∂B. Then ∂B is the m-fold connected sum $\partial B = L\# \cdots \# L$ and $\pi_1(\partial B)$ is the m-fold free product $\pi_1(L) * \cdots * \pi_1(L)$. It follows that $\pi_1^\infty(\tilde{P}_L)$, the inverse limits of these free products, is not trivial.

3.2.3 NPC Manifolds Not Covered by Euclidean Space

If L is a simply connected GHS^{n-1}, then, by Theorem 3.36, the CAT(0) cube complex $\tilde{P}_L$ is a topological manifold. Can we choose L so that $\tilde{P}_L$ will not be simply connected at infinity? For $n \geq 5$, if we choose L as in (3.19) below, then this is indeed the case.

Let A^{n-1} be a compact acyclic smooth manifold with boundary such that

(a) $\pi_1(\partial A^{n-1}) \to \pi_1(A^{n-1})$ is onto, and
(b) $A \cup_{\partial A} A$, the double of A along ∂A, is not simply connected.

As shown in [91, p. 383], it is easy to find examples of such A. Triangulate A as a flag complex so that ∂A is a full subcomplex and let L be the GHS^{n-1} defined by

$$L := A \cup_{\partial A} \text{Cone}(\partial A). \tag{3.19}$$

By (a), L is simply connected.

Gromov asked if a topological manifold is CAT(0) must it then be homeomorphic to euclidean space? From the above discussion we get the following theorem that answers Gromov's question in the negative. A different proof of this theorem using hyperbolization techniques together with methods from [7] will be given in the proof of Theorem 6.10 of Sect. 6.2.3.

Theorem 3.41 ([91, Theorem 5b.1]) *Suppose L is the flag complex defined by* (3.19). *Then the NPC cube complex P_L is a topological manifold whose universal cover $\tilde{P}_L$ is not simply connected at infinity. Hence, in each dimension $n \geq 5$, there is a* CAT(0) *cube complex $\tilde{P}_L$ which*

(i) *is a topological n-manifold and*
(ii) *is not homeomorphic to euclidean n-space.*

So, there exist NPC polyhedra that are topological manifolds and are not covered by euclidean space. (The improvement over Theorem 3.40 is that the manifold M_L can be taken to be an NPC cube complex P_L.)

Proof Since L is simply connected it follows from Theorem 3.33 that P_L is a topological manifold. There is a discrete set of edges $\mathcal{E}$ in the cubical structure on $\tilde{P}_L$ so that $\tilde{P}_L$ has PL singularities only along the edges in $\mathcal{E}$. This set of edges can be described as follows. Let $v \in L$ be the vertex corresponding to the cone point of $\text{Cone}(\partial A)$. Let $s_v \in S$ be fundamental reflection corresponding to v. There is an edge e_v in $\tilde{P}_L$ running from the cone point c of K_L (= Cone(L)) to $s_v c$. Then $\mathcal{E}$ is the set of all W_L translates of e_v. A regular neighborhood R of e_v in $\tilde{P}_L$ is the union of two copies of Cone(L) glued together along Cone(∂A). Thus, R is a contractible manifold with boundary where its boundary is the double of A along ∂A, i.e., $\partial R = A \cup_{\partial A} A$. By property (b), ∂R is not simply connected. The proof is completed by using the argument in the proof of Theorem 3.40. Let B be a metric ball in $\tilde{P}_L$ chosen so that it contains m copies of R in its interior and ∂B intersects no other copies of R. Then ∂B is the m-fold connected sum $\partial R \# \cdots \# \partial R$ and $\pi_1(\partial B)$ is the m-fold free product $\pi_1(\partial R) * \cdots * \pi_1(\partial R)$. As before, this implies the inverse limit, $\pi_1^{\infty}(\tilde{P}_L)$, is not trivial.

The manifold P_L is smoothable. (Since its fundamental domain is homeomorphic to a contractible manifold with boundary, the stable topological tangent bundle of P_L is trivial and hence, lifts to a vector bundle.) This gives the following corollary to Theorem 3.41.

Corollary 3.42 *In each dimension $\geq$ 5, there exists a smooth n-manifold with an NPC polyhedral metric that does not admit a smooth Riemannian metric of* *nonpositive sectional curvature (because its universal cover is not homeomorphic to euclidean space).*

Remark 3.43 (Strict Negative Curvature) By using the strict hyperbolization procedure of Sect. 6.5 we can even arrange for the n-manifold in this corollary to be locally CAT(-1).

Remark 3.44 (Dimension 4**)** Corollary 3.42 also holds in dimension 4 (cf. [90]). The proof uses the Flat Torus Theorem 2.37 rather than the fundamental group at infinity. In [90], by using a suitable right-angled reflection group, we get an NPC cube complex containing a totally geodesic, 2-dimensional subcomplex which is metrically a flat 2-torus and which is locally knotted. Moreover, this cube complex is homeomorphic to a smooth 4-manifold M^4. The metric on the complement of this 2-torus is CAT(-1) so that the RACG is relatively hyperbolic (see Sect. 4.2.5). The universal cover of M^4 has a 2-flat whose limit set is knotted in the visual boundary of the universal cover. (The visual boundary is necessarily S^3.) With a little more care (such as insuring the 2-flat is isolated) one sees that the Flat Torus Theorem implies that this situation is impossible for a nonpositively curved Riemannian metric on any smooth 4-manifold homotopy equivalent to M^4.

3.2.4 Real Toric Manifolds

The polyhedral product P_L is also known as a (real) *generalized moment complex* (or *generalized moment angle manifold* when L is a simplicial polytope). It occurs in nature in several different contexts, e.g., as the solution to a system of real quadratic equations (cf. [132] or [27]) or in the theory of real toric manifolds (cf. [92]).

Davis–Januskiewicz Space Given a simplicial complex L with vertex set $I = \{1, \dots, m\}$, define the (real) *Davis–Januskiewicz space*, $B_{\mathbb{R}}(K_L)$, to be the Borel construction on P_L with respect to the action of the group $H := (\mathbf{C}_2)^I$. Recalling that the RACG, W_L, is the group of all lifts of the H-action to the universal cover $\tilde{P}_L$, we see that $B_{\mathbb{R}}K_L$ is also the Borel construction on $\tilde{P}_L$ with respect to the W_L-action, i.e.,

$$B_{\mathbb{R}}K_L = P_L \times_H EH = \tilde{P}_L \times_{W_L} EW_L. \tag{3.20}$$

Remark 3.45 Omitting the adjective "real," we get the notion of a *generalized moment angle manifold* with respect to $\mathbb{C}$: it is a polyhedral product of the form $(D^2, S^1)^L$. It admits a $\mathbb{T}^I$-action with orbit space K_L. The "real part" of this is the polyhedral product $P_L = ([-1, 1], \{\pm 1\})^L$, which supports a $(\mathbf{C}_2)^I$-action with the same orbit space. Similarly, a *toric manifold* over K_L essentially is a smooth $2d$-manifold M^{2d} with a $\mathbb{T}^d$-action and $M^{2d}/T^d = K_L$. Its real part is a smooth d-manifold M^d so that $(\mathbf{C}_2)^d \curvearrowright M^d$ with strict fundamental domain K_L. (This is the justification for the use of the words "real" and"toric.")

Definition 3.46 Suppose L is a simplicial complex with vertex set $\{1, \dots, m\}$. Let $\mathbb{F}$ be a field and let $\{v_1, \dots v_m\}$ be the standard basis for $\mathbb{F}^{\{1,\dots,m\}}$. The Stanley–Reisner *face ring*, $\mathbb{F}[L]$, is the quotient of the polynomial ring $\mathbb{F}[v_1, \dots v_m]$ by the ideal I generated by all square free monomials of the form $v_{i_1} \cdots v_{i_k}$, where $\{v_{i_1}, \dots, v_{i_k}\}$ does not span a positive dimensional simplex of L.

Theorem 3.47 (Properties of Real Davis–Januszkiewicz Complexes, cf. [92]) *Let L be a simplicial complex.*

(i) $\pi_1(B_{\mathbb{R}} K_L) = W_L$.
(ii) *If L is a flag complex, then* $B_{\mathbb{R}} K_L = BW_L$.
(iii) *Let* $\mathbb{F}_2$ *denote the field with two elements and let* $\mathbb{F}_2[L]$ *denote the Stanley–Reisner face ring of L over* $\mathbb{F}_2$. *Then* $H^*(B_{\mathbb{R}}(K_L); \mathbb{F}_2) = \mathbb{F}_2[L]$.
(iv) $H^*(W_L; \mathbb{F}_2) = \mathbb{F}_2[L]$.

Proof The space $\tilde{P}_L \times EW_L$ is simply connected and W_L acts freely on it. So, W_L is the fundamental group of the quotient. Hence, using (3.20) it is the fundamental group of $\pi_1(B_{\mathbb{R}} K_L)$. This proves (i). If L is a flag complex, then $\tilde{P}_L$ is contractible and consequently, so is $\tilde{P}_L \times EW_L$. This proves (ii). Statement (iii) is proved in [92, Thm. 4.11]. Formula (iv) follows from (i), (ii) and (iii).

Definition 3.48 (Cohen–Macaulay Algebras) An $\mathbb{N}$-graded algebra R over a field $\mathbb{F}$ is *Cohen–Macaulay* if there exists a sequence $(\lambda^1, \dots, \lambda^d)$ of algebraically independent elements of degree one in R such that R is a free $\mathbb{F}[\lambda^1, \dots \lambda^d]$-module ($(\lambda^1, \dots, \lambda^d)$ is called a *regular sequence*.

Definition 3.49 (Cohen–Macaualy Complexes) A finite simplicial complex L of dimension $d-1$ is *Cohen–Macaulay* over $\mathbb{F}$ if for each i-simplex $\sigma \in \mathcal{S}(L)$, $\overline{H}_i(\text{Lk}(\sigma, L); \mathbb{F})$ is nonzero only in degree $d-i-2$. (For $\sigma = \emptyset$ this means that $\overline{H}_i(L; \mathbb{F}) = 0$ for $i < d-1$.)

For example, if L is a simplicial $(d-1)$-sphere (or even a GHS^{d-1}), then L is a Cohen–Macaulay complex. Other examples include spherical buildings. The connection between the two concepts in Definitions 3.48 and 3.49 for the face ring $\mathbb{F}[L]$ is provided by the following theorem of Reisner.

Theorem 3.50 (Reisner [193]) *The following conditions are equivalent.*

(i) *The simplicial complex L is Cohen–Macaulay.*
(ii) *The face ring* $\mathbb{F}[L]$ *is Cohen–Macaulay.*

Let $H_L(t) = \sum_{i=0}^{\infty} H_k t^k$ be the Hilbert series of $\mathbb{F}_2[L]$, i.e., H_k is the dimension of the degree k part of $\mathbb{F}_2[L]$ as an $\mathbb{F}_2$-vector space. By Theorem 3.47 (iii), $H_L(t)$ is equal to the Poincaré series of Davis–Januszkiewicz space, i.e., $H_k = b_k(B_{\mathbb{R}}(L); \mathbb{F}_2)$, the kth Betti number of $B_{\mathbb{R}}(L)$ with coefficients in $\mathbb{F}_2$.

When the $(d-1)$-dimensional simplicial complex L is Cohen–Macaulay, the fact that $\mathbb{F}_2[L]$ is a Cohen–Macaulay ring implies that the Hilbert series for $\mathbb{F}_2[L]$ is the product of the Hilbert series for $\mathbb{F}_2[\lambda^1, \dots, \lambda^d]$ with a polynomial $h(t)$ of degree d, i.e.,

$$\frac{H_L(t)}{(1-t)^d} = h(t) = h_0 + h_1 t + \cdots + h_d t^d. \tag{3.21}$$

When L is the boundary simplicial convex polytope, $h(t)$ is called its *h-polynomial* (or the h-polynomial of the dual polytope K_L). It has been much studied by combinatorialists in connection with the theory of the number of faces in such polytopes, e.g. in [137, 210, 235]. There is a simple expression for the coefficients h_i in terms of the numbers f_i of i-simplices of L,

$$\sum_{i=0}^{i=d} h_i t^{m-i} = \sum_{i=0}^{i=d} f_{i-1}(t-1)^{m-i}$$

Intersections of Quadric Hypersurfaces A d-dimensional polytope K_L can be defined as the intersection of the standard simplex Δ^{m-1} defined by $\sum_{i=1}^m r_i = 1$, with $r_i \geq 0$, and a collection of k linear hyperplanes defined by

$$\sum_{i=1}^m A_i r_i = 0,$$

where the A_i are the column vectors of some $(k \times m)$-matrix. If these hyperplanes intersect transversely, then K_L is a simple polytope of dimension $d := m - 1 - k$ meaning that the boundary complex of its dual polytope L is a triangulation of S^{d-1}. Let P_L denote the intersection of the unit $(d-1)$-sphere defined by $\sum_{i=1}^{i=m} x_i^2 = 1$ with the k quadric hypersurfaces surfaces defined by the equations:

$$\sum_{i=1}^m A_i x_i^2 = 0.$$

The group $(\mathbf{C}_2)^m$ acts on the intersection P_L, which can be identified with the polyhedral product (or generalized moment angle polytope) defined by equation (3.3) or (3.6) in Sect. 3.1 (cf. [132], [172] or [27]). Similarly, the intersection of the complex quadric hypersurfaces $\sum_{i=1}^{i=m} A_i |z_i|^2 = 0$ in $\mathbf{C}^m$ with the unit sphere, $|z_1|^2 + \cdots + |z_m|^2 = 1$ in $\mathbf{C}^m$ is the polyhedral product, $(D^2, S^1)^L$.

Small Covers (or Real Toric Manifolds) Recall that if L^{d-1} is the boundary complex of a simplicial d-polytope, then its dual K_L is a simple convex polytope. Similarly, if L^{d-1} is nearly a PL-sphere, then K_L is a "simple cell" (meaning that each dual cone is a cell). In either case, P_L $(= D((\mathbf{C}_2)^I, K_L))$ is a closed d-manifold. For the remainder of this subsection we suppose this to be the case.

Definition 3.51 (cf. [92]) Suppose L is nearly a PL sphere of dimension $(d-1)$ and that $I = \mathrm{Vert}(L)$. Let $\mathbb{F}_2$ be the field with 2 elements. A *characteristic function* for P_L is an I-tuple $(\lambda_i)_{i\in I}$ of nonzero vectors in $\mathbb{F}_2^d$ so that for each $(d-1)$-simplex σ in L, $\{\lambda_i\}_{i\in I(\sigma)}$ is a basis for $\mathbb{F}_2^d$. (As usual $I(\sigma) = \{i \in I \mid i \text{ is a vertex of } \sigma\}$.) We can regard the λ_i as defining a homomorphism, called the *characteristic homomorphism*, $\lambda = (\lambda_1, \dots, \lambda_m) : \mathbb{F}_2^m \to \mathbb{F}_2^d$, where $m = \mathrm{Card}\, I$ and where λ takes the standard basis element of $\mathbb{F}_2^m$ corresponding to $i \in I$ to λ_i.

If we identify $\mathbb{F}_2 = \{0, 1\}$ with the multiplicative group $\mathbf{C}_2 = \{\pm 1\}$, then we can also regard λ as a homomorphism $(\mathbf{C}_2)^m \to (\mathbf{C}_2)^d$. The vectors $\lambda_1, \dots, \lambda_m$ are the column vectors of the matrix (a_i^j), $1 \le j \le m$, $1 \le i \le d$ that represents λ. Each λ_i is a linear function on $\mathbb{F}_2^m$, hence, a linear combination of $\{v_1, \dots, v_m\}$. The row vectors of (a_i^j) are denoted $\lambda^1, \dots, \lambda^d$.

When L is the boundary complex of a simplicial polytope, K_L is a simple polytope. Similarly, when L is nearly a PL-sphere, K_L has the structure of a manifold with corners so that each stratum is a disk. A codimension k-stratum of K_L is dual to a $(k-1)$-simplex of L. For example, a vertex of L is dual to a codimension-one face (or *facet* of K_L). So each vertex $i \in I$ corresponds to a facet K_i of K_L. A collection of facets corresponding to a subset σ of I has nonempty intersection K_σ if and only if σ is a simplex of L. Thus, a characteristic function λ can be regarded as a function $\mathrm{Facet}(K_L) \to \mathbb{F}_2^d$, where $\mathrm{Facet}(K_L)$ ($\cong I$) is the set of facets of K_L and $\mathbb{F}_2^d = (\mathbf{C}_2)^d$.

Definition 3.52 Let $\lambda : \mathrm{Facet}(K_L) \to \mathbb{F}_2^d$ be a characteristic function and $\lambda : (\mathbf{C})^m \to (\mathbf{C}_2)^d$ the associated characteristic homomorphism. The associated *small cover* of K_L is defined by

$$M(\lambda) = P_L \,/\, \ker \lambda. \tag{3.22}$$

We view λ as a m-tuple of vectors $(\lambda_1, \dots, \lambda_m)$, where $\lambda_i \in (\mathbf{C}_2)^d$. The manifold $M(\lambda)$ is also called a *(real) toric manifold* over K_L.

The toric manifold $M(\lambda)$ is d-dimensional; it comes equipped with a $(\mathbf{C}_2)^d$-action with strict fundamental domain K_L. The orbifold K_L is a simple d-cell.

Remark 3.53 There is a similar notion of a (complex) toric manifold $M^{2d}(\lambda)$ corresponding to a "characteristic function" $\lambda : \mathbb{Z}^I \to \mathbb{Z}^d$. The manifold $M^{2d}(\lambda)$ admits an action of the d-dimensional torus $\mathbb{T}^d$; the orbit space is again K_L. The vector $\lambda_i \in \mathbb{Z}^d$ specifies the codimension-one subtorus which fixes the ith facet. We call it a "(complex) toric manifold." This notion is related to the concept in algebraic geometry of a type of projective variety called a "nonsingular toric variety." A toric variety is a special case of a toric manifold; it is a compact complex manifold M of real dimension $2d$; $\mathbb{T}^d \curvearrowright M$; and $M/\mathbb{T}^d$ is a simple convex polytope. However, most characteristic functions λ do not correspond to nonsingular toric varieties. (To distinguish between these two notions Buchstaber–Panov [41] changed the name from "toric manifold" to "quasitoric manifold;" however, we will stick with the original definition of toric manifold in [92].) The set of real points of a toric variety is a "real toric variety;" so, a real toric variety is a special case of a real toric manifold.

Example 3.54 (Projective Spaces) Suppose L is the boundary complex of a d-simplex Δ^d and that $I = \{0, 1, \dots, d\}$ is its vertex set. Then K_L is the dual simplex Δ^* and the polyhedral product P_L is the boundary complex of the $(d+1)$-cube $[-1,1]^I$, i.e., $P_L = \partial([-1,1]^{d+1}) = S^d$. The group $(\mathbf{C}_2)^{d+1}$ is a reflection group on S^d and the fundamental chamber is Δ^*. Up to an automorphism of $(\mathbf{C}_2)^d$ (i.e., up to an element of $GL_2(\mathbb{F}_2)$), there is only one possibility for the characteristic function, $\lambda : \{0, 1, \dots, d\} \to \mathbb{F}_2^d$: it is defined by $i \mapsto e_i$, for $1 \leq i \leq d$, and for $i = 0$ by $0 \mapsto e_1 + \cdots + e_d$, where $e_1, \dots, e_d$ is the standard basis for $\mathbb{F}_2^d$. Then $\ker \lambda$ is the cyclic group of order 2 with nontrivial element $(-1, \dots, -1)$. Hence, $\ker \lambda$ acts on S^d via the antipodal map and

$$M^d(\lambda) = S^d/\{\pm 1\} = \mathbb{R}P^d.$$

Example 3.55 (Dimension 2) We expand the discussion from Example 3.2 (iii). Suppose L is an m-circuit so that K_L is an m-gon. Then P_L is an orientable surface tiled by 2^m m-gons with 4 meeting at each vertex. Its Euler characteristic is $2^m(1 - m/2 + m/4) = 2^m(1 - m/4)$ (cf. [68, p. 57]). The facets of K_L are its edges. Suppose m is even. Then we can define $\lambda : \mathrm{Facet}(K_L) \to (\mathbb{F}_2)^2$ by alternately labeling the edges by one of the two standard basis elements e_1, e_2 of $(\mathbb{F}_2)^2$. One sees that $M(\lambda) = P_L/\ker \lambda$ is an orientable surface of genus $(m-2)/2$ (cf. [92, Ex. 1.20]). Independent of whether m is even or odd, one can always find a characteristic function λ so that $M(\lambda)$ is the connected sum of $m-2$ copies of $\mathbb{R}P^2$, i.e., it is a nonorientable surface of genus $m-2$. For example, if $m = 3$, then $M(\lambda)$ must be $\mathbb{R}P^2$.

Before continuing with more examples of toric manifolds, we explain the beautiful connection between the cohomology of the toric manifold $M(\lambda)$ and the cohomology of Davis–Januszkiewicz space $B_{\mathbb{R}}(L)$. By (3.20), $B_{\mathbb{R}}(L)$ is the Borel construction on P_L: $B_{\mathbb{R}}(L) = P_L \times_H EH$, where $H = (\mathbf{C}_2)^m$, where $m = \mathrm{Card}\, I$. Then P_L is the covering space of $M(\lambda)$ corresponding to the characteristic function $\lambda : (\mathbf{C}_2)^m \to (\mathbf{C}_2)^d$ (or $\mathbb{F}_2^m \to \mathbb{F}_2^d$). Regarding the homomorphism $\lambda : \mathbb{F}_2^m \to \mathbb{F}_2^d$ as a $(d \times m)$-matrix with column vectors $\lambda_1, \dots, \lambda_m$ and row vectors $\lambda^1, \dots \lambda^d$, we see that it defines a homomorphism $H_1(B_{\mathbb{R}}(L); \mathbb{F}_2) \to H_1((B\mathbf{C}_2)^d; \mathbb{F}_2) = H_1((\mathbb{R}P^\infty)^d; \mathbb{F}_2)$ and hence, the homotopy class of a map

$$\lambda : B_{\mathbb{R}}(L) \to (B\mathbf{C}_2)^d, \tag{3.23}$$

which we continue to denote by λ. This map is the projection map of fiber bundle with fiber $M(\lambda)$. In the following theorem all cohomology groups are with coefficients in $\mathbb{F}_2$.

Theorem 3.56 ([92, Theorems 5.9 and 5.12]) *Suppose $M(\lambda)$ is a toric manifold over K_L where L is a* GHS^{d-1}.

(i) *The sequence* $(\lambda^1, \cdots, \lambda^d)$ *is a regular sequence (see Definition 3.48) of degree one elements in the face ring* $\mathbb{F}_2[L] = H^*(B_{\mathbb{R}}(L); \mathbb{F}_2)$.
(ii) *The Serre spectral sequence for the fiber bundle* (3.23) *degenerates at* E_2*:*

$$E_\infty^{p,q} = E_2^{p,q} = H^p((B\mathbf{C}_2)^d) \otimes H^q(M(\lambda)).$$

(iii) *The Poincaré polynomial of* $M(\lambda)$ *is the h-polynomial,* $h(t)$ *of* (3.21), *i.e.,* h_k *is the* k^{th} *Betti number* $b_k(M(\lambda); \mathbb{F}_2)$.
(iv) *The cohomology ring* $H^*(M(\lambda))$ *is the quotient* $\mathbb{F}_2[v_1, \ldots, v_m]/(I + J)$ *of the polynomial ring by the ideal* $I + J$*, where* I *is the Stanley–Reisner ideal that defines the face ring (cf. Definition 3.46) and* J *is the ideal generated by* $\lambda_1, \ldots, \lambda_m$ *(where the* λ_i *are regarded as linear combinations of the* v_j*).*

Proof When K_L is a simple convex polytope one can prove (iii) as in [92, Thm. 3.1] by finding a perfect Morse function on $M(\lambda)$ with the number of critical points of index k equal to h_k. To do this, start with a generic height function f on the convex polytope K_L (i.e., the level sets of f are never tangent to the edges of K_L). The critical points of f are the vertices and the index of f at a vertex c is the number of downward pointing edges at c (i.e., the number of edges incident to c on which f is decreasing). One of the equivalent definitions of h_k is that it is equal to the number of vertices of index k. If $p : M(\lambda) \to K_L$ is the orbit projection, then fp is a Morse function with all its critical points at a vertices. The closure of a descending k-cell is the fixed point set of a subgroup of $(\mathbf{C}_2)^d$. This means that this closure is a closed k-dimensional submanifold of $M(\lambda)$. This implies that the cell structure defined by fp is perfect with respect to homology with coefficients in $\mathbb{F}_2$ and therefore, proves (iii).

The E_2 page of the Serre spectral sequence is $H^*((B\mathbf{C}_2)^d) \otimes H^*(M(\lambda))$ which has Hilbert series $h(t)/(1 - t)^d$. Since $\mathbb{F}_2[L]$ is Cohen–Macaulay, this is the Hilbert series of $H^*(B_{\mathbb{R}}(L))$. It follows that $E_2 = E_\infty$, which proves (ii). We can identify each row vector λ^i with an element of $H^1(B_{\mathbb{R}}(L))$. It follows from (ii) that $H^1((B_{\mathbb{R}}(L))$ is a free $\mathbb{F}_2[\lambda^1, \ldots, \lambda^d]$-module and hence, that $\lambda^1, \ldots \lambda^d$ is a regular sequence. This proves (i) and (iv).

Here are some more examples of real toric manifolds.

Example 3.57 (Real Points of Toric Varieties) One way to understand a (complex) toric variety is as the closure of a $(\mathbb{C}^*)^d$-orbit on a complex d-manifold (where $\mathbb{C}^* = \mathbb{C} - \{0\}$) Similarly, a real toric variety can be understood as the closure of a $(\mathbb{R}^*)^d$-orbit. Regard $(\mathbb{R}^*)^d$ as the group of diagonal $(d + 1) \times (d + 1)$ matrices divided out by the subgroup of scalar matrices $(= \{(t, \ldots, t)\}_{t\in\mathbb{R}^*})$. A prototypical example of a real toric variety comes from considering the $(\mathbb{R}^*)^d$-action on $\mathbb{R}P^d$, given by $(t_0, t_1, \ldots, t_d) \cdot [x_0, \ldots x_d] = [t_0x_0, \ldots, t_dx_d]$. The complement of the union of coordinate hyperplanes in $\mathbb{R}P^d$ is a single $(\mathbb{R}^*)^d$-orbit consisting of 2^d copies of an open d-simplex, i.e., $\{\pm 1\}^d \times \mathrm{int}(\Delta^*)$.

Similarly, the group $(\mathbb{R}^*)^d$ acts on the real flag manifold $GL_{d+1}(\mathbb{R})/B$ where B is the group of upper triangular matrices whose entries along the diagonal are some constant multiple of the identity. (Here we are thinking of $(\mathbb{R}^*)^d$ as the group of diagonal matrices of determinant 1.) Consider a generic $(\mathbb{R}^*)^d$-orbit, i.e., the orbit of a diagonal matrix with distinct entries. The closure of such an orbit is a real toric variety M^d. It can be described as follows. The group $(\mathbf{C}_2)^d$ acts on it with strict fundamental domain a permutohedron. This is the d-dimensional "Coxeter zonotope" that can be described as the convex hull of a generic point in the reflection representation of the symmetric group S_{d+1} on $\mathbb{R}^d$ (cf. Examples 3.73 and 3.79 in Sects. 3.3.1 and 3.3.2).

Example 3.58 (Tomei Manifolds) In [226] Tomei described another interesting way that real toric manifolds occur in nature. Their description here is taken from [76]. Let Y be the vector space of real $(d + 1) \times (d + 1)$ symmetric tridiagonal matrices. The coordinates of the diagonal entries and are denoted by $a_0, \dots, a_d$. The entries that lie on the diagonal one above the main diagonal (which are equal to the entries on the diagonal one below the main diagonal) are denoted $b_1, \dots, b_d$. Thus, Y is the $(2d + 1)$-dimensional vector space consisting of matrices of the following form:

$$\begin{pmatrix} a_0 & b_1 & 0 & \dots & & 0 \\ b_1 & a_1 & b_2 & \dots & & 0 \\ \vdots & \vdots & \ddots & \ddots & & \vdots \\ 0 & 0 & \dots & \dots & a_{d-1} & b_d \\ 0 & 0 & & \dots & b_d & a_d \end{pmatrix}$$

Let Υ be a set of $d + 1$ real numbers and let $M^d(\Upsilon)$ be the set of matrices in Y with spectrum equal to Υ. If the numbers in Υ are distinct, then $M^d(\Upsilon)$ is a smooth, closed d-manifold. (Let Y_{reg} be the set of matrices in Y whose eigenvalues are distinct. Then the map from Y_{reg} to the space of unordered distinct $(d + 1)$-tuples of real numbers is the projection of a smooth fiber bundle; the fiber over Υ is $M^d(\Upsilon)$.) The manifold $M^d(\Upsilon)$ is the *Tomei manifold* of isospectral tridiagonal matrices. At this point, one might ask if $M^d(\Upsilon)$ is diffeomorphic to some other well-known manifold. It is not hard to see that $M^1(\Upsilon)$ is a circle. We shall see below $M^2(\Upsilon)$ is an orientable surface of genus 2 while $M^3(\Upsilon)$ is the double of a certain finite-volume hyperbolic 3-manifold along its cusps (it has 6 cusps, each of which is a torus). So, $M^d(\Upsilon)$ is generally not a well-known manifold from classical topology.

The orthogonal group $O(d + 1)$ acts by conjugation on the vector space of symmetric $((d + 1) \times (d + 1))$-matrices, preserving their spectra. The kernel of this action is $\{\pm \mathrm{id}\}$, where id is the identity matrix. Let D be the set of diagonal matrices in $O(d + 1)$. The diagonal entries of a matrix in D are numbers in $\{\pm 1\}$. The subgroup D acts on Y as sign changes of the off-diagonal entries and it acts trivially on the diagonal entries. Thus, $D/\{\pm \mathrm{id}\} = (\mathbf{C}_2)^d$ acts on Y as a reflection group and since it preserves spectra, it is also a reflection group on $M^d(\Upsilon)$. The fundamental chamber Q is the intersection of $M^d(\Upsilon)$ with the subset of Y defined by the inequalities: $b_1 \geq 0, \ldots, b_d \geq 0$. Thus, Q is a smooth d-manifold with corners that can be regarded as a "Coxeter orbifold." It follows that $M^d(\Upsilon)$ is a real toric manifold over Q. It turns out that Q is combinatorially isomorphic to a simple convex polytope. In fact, this polytope is the same permutohedron that occurred in Example 3.57. In other words, ∂Q is the dual to the barycentric subdivision of $\partial \Delta^{d+1}$ (which is equal to Coxeter complex $L(\mathbf{A}_d)$). The proof that Q is combinatorially equivalent to a convex polytope depends on the fact that there is a natural $\mathbb{R}$-action on the interior of Q called the "Toda flow." (A definition of this flow can be found in [76, §11].) The Toda flow is part of a completely integrable system, that is, it is part of a free action of $\mathbb{R}^d$ on $\mathrm{int}(Q)$. Since the faces of Q are also $\mathbb{R}^d$-orbits, one can show that Q is a permutohedron. Thus, $M^d(\Upsilon)$ is tiled by 2^d permutohedra. If $d = 1$, Q^1 is an interval and hence, $M^1(\Upsilon)$ is a circle. If $d = 2$, then Q^2 is a hexagon. If we double Q^2 along 3 nonadjacent edges we get a pair of pants; doubling again along the 3 boundary components we get a surface of genus 2. When $d = 3$, Q^3 is a 3-dimensional permutohedron with 6 square faces (corresponding to cusps) and 8 hexagonal faces. It follows from Andreev's Theorem (cf. [8, 9], or [104, Thm. 10.3.1]) that Q^3 can be realized as a right-angled ideal polytope in hyperbolic 3-space with the squares corresponding to the 6 ideal vertices. (See also Sect. 4.2.4 and Remark 4.22.) There are two groups of four hexagonal facets of Q^3, each group corresponds to a mirror of a $(\mathbf{C}_2)^2$ reflection group. Gluing together 4 copies of Q^3 using this action we get a finite volume hyperbolic 3-manifold with 6 cusps. Doubling along the cusps we get $M^3(\Upsilon)$. It follows that in the associated generalized moment manifold P_L is the same as the one for the real toric variety M^d of Example 3.57. However, $M^3(\Upsilon)$ is *not* the set of real points of a toric variety. Moreover, M^d and $M^d(\Upsilon)$ are not diffeomorphic (since $M^d(\Upsilon)$ is stably parallelizable while M^d is not). As real toric manifolds they correspond to completely different characteristic functions $\lambda : \mathrm{Facet}(Q) \to \mathbb{F}_2^d$.

Remark 3.59 (Gaifullin's Results on Steenrod Representability) An unexpected use for Tomei manifolds was discovered by Gaifullin [123], [124] in connection with Steenrod's representability problem. In the 1940s Steenrod asked if, given an arbitrary space X and homology class $z \in H_d(X; \mathbb{Z})$, is it possible to find a closed manifold N^d and a map $f : N^d \to X$ such that $f_*([N^d]) = z$? René Thom found that although the answer to this question was negative in general, after replacing z by a nonzero multiple the answer is "yes." As a consequence of his computation of the oriented bordism group $\Omega_d(X) \otimes \mathbb{Q}$, Thom showed in 1954 that there is an integer k $(= k(d))$ so that for any homology class $z \in H_d(X; \mathbb{Z})$, kz can be represented by a manifold $f_*([N^d])$. Gaifullin showed that one always can take N^d to be a finite cover of the Tomei manifold $M^d(\Upsilon)$. More precisely, there is a finite cover $\tilde{M}$ of $M^d(\Upsilon)$ and a map $f : \tilde{M} \to X$ so that $f_*(\tilde{M}) = qz$ for some nonzero integer q (depending on z).

3.2.5 Haken Manifolds

The notion of "Haken manifolds" in dimensions greater than 3, was defined in Foozwell's PhD thesis [118], as well as in Foozwell-Rubinstein [119]. The theory was developed further in [89]. The manifolds P_L, where L is a PL triangulation of a sphere, are prototypical examples of such Haken manifolds. However, the theory of Haken manifolds gives a more general method of cutting open a manifold along hyperplanes, which covers many more examples than any method coming from right-angled reflection groups.

A *boundary pattern* for an n-manifold with boundary is a decomposition of part of its boundary into connected $(n-1)$-manifolds (with boundary) such that the intersection of any k of them is either empty or a connected $(n-k)$-manifold. An element of the boundary pattern is a *facet*; a component of an intersection of facets is a *stratum*. The boundary pattern is *complete* if the union of the facets is the entire boundary of the n-manifold.

As in Sect. 3.2.1, a *simple n-cell* is a compact manifold with boundary pattern such that each stratum is homeomorphic to D^{n-k} and the stratification is dual to a simplicial complex L^{n-1} that is nearly a PL $(n-1)$-sphere (cf. Sect. 3.2.1). Here $k-1$ is the dimension of the simplex dual to the stratum. A *simple homotopy n-cell* is a compact n-manifold C with complete boundary pattern so that each stratum is a compact contractible $(n-k)$-manifold and so that the stratification is dual to a GHS^{n-1} $(= L)$. So, as in Sect. 3.2.1, C is a resolution of L.

Let M^n be a manifold with boundary pattern. A *hypersurface* in M is a codimension-one submanifold which meets each stratum transversely. Any hypersurface $F < M$ has a normal S^0-bundle; it is a trivial S^0-bundle if and only if F is 2-sided in M. To "cut M open" along a hypersurface F we remove F from M and then adjoin to $M-F$ the normal S^0-bundle of F to obtain a new manifold with boundary pattern, denoted by $M \odot F$. (So, $M \odot F$) is diffeomorphic to the complement of an open interval bundle around F.) A *prehierarchy* $(M_0, F_0), \dots, (M_m, F_m)$ for a compact manifold M with complete boundary pattern is a sequence of manifolds M_k with complete boundary patterns together with hypersurfaces $F_k < M_k$ so that $M_0 = M$, $M_{k+1} = M_k \odot F_k$ and the prehierarchy terminates at $M_{m+1} := M_m \odot F_m$, which is required to be a disjoint union of simple homotopy cells. The boundary pattern is called *useful* if

(a) Each facet is simply connected.
(b) The intersection of any two facets is connected.
(c) If three facets have pairwise nonempty intersections, then the intersection of all three is nonempty.

To define "Haken manifold" we need to say what it means for a hypersurface to be "essential." (When $\dim M = 3$, it means that F is an incompressible and boundary incompressible surface.) In general, it means that the inclusion of each hypersurface $F_k \hookrightarrow M_k$ induces injections $\pi_1(F_k) \to \pi_1(M_k)$ and $\pi_1(F_k, \partial F_k \cap G) \to \pi_1(M_k, G)$, for any facet G of M_k.

A *Haken homotopy n-cell* is defined inductively as a homotopy n-cell with a complete useful boundary pattern so that each facet is itself a Haken homotopy $(n-1)$-cell. In particular, a Haken homotopy n-cell is a simple homotopy cell. The manifold is a *Haken n-cell* if each of its strata is actually a disk. In dimension two, a p-gon is a Haken 2-cell if and only if $p \geq 4$. From this and properties (a), (b) and (c) above, we get the following characterization of Haken homotopy n-cells.

Theorem 3.60 (cf. [89, Thm. 3.5]) *A simple homotopy n-cell C is a Haken homotopy cell if and only if its boundary complex is dual to a flag complex.*

For example, a 3-dimensional simple convex polytope is a Haken 3-cell if it satisfies (i) each facet is a polygon with at least 4 sides, and (ii) whenever three facets intersect pairwise the triple intersection is nonempty and equal to a vertex (cf. (c), above). (Note that these conditions are implied by Andreev's Condition for right-angled simple polytopes in $\mathbb{H}^3$; see Remark 4.22 and Theorem 4.23 in Sect. 4.2.4.)

Definition 3.61 A *hierarchy* on a manifold M with boundary pattern is a prehierarchy, $(M_0, F_0), \dots, (M_m, F_m)$, where each hypersurface $F_k < M_k$ is essential and where M_{m+1} is a disjoint union of Haken homotopy n-cells.

Definition 3.62 (cf. [89]). A manifold with complete boundary pattern is a *Haken manifold* (respectively, a *generalized Haken manifold*) if it admits a hierarchy $(M_0, F_0), \dots, (M_m, F_m)$, where the end of the hierarchy, M_{m+1}, is a disjoint union of Haken n-cells (respectively, Haken homotopy n-cells).

Note that if $F_k < M_k$ is any hypersurface in the hierarchy, then F_k is a Haken $(n-1)$-manifold.

Theorem 3.63 (Foozwell [118], also cf. [119] and [89, Prop. 3.1]) *Each connected component of a Haken manifold is aspherical.*

Sketch of Proof One shows by induction on dimension that each connected component of a hypersurface is aspherical. Next, Whitehead's Lemma asserts that the result of gluing aspherical spaces together along π_1-injective aspherical subspaces is again aspherical. (See Corollary A.23 in Appendix A.4 or [146, Thm. 1B.11].) Since M_k is constructed by gluing together M_{k+1} along F_{k+1}, Theorem 3.63 follows.

Example 3.64 (Right-Angled Reflection Manifolds) Suppose L is a flag complex with vertex set $I = \{0, 1, \dots, m\}$. As before, we get P_L, a subcomplex of the cube $[-1, 1]^I$. Cutting the cube open along the hyperplanes $\{x_i = 0\}$, $i \in I$, we get a decomposition of the cube into 2^m components, each isomorphic to a subcube of the form $[0, 1]^I$. This induces a decomposition of P_L into 2^m pieces each isomorphic to a chamber for the action of $(\mathbf{C}_2)^I$ as a reflection group on P_L. Now suppose L is nearly a PL triangulation of S^{n-1} so that $M = P_L$ is an n-manifold and each of the hypersurfaces $F_i' := P_L \cap \{x_i = 0\}$ is a (not necessarily connected) codimension-one submanifold. Then M has the structure of a Haken manifold. The hierarchy, $(M_0, F_0), \dots (M_m, F_m)$ is defined explicitly by putting $M = M_0$, $F_0 = F_0'$ and F_i to be the closure of cut open copies of F_i' (cut open along $F_{i-1} \cap F_i'$). So, the number of components of F_i' is doubled each time we cut along a 2-sided hypersurface, F_j, with $F_j \cap F_i' \neq \emptyset$ and $j < i$. Similarly, suppose L is only required to be a generalized homology $(n-1)$-sphere (i.e., a GHS^{n-1}). Then, after resolving the fundamental domain to a simple homotopy n-cell, we get a generalized Haken n-manifold. (See Sect. 3.2.1 for the definition of a *generalized homology sphere* and its resolution.) Another mild generalization of these examples is to the case where M is a NPC cube complex, the link of each vertex is nearly a PL-sphere and each hyperplane is embedded. ("Hyperplanes" are defined using midcubes as in Sect. 5.1.1.) Then the resulting decomposition of M into Haken n-cells is just the tiling of M dual to its tiling by cubes.

It is remarkable that Theorem 3.63 can be proved without resorting to nonpositive curvature. The decomposition of M into Haken n-cells need not be dual to any cubical structure. Indeed, it might not give a tiling of M. (In other words, the intersection of two cells in the decomposition need not be a union of faces.) One way to construct such examples is to begin with an example coming from a reflection group as the first term in a hierarchy (M_0, F_0) and then choose a homeomorphism $f : F_0 \to F_0$ which does not preserve the cell structure on F_0. Then use f to reglue $M_1 = M_0 \odot F_0$ to obtain a new manifold M_0'. The new manifold is still Haken since the remaining terms in the hierarchy, $(M_1, F_1), \ldots, (M_m, F_m)$, are not changed. If F_0 is a manifold whose fundamental group has a fairly large outer automorphism group, such as a torus or a product of surfaces, then we can choose the homeomorphism f that is not isotopic to one that preserves any cell structure on F_0; hence, we will not be able to find a hierarchy with initial hyperplane F_0 so that the resulting decomposition into cells is a tiling. N.B.: Strictly speaking, this means that the universal cover of M might not be a polyhedron in the classical sense. Although it decomposes into Haken cells, two such cells need not intersect in a common face.

Example 3.65 (Fiber Bundles) Suppose $\pi : M \to B$ is the projection map of a smooth fiber bundle with fiber Σ. If the base and fiber are both generalized Haken manifolds, then so is M. To see this suppose $F < B$ is an incompressible hypersurface in B. Then $\pi^{-1}(F)$ is an incompressible hypersurface in M and $M \odot \pi^{-1}(F)$ is naturally a fiber bundle over $B \odot F$. Hence, the inverse image of a hierarchy for B yields the beginning of a hierarchy for M which cuts M into a disjoint union of manifolds of the form $\Sigma \times c$, where c is a homotopy Haken cell (since c is contractible any bundle over it is trivial). A hierarchy for Σ then gives a hierarchy for M. Many such fiber bundles do not admit any NPC metric. For example, if M is a S^1-bundle over B and its Euler class in $H^2(B; \mathbb{Z})$ is not of finite order, then M does not admit any locally CAT(0) metric (cf. [35, Thm. II.6.12]). Another class of examples are solvmanifolds. Since any solvmanifold can be constructed as an iterated sequence of torus bundles, starting from a torus, solvmanifolds are Haken. However, if the fundamental group of a solvmanifold is not virtually abelian then it does not admit an NPC metric by the Solvable Subgroup Theorem (cf. [35, Thm. II.7.8]).

Remark 3.66 The Euler Characteristic Conjecture states that if M^{2n} is a closed aspherical manifold of even dimension, then its Euler characteristic satisfies:

$$(-1)^n \chi(M^{2n}) \geq 0.$$

If C is a Haken homotopy cell, then its "orbifoldal Euler characteristic" is defined by assigning each face of codimension k a weight of $1/2^k$ and then taking the weighted alternating sum of the number of cells. The Charney–Davis Conjecture states that the sign of the orbifodal Euler characteristic of a $2n$-dimensional Haken homotopy cell is $(-1)^n$. It is shown in [58] that the Charney–Davis Conjecture is equivalent to the Euler Characteristic Conjecture for $2n$-manifolds with the structure of a NPC cube complex. The main result of [89] is that the Euler characteristic of a closed Haken manifold M^{2n} is equal to the sum of the orbifoldal Euler characteristics of the homotopy cells at the end of the hierarchy. Hence, the Euler Characteristic for Haken manifolds is equivalent to the Charney–Davis Conjecture for Haken homotopy cells. By using L^2-homology it is proved in [104] that the Charney–Davis Conjecture holds in dimension 4. Hence, not only is the Euler Characteristic Conjecture true for closed 4-manifolds that are NPC cube complexes, it is also true for all Haken manifolds of dimension 4.

3.2.6 *The Reflection Group Trick*

Given a generalized homology $(n-1)$-sphere L, we explained in Sect. 3.2.2 how to replace the fundamental chamber K_L by a contractible manifold with corners C so that the resulting manifold $D((\mathbf{C}_2)^I, C)$ is homotopy equivalent to P_L. (If L is a PL homology sphere, then ∂C is homeomorphic to L). The idea behind the Reflection Group Trick is that when L is a triangulation of a PL manifold, not necessarily a sphere, then we can replace the fundamental chamber K_L ($=$ Cone L) by a manifold X cobounding L. The result of the basic construction is the manifold,

$$M := D((\mathbf{C}_2)^I, X) = ((\mathbf{C}_2)^I \times X)/\sim, \tag{3.24}$$

where $\sim$ was defined in (3.9). Since X is a strict fundamental domain for $(\mathbf{C}_2)^I$ on M, it follows that X is a retract of M. The retraction $r : M \to X$ is given by $[g, x] \mapsto x$ (see Lemma 3.7 (iv)). We will see in Theorem 3.68 that if X is aspherical, then so is M.

A group G is *type F* if its classifying space BG is homotopy equivalent to a finite CW complex. (See Sect. 7.2 for a discussion of groups which are type F or satisfy other finiteness properties.) So, given a group G of type F, the goal of this subsection is to explain a method of [82, Chapter 11] for constructing a closed aspherical manifold M together with a retraction $r : M \to BG$. The upshot is that fundamental groups of aspherical manifolds cannot be expected to enjoy properties that are stronger than those enjoyed by general groups of type F. First applications of this trick appear in [74] and in Mess [178]. More applications are listed in Proposition 3.69 below.

First, realize BG as a finite CW complex and then find a thickening of it to X^n, a compact n-manifold with boundary. (To say X is a *thickening* of BG means only that X is homotopy equivalent to BG.) Next, take a PL triangulation of ∂X as a flag complex L. (The hypothesis that the triangulation is PL can be weakened: as in Sect. 3.2.1 all that is needed is that the dual cones can be resolved to simple homotopy cells.) Cellulate ∂X by the dual cells to the simplices of L. These dual cells define a boundary pattern on X as in Sect. 3.2.5. Equivalently, the dual cells give X the structure of an orbifold locally modeled on finite right-angled reflection groups: the local group at a cell dual to a simplex σ is $(\mathbf{C}_2)^{\mathrm{Vert}(\sigma)}$. Finally, apply the basic construction of Sect. 3.1.1 where the group is $(\mathbf{C}_2)^I$ and the strict fundamental domain is X. Let M^n be the resulting closed manifold defined by (3.24).

Remark 3.67 A variation of the Reflection Group Trick shows that given any Haken manifold X with complete boundary pattern, there is a closed Haken manifold that retracts onto it. The point is that the boundary pattern gives X the structure of a locally smooth orbifold with local groups of the form $(\mathbf{C}_2)^k$, $1 \leq k \leq n$. Using (3.24) we can then develop X to a closed manifold M on which $(\mathbf{C}_2)^I$ acts as a reflection group. Since X is Haken, each component of the fixed point set of any reflection on M will be an essential hypersurface in M. Cutting M open along these hypersurfaces we get back X.

One can show that the manifold M resulting from the Reflection Group Trick is aspherical by using the argument in the proof of Theorem 3.63. In other words, first show that M is constructed from copies of X by gluings corresponding to a sequence of amalgams and HNN extensions; then apply Whitehead's Lemma (Corollary A.23 in Appendix A.4). A more careful proof of the asphericity of M using the wreath-graph product construction will be given below. This gives us the following theorem (cf. [73, Remark 15.9] or [82, Thm. 11.1.1]).

Theorem 3.68 (The Reflection Group Trick) *Given a group G of type F, there is a closed aspherical manifold M that retracts onto BG.*

As a consequence we get a number of interesting examples of fundamental groups of closed aspherical manifolds listed in the next proposition.

Proposition 3.69 *The fundamental group of a closed aspherical manifold*

(i) *need not have solvable word problem* (Weinberger [231, p. 106]).
(ii) *need not be residually finite* (Mess [178]).
(iii) *need not coarsely contain expanders in its Cayley graph* (Sapir [204, Cor. 1.3]).
(iv) *can contain a torsion-free "Tarski monster," i.e., a noncyclic finitely generated group such that all its nontrivial proper subgroups are infinite cyclic* (Sapir [204, Cor. 1.3]).

As a first example that can be used to prove part (ii) of the proposition, take G to be one of the Burger–Mozes examples from Sect. 2.4.4, i.e., an infinite simple group that is the fundamental group of one of the square complexes from Theorem 2.50. Thicken BG to a manifold X. Since G is simple, it is not residually finite; hence, neither is any group which retracts onto it. This proves (ii). Similarly, for (i) there is a group G of type F with unsolvable word problem and with $\dim BG = 2$. The Reflection Group Trick then provides an aspherical M such that $\pi_1(M)$ retracts onto G. Hence, the word problem for $\pi_1(M)$ is also not solvable. The reasoning for (iii) and (iv) is similar. Part (iii) means that $\pi_1(M)$ cannot coarsely embed in a Hilbert space and hence, does not have finite asymptotic dimension.

Remark 3.70 In all the examples in Proposition 3.69 one can take BG to be a 2-dimensional CW complex. Since any 2-dimensional CW complex can be thickened to a 4-manifold X, M can be taken to be a closed 4-manifold.

Proof of Theorem 3.68 Let $\tilde{X}$ be the universal cover of X and let $\pi = \pi_1(X)$. Then $\partial\tilde{X} \to \partial X$ is a (not necessarily connected) covering space. Let $\tilde{L}$ be the induced covering of L; so, $\tilde{L}$ is a triangulation of $\partial\tilde{X}$. Let $\tilde{W} := W_{\tilde{L}}$ be the RACG associated to the 1-skeleton of $\tilde{L}$ and put

$$\tilde{M} = D(\tilde{W}, \tilde{X}).$$

Then $\tilde{W} \curvearrowright \tilde{M}$ with $\tilde{X}$ as strict fundamental domain. Since each stratum of $\tilde{X}$ is contractible, it follows from [82, Thm. 9.1.4] that $\tilde{M}$ is contractible. We also have that $\pi \curvearrowright \tilde{L}$ via simplicial automorphisms. Since $\mathrm{Vert}(\tilde{L})$ corresponds to a fundamental set of generators for $\tilde{W}$, this exhibits π as a group of automorphisms of $\tilde{W}$. Form the wreath-graph product, $\tilde{W} \rtimes \pi$. As in Sect. 3.1.3, $\tilde{W} \rtimes \pi$ acts on $\tilde{M}$ with quotient space X $(= \tilde{X}/\pi)$. Finally, we claim that $\tilde{M}$ is a covering space of the manifold M defined in (3.24). To see this, first note that $D(W_L, X) \to M$ is a regular covering space with group of deck transformations $\Gamma_L := \ker(W_L \to \sum \mathbf{C}_2)$. Taking the pull back with respect to $\tilde{X} \to X$ we see that $\tilde{M} \to M$ is a covering space with group of deck transformations $\tilde{\Gamma} \rtimes \pi$ where $\tilde{\Gamma}$ is the kernel of the obvious homomorphism $\tilde{W} \to \sum \mathbf{C}_2$. Since $\tilde{M}$ is homotopy equivalent to the Davis–Moussong complex for $\tilde{W}$, it is contractible. Hence, M is aspherical.

3.3 Blowing Up Zonotopal Cell Complexes

In this section we explain another method for constructing manifolds with the structure of NPC cube complexes. Although there are a number of similarities, this method is different from techniques using right-angled reflection groups. It is the method of [136] and [94, 95] of blowing up zonotopal cell complexes in order to produce NPC cube complexes. The simplest case of this technique, known as Gromov's "Möbius band hyperbolization procedure," is described below and in

[136, §3.4]. Manifolds that result from this process occur in nature, for example, as real toric manifolds (cf. Example 3.57) or as compactifications of the moduli space of points in $\mathbb{R}P^1$.

3.3.1 Zonotopes

Any affine map from $\mathbb{R}^n \to \mathbb{R}^d$ has the form $x \mapsto Vx + y_0$, where V is a linear and $y_0 \in \mathbb{R}^d$. Thus, V can be represented as a $(d \times n)$-matrix $V = [v_1, \dots, v_n]$ consisting of n column vectors $v_i \in \mathbb{R}^d$, $1 \le i \le n$.

Definition 3.71 A *zonotope* in $\mathbb{R}^d$ is the image of the n-cube $[-1, 1]^n$ under an affine map $\mathbb{R}^n \to \mathbb{R}^d$.

Most often we will assume that the translation vector y_0 is 0 so that the zonotope is just the image of a cube under the linear map defined by V. In this case we will say that the zonotope is *centered* (at the origin). Equivalently, a zonotope is a translate of a Minkowski sum $[-v_1, v_1] + \cdots + [-v_n, v_n]$ of line segments in $\mathbb{R}^d$, i.e., it is the convex polytope consisting of all points y in $\mathbb{R}^d$ of the form:

$$y = \sum_{i=1}^{n} x_i v_i \, , \quad \text{with } -1 \le x_i \le 1.$$

A third way is to define a zonotope as the dual to a central hyperplane arrangement, $\{H_i\}_{1 \le i \le n}$, in $(\mathbb{R}^d)^*$. Once we use the inner product to identify the dual space $(\mathbb{R}^d)^*$ with $\mathbb{R}^d$, the hyperplane H_i is defined by the equation $y \cdot v_i = 0$. If Z is the zonotope dual to a hyperplane arrangement $\mathcal{A}$, then ∂Z is dual to the "spherical fan," $\mathbb{S}\text{Fan}(\mathcal{A})$, determined by chambers cut out by the hyperplane arrangement $\mathcal{A}$. (See Sect. 4.2.1 below for the definition of a "spherical fan.") If each chamber of $\mathcal{A}$ is a simplicial cone, then $\mathcal{A}$ is said to be a *simplicial arrangement* and the dual zonotope Z is a *simple zonotope*. (In particular, Z is a simple polytope since ∂Z is dual to a simplicial triangulation of a sphere.) A central hyperplane arrangement is *reducible* if it is combinatorially isomorphic to a nontrivial direct sum of hyperplane arrangements; otherwise, it is *irreducible*. Similarly, a subspace of a hyperplane arrangement is either irreducible or reducible if the normal hyperplane arrangement to the subspace has the respective property. (A *subspace* of the arrangement is a nonempty intersection of hyperplanes in $\mathcal{A}$.) Finally, a face of a zonotope is either *irreducible* or *reducible* as the normal arrangement to the dual subspace has the respective property.

A basic reference for the material in this subsection is Chapter 7 of Ziegler's book [235]. In the next proposition we list some of the basic properties of zonotopes.

Proposition 3.72 (Some Properties of Zonotopes)

(i) *Any zonotope is centrally symmetric. (When Z is centered, the central symmetry is the antipodal map $a : y \to -y$.)*
(ii) *Any face of a zonotope is itself a zonotope.*
(iii) *A polygon is combinatorially isomorphic to a zonotope if and only if it has an even number of edges.*
(iv) *A convex polytope is a zonotope if and only if each of its 2-dimensional faces is centrally symmetric. (See* [235, p. 200] *and the references given there.)*

Example 3.73 (Coxeter Zonotopes) These will be discussed in more detail in Sect. 4.2.2. Suppose $W \curvearrowright \mathbb{R}^d$ as a finite linear reflection group. The collection of hyperplanes which are fixed by some reflection in W is a *reflection arrangement*. If W fixes only the origin, then the hyperplane arrangement is simplicial and the associated zonotope is a *Coxeter zonotope*. Equivalently, a Coxeter zonotope is the convex hull of the W-orbit of a point y which lies in the complement of the hyperplane arrangement. For example, if W has diagram $\mathbf{A}_1 \times \cdots \times \mathbf{A}_d$ (i.e., if $W = (CbC_2)^d$), then the zonotope Z is the *d-cube* $\square^d$. In this case the column vectors of the linear map $V = [e_1, \ldots, e_d]$ form the standard basis of $\mathbb{R}^d$. As another example, if the diagram is $\mathbf{A}_d$, then W is the symmetric group on $d + 1$ letters and Z is the d-dimensional *permutohedron*. (See Fig. 4.1 in Sect. 4.2.2.)

Definition 3.74 A cell complex X is a *zonotopal complex* if each of its cells is combinatorially isomorphic to a zonotope.

For example, a cube complex is a zonotopal complex.

Zonotopal complexes will play a big role in Chap. 4. Given a Coxeter system (W, S), its Davis–Moussong complex $\Sigma(W, S)$ as well as the Salvetti complex $X(W, S)$ for the corresponding Artin group are zonotopal complexes.

3.3.2 *Blowing Up Zonotopes to Cube Complexes*

Next we describe a procedures of [91, 136] and [94] for converting a zonotopal complex X into an cube complex $X_\#$. In most cases of interest cases the cube complex will be nonpositively curved. These blowup constructions are examples of "hyperbolization procedures," which will be discussed later in Chap. 6.

The data needed to define a blowup of a zonotope Z that is dual to a simplicial hyperplane arrangement $\mathcal{A}$ are encapsulated in the notion of a *building set*. This means a certain collection $\mathcal{B}$ of subspaces of the hyperplane arrangement in $(\mathbb{R}^d)^*$. Equivalently, a building set for a zonotope is a certain collection (of parallel classes) of faces of the zonotope Z. (The term "building set" comes from the paper [106] of De Concini and Procesi.) There are two extreme examples of a building set: the *maximal building set* consisting of all proper subspaces of $\mathcal{A}$ and the *minimal building set* consisting of all proper irreducible subspaces. The dual notion is that the maximal building set for the zonotope Z is the collection of all positive dimensional faces, while the minimal building set is the collection of such faces which are irreducible, i.e., which do not have a nontrivial splitting as a product. For any building set $\mathcal{B}$, the blowup of Z will always be a cube complex $Z_{\#\mathcal{B}}$. When $\mathcal{B}$ is the maximal building set, $Z_{\#\mathcal{B}}$ will be NPC. It will also be NPC in the case of the minimal building set provided Z is irreducible.

Given a zonotopal complex X, the correspondence $X \mapsto X_{\#\mathcal{B}}$ is functorial in the following sense: if $f : X \to Y$ is a cellular immersion from X to Y, then there is an induced locally isometric immersion of cube complexes, $X_{\#\mathcal{B}} \to Y_{\#\mathcal{B}}$. (To say that f is a "cellular immersion" we mean that it takes each cell of X isomorphically onto a cell of Y. For example, if f is a combinatorial isomorphism, a covering projection, an inclusion of a subcomplex, or a composition of such maps, then it is a cellular immersion.) In case the building set is maximal (resp., minimal) we shall write $\mathcal{H}_{\#\max}$ (resp., $\mathcal{H}_{\#\min}$). To emphasize the functorial nature of the construction we will use the notation $\mathcal{H}_{\#}(X)$ for $X_{\#}$ even when the blowup set is not specified. In the next proposition we list a few properties of this construction.

Proposition 3.75 (Properties of Blowups, cf. [94]) *Suppose X is a zonotopal cell complex and that $\mathcal{H}_{\#}(X)$ is its blowup with respect to either the maximal or minimal building set.*

(i) *There is a cube complex $\mathcal{H}_{\#}(X)$ of the same dimension as X and with the same vertex set as X.*
(ii) *If Z is a simple zonotope of dimension $d+1$, then $\mathcal{H}_{\#}(Z)$ is $(d+1)$-manifold with boundary*
(iii) *(Preservation of local structure). The link of any vertex of $\mathcal{H}_{\#}(X)$ is a subdivision of the link of the corresponding vertex of X. In particular, if X is a d-manifold, then so is $\mathcal{H}_{\#}(X)$.*
(iv) *The maximal blowup $\mathcal{H}_{\#\max}(X)$ is always an NPC cube complex.*

The cube complex $\mathcal{H}_{\#}(X)$ is defined by induction on the dimension of X. For simplicity we will only deal with the maximal blowup, putting $\mathcal{H}_{\#}(X) = \mathcal{H}_{\#\max}(X)$. Basically, one defines $\mathcal{H}_{\#}(Z)$ for a d-dimensional zonotope Z after assuming that $\mathcal{H}_{\#}(\partial Z)$ has already been defined. (In the case of the minimal blowup, when Z is reducible and has a nontrivial decomposition as $Z = Z_1 \times Z_2$, one defines $\mathcal{H}_{\#\min}(Z)$ to be $\mathcal{H}_{\#\min}(Z_2) \times \mathcal{H}_{\#\min}(Z_2)$.) The induction starts with $\dim X = 1$, in which case $\mathcal{H}_{\#}(X) = X$. The case where Z is a 2-dimensional zonotope is considered in the next paragraph.

The Case When dim Z = 2 By Proposition 3.72 (iii), Z is a polygon whose number of sides is even, say $2m$, and the central symmetry a acts cellularly on it. So, ∂Z and $\partial Z/a$ are cellulations of S^1 by 1-cubes and the natural projection $p : \partial Z \to \partial Z/a$ is the 2-fold covering. Define $\mathcal{H}_{\#}(Z)$ to be the associated $[-1, 1]$-bundle. In other words, $\mathcal{H}_{\#}(Z)$ is a Möbius band. (For this reason, this construction is called the "Möbius band hyperbolization procedure" in [91].) The structure on $\partial Z/a$ as a 1-cube complex (i.e., an m-gon) induces the structure of a square complex on $\mathcal{H}_{\#}(Z)$: each edge in $\partial Z/a$ gives rise to a twisted square connecting a pair of opposite edges in ∂Z. So, if ∂Z is a $2m$-gon, then $\partial Z/a$ is an m-gon and $\mathcal{H}_{\#}(Z)$ is a subdivision of the Möbius band into m squares. Notice that the boundary of $\mathcal{H}_{\#}(Z)$ is canonically identified with ∂Z. In general, if X is any 2-dimensional zonotopal complex, then for each 2-face Z of X one glues $\mathcal{H}(Z)$ onto X^1 via the characteristic map $\partial Z \to X^1$.

The Case When dim $Z = d$ We continue the induction by replacing 2 by d and repeating the previous paragraph. Suppose that $\mathcal{H}_{\#}(X)$ has been functorially defined for all zonotopal complexes of dimension $< d$ and that Z is a d-dimensional zonotope. Then, by induction, $\mathcal{H}_{\#}(\partial Z)$ is defined and, by functorality, the central symmetry a induces a free involution $\mathcal{H}_{\#}(a)$ on $\mathcal{H}_{\#}(\partial Z)$ so that $\mathcal{H}_{\#}(\partial Z/a) = \mathcal{H}_{\#}(\partial Z)/\mathcal{H}_{\#}(a)$ and $\mathcal{H}_{\#}(\partial Z) \to \mathcal{H}_{\#}(\partial Z/a)$ is a 2-fold cover. Define $\mathcal{H}_{\#}(Z)$ to be the associated $[-1, 1]$-bundle. The cubical structure on $\mathcal{H}_{\#}(\partial Z/a)$ induces a cubical structure on $\mathcal{H}_{\#}(Z)$: by taking the product with $[-1, 1]$, each k-cube $\mathcal{H}_{\#}(\partial Z/a)$ gives rise to a $(k + 1)$-cube in $\mathcal{H}_{\#}(Z)$. Finally, if $\dim X = d$, then, by induction, $\mathcal{H}_{\#}(X^{d-1})$ is a well-defined cube complex and one glues each blown up d-face Z to $\mathcal{H}_{\#}(X^{d-1})$ via the functor applied to the characteristic map $\partial Z \to X^{d-1}$ to get the gluing map $\mathcal{H}_{\#}(\partial Z) \to \mathcal{H}_{\#}(X^{d-1})$, as before.

Remarks on the Proof of Proposition 3.75 By induction on dimension, we can assume that the result of applying the construction to the $(d - 1)$-skeleton, $\mathcal{H}_{\#}(X^{d-1})$, is a $(d - 1)$-dimensional cube complex. Since the restriction of the $[-1, 1]$-bundle to any cube in $\mathcal{H}_{\#}(X^{d-1})$ is a trivial interval bundle, we see that $\mathcal{H}_{\#}(X^d)$ inherits the structure of a d-dimensional cube complex. By similar easy arguments using induction one sees that $X \mapsto \mathcal{H}_{\#}(X)$ is functor with respect to cellular immersions and that properties (i), (ii) and (iii) hold. Why is the cube complex $\mathcal{H}_{\#\max}(X)$ nonpositively curved? The point is that if v is a vertex of X, then, by induction, one proves that the link of v in $\mathcal{H}_{\#\max}(X)$ is the barycentric subdivision of the corresponding link of v in X, i.e., $\mathrm{Lk}(v, \mathcal{H}_{\#\max}(X)) = b\,\mathrm{Lk}(v, X)$ (see [94, §3.3 and Thm. 4.4.1]). Since the barycentric subdivision of any cell complex is a simplicial flag complex, it follows from Gromov's Lemma (i.e., Corollary 2.31) that $\mathcal{H}_{\#\max}(X)$ is NPC. This proves (iv). (When X is a cube complex this result is stated in [136, §3.4 and §4.3].)

Remark 3.76 If the zonotope Z is dual to a central hyperplane arrangement in $(\mathbb{R}^d)^*$, the $(\partial Z)/a$ is dual to a triangulation of $\mathbb{R}P^{d-1}$. Since $(\partial Z)/a$ is the 0-section of the canonical $[-1, 1]$-bundle, we see that $\mathcal{H}_{\#}((\partial Z)/a)$ is a codimension-one submanifold of the cube complex $\mathcal{H}_{\#}(Z)$. It is a "hyperplane" of the cube complex in the sense Definition 5.1 in Chap. 5, i.e., it is a union of "midcubes." For example, if Z is a $2m$-gon, then the Möbius band $\mathcal{H}_{\#}(Z)$ is a union of m squares whose 0-sections tessellate the core $\mathbb{R}P^1$ into m intervals. (In the case of a complex blowup as in [106], $\mathcal{H}_{\#}((\partial Z)/a)$ corresponds to the "exceptional divisor.") Thus, for each face of a zonotopal complex X, there is unique exceptional divisor in the interior of the blowup of that face.

3.3.3 Mock Reflection Groups

The theme of [95] and Scott [206] is to develop some examples of cubical structures on manifolds that are analogous to the examples given by right-angled reflection groups in Sect. 3.2. These examples are constructed by blowing up hyperplane arrangements associated to the action of a finite reflection group W on euclidean space. A detailed description of finite Coxeter groups and their associated zonotopes will be given in Sect. 4.2 of the next chapter. Although it might have been preferable to postpone the discussion of these blowups to that section, because of the close analogy with RACGs and cube complexes, we will, instead, discuss them in this subsection.

Suppose the zonotope Z^{d+1} is a "Coxeter zonotope" dual to a hyperplane arrangement corresponding to a finite Coxeter group W acting as a reflection group on $(\mathbb{R}^{d+1})^*$. Let S be a set of fundamental reflections and let $\sigma^d < \mathbb{S}^d$ be the corresponding fundamental spherical simplex (cf. Sect. 4.2.1). Then $\mathcal{H}_\#(Z)$ $(= Z_\#)$ is a $(d+1)$-manifold with boundary with the structure of a cube complex; its boundary M^d is given by

$$M^d := \mathcal{H}_\#(\partial Z).$$

Since all vertices of Z lie in ∂Z, the vertices of the cube complex $\mathcal{H}_\#(Z)$ all lie in M^d. By functorality, W acts on the cube complex. Since W is simply transitively on the vertex set of ∂Z, it is also simply transitive on the vertices of M^d. The simplex σ is a fundamental domain for W on $\mathbb{S}^d$ and it turns out that a fundamental domain for W on M is the manifold with corners $\sigma_\#$ obtained by truncating all the faces of σ that lie in the relevant building set. (N.B.: Although σ is a strict fundamental domain for W on $\mathbb{S}^d$, its truncation $\sigma_\#$ is generally *not* a strict fundamental domain on M^d. In order to construct M/W one needs to make some identifications on the boundary of $\sigma_\#$.) In the case of the maximal building set, we truncate all faces of σ so that the result $\sigma_{\#\max}$ has a codimension-one face for each proper nonempty subset $T < S$. For example, when σ is a 2-simplex, $\sigma_{\#\max}$ is a hexagon; when it is a d-simplex, $\sigma_{\#\max}$ is a d-dimensional permutohedron. When the reflection arrangement is irreducible and we are taking the the minimal building set, then $\sigma_{\#\min}$ an associahedron (if the Coxeter diagram of (W, S) is a connected line segment) or something similar. For example, when $d = 2$, $\sigma_{\#\min}$ is a pentagon. So, in the maximal case, M is a manifold tiled by permutohedra while in the minimal case, at least when (W, S) is irreducible, the tiles are associahedra or their cousins. Further details can be found in [95, §3.1, §3.2].

Hyperplanes in M^d Let w_S denote the element of longest length in W. It turns out that w_S is an involution and that conjugation by w_S takes S to itself. (See [28, Ex. 22, p. 40] or [82]*§4.6 for the definition and properties of the element of longest length in a finite Coxeter group.) Denote the resulting permutation by $j_S : S \to S$. Let $\sigma_\#$ denote the truncated simplex. To simplify the discussion suppose we are dealing with the maximal building set so that $\sigma_\#$ is a permutohedron. Then $\sigma_\#$ has a codimension-one face $(\sigma_\#)_T$ for each proper nonempty subset $T < S$. This face extends to a codimension-one submanifold $M_T < M$, called a "hyperplane" (cf. Remark 3.76). This hyperplane is fixed by the antipodal map a_T in the normal representation to the subspace corresponding to T in $(\mathbb{R}^{(d+1)})^*$. The translates of $\sigma_\#$ by elements of W give a tessellation of M by copies of $\sigma_\#$. For each subset T, let W_T denote the subgroup generated by T and let $w_T \in W_T$ be its element of longest length. The involution w_T acts on M stabilizing the face $(\sigma_\#)_T$, as well as, the codimension-one manifold M_T. If $j_T : T \to T$ is the identity permutation, then W_T fixes M_T pointwise and hence, acts locally as a reflection. Two hyperplanes M_T and $M_{T'}$ intersect if and only if the subsets T and T' are nested, i.e., if $T' < T$ or $T < T'$. (In the case of a minimal building set, M_T and $M_{T'}$ also intersect if T and T' correspond to disjoint commuting subsets of the Coxeter diagram.) Moreover, $\{M_T\}_{T<S}$ is a family of divisors with "normal crossings," in that M_T and $M_{T'}$ are the only hyperplanes containing the codimension-two submanifold $M_T \cap M_{T'}$ (so it looks like they intersect orthogonally). Thus, $\{M_T\}_{T<S}$ closely resembles the family of reflecting hyperplanes $\{x_i = 0\}$ in the $(\mathbf{C}_2)^I$-action on the polyhedral product P_L. The difference is that the family $\{w_T\}_{T<S}$ do not pairwise commute. In fact, if $T' < T$ and $T'' = j_T(T')$, we have the relation,

$$w_T w_{T'} w_T = w_{T''}.$$

In [95], the w_T are said to act on M as *mock reflections*.

Lifting to the Universal Cover of M^d As we did with the $(\mathbf{C}_2)^I$-action on P_L, we can lift the W-action on M to the action of a group G on the universal cover $\tilde{M}$. Fix a lift $\tilde{\sigma}_\#$ of the truncated simplex $\sigma_\#$ and let α_T be the lift of w_T which takes $\tilde{\sigma}_\#$ to the adjacent truncated simplex across the face indexed by T. It can be is shown in [95] that G is generated by the set of involutions $\{\alpha_T\}_{T<S}$ and that it has a presentation of the form:

$$\langle \{\alpha_T\}_{T<S} \mid \{(\alpha_T)^2\}_{T<S};\, \alpha_T\alpha_{T'}\alpha_T\alpha_{T''}, \quad \text{when } T' < T \text{ and } T'' = j_T(T")\rangle. \tag{3.25}$$

We give a proof that this is, in fact, a presentation for G in a subsequent paragraph.

Next we discuss the cube complex structures on M and $\tilde{M}$. There is a spherical simplex σ centered at each vertex of ∂Z. The group W acts simply transitively on the vertex set of M $(= \mathrm{Vert}(\partial Z))$. Hence, M $(= \mathcal{H}_\#(\partial Z))$ is tiled by copies of $\sigma_\#$ each of which is centered at a vertex (one for each element of W). Fix the vertex $v \in \partial Z$ at the center of the fundamental simplex σ. Then v and another vertex $wv \in M$ span an edge of the cube complex if and only if they span one of the original edges in ∂Z (i.e., if $w \in S$) or if v and wv are antipodal vertices across the face corresponding to a subset $T < S$ (i.e., if $wv = w_T v$). Next consider $\tilde{M}$. We have a short exact sequence,

$$1 \to \pi_1(M) \to G \to W \to 1$$

(cf. (3.13) in Sect. 3.1.1). As was the case for M, $\tilde{M}$ is tiled by copies of the truncated simplex $\sigma_\#$, one copy for each element of G. The cube complex structure on M lifts to a cube complex structure on $\tilde{M}$ and G acts simply transitively on $\mathrm{Vert}(\tilde{M})$. It follows that the 1-skeleton of the cube complex for $\tilde{M}$ is the Cayley graph of G and that its 2-skeleton is the Cayley 2-complex. In detail, let $\tilde{v}$ be a lift of v. There is an edge emanating from $\tilde{v}$ for the lift of each edge in M corresponding to a nested pair $\{T', T\}$. Let α_T be the lift of the involution $w_T \in W$ that stabilizes this edge. It follows that $\{\alpha_T\}_{T<S}$ is a set of generators for G and that G has a presentation given by (3.25). So, the situation is completely analogous to the discussion of RACGs in the last paragraph of Sect. 3.1.1—the only difference being that rather than having commutation relations, $s_i s_j s_i s_j$, of length 4 between fundamental generators we now have relations of length 4 of the form $\alpha_T \alpha_{T'} \alpha_T \alpha_{T''}$ where T, T', T'' are as in (3.25) (and where T'' is not necessarily equal to T').

Remark 3.77 The orbit space M/G is usually not isomorphic to $\sigma_\#$; so, $\sigma_\#$ is *not* a strict fundamental domain and the complex of groups structure over $\sigma_\#$ is generally *not* that of a simple complex of groups.

Theorem 3.78 (Maximal Blowups (Davis-Januszkiewicz-Scott [94, 95])) *Suppose that (W, S) is a finite Coxeter system of rank $(d+1)$, that Z is the corresponding zonotope and that $M^d = \mathcal{H}_{\#\max}(\partial Z)$ is the maximal blowup. Then M^d is tiled by permutohedra and the dual cubical structure is NPC. The mock reflection group G defined by the presentation* (3.25) *acts simply transitively on the vertices of the* CAT(0) *cube complex $\tilde{M}^d$.*

Example 3.79 (The Maximal Blowup of the Boundary of a Cube) Consider the case where the zonotope Z is a $(d+1)$-cube, the Coxeter group is $(\mathbf{C}_2)^{d+1}$, and where we are using the maximal building set. Then ∂Z is an octahedron. The truncated fundamental simplex $\sigma_\#$ is a permutohedron. The group $(\mathbf{C}_2)^{d+1}$ acts on $M = \mathcal{H}_\#(\partial Z)$. For each $T < S$ the permutation j_T is $\mathbb{I}_T$, the identity permutation of T and the involution w_T acts as the identity on the codimension-one submanifold M_T. So, each mock reflection is an actual reflection and $(\mathbf{C}_2)^{d+1}$ is a reflection group on M with fundamental chamber the right-angled permutohedron, $\sigma_\#$. It turns out that M is the real toric variety that was described in Example 3.57.

Theorem 3.80 (Minimal Blowups, cf. Davis-Januszkiewicz-Scott [94, 95]) *Suppose that (W, S) is a finite irreducible Coxeter system of rank $(d+1)$, that Z is the corresponding zonotope and that $M^d = \mathcal{H}_{\#\min}(\partial Z)$ is the minimal blowup. (In other words, only the irreducible faces of ∂Z are blown up.) The cubical structure on M^d is NPC. If the Coxeter diagram of (W, S) is a connected line segment, then M^d is tiled by copies of a d-dimensional associahedron (otherwise, it is tiled by graph associahedra). There is a mock reflection group G defined by a presentation similar to* (3.25) *but it uses only pairs $\{T, T'\}$ which are nested with respect to the minimal building set. The group G is simply transitive on the vertex set of $\tilde{M}^d$.*

Example 3.81 (The Minimal Blowup of the Boundary of a Permutohedron) Suppose Z^{d+1} is the permutohedron associated to the Coxeter group with diagram

$\mathbf{A}_{d+2}$ (i.e., W is the symmetric group on $d+2$ letters). Consider the minimal blowup $M^d := \mathcal{H}_{\#\min}((\partial Z)/a)$. It has as a (nonstrict) fundamental domain the truncated d-simplex $\sigma_{\#\min}$, and it turns out that this is a d-dimensional associahedron. (See [95, §8.1] for the definition of associahedron.) As is the case with all of these blowups of zonotopes via building sets, the manifold M is naturally a cube complex. Since we are taking a minimal blowup and the Coxeter system is irreducible, this cube complex is NPC. Kapranov [165] proved that M^d has several other equivalent descriptions. In particular, he showed that the following varieties are isomorphic (also see [94, Thm. 0.3.1]):

(i) $\mathcal{H}_{\#\min}((\partial Z)/a)$, the minimal blowup of the braid arrangement;
(ii) the Chow quotient $G(2, d+3)//(\mathbb{R}^*)^{d+3}$, where $G(2, d+3)$ means the Grassmann of 2-planes in $\mathbb{R}^{d+3}$;
(iii) the Chow quotient of $(d+3)$ points in the projective line, $(\mathbb{R}P^1)^{d+3}//GL(2, \mathbb{R})$;
(iv) The real points of the Grothendieck-Knudsen moduli space for stable $(d+3)$-pointed curves of genus 0, $\overline{\mathcal{M}}_{0,d+3}$.

We note that descriptions (iii) and (iv) are both compactifications of $\mathcal{M}_{0,d+3}$, the moduli space of $(d+3)$ distinct points in the projective line.

The associated mock reflection group G of Theorem 3.80 acting on the universal cover $\tilde{M}$ was been called the *cactus group* in [113], [129]. (I am not sure of the reason for this terminology.) The kernel of the natural homomorphism to the symmetric group is called the *pure cactus group*; it can be identified with $\pi_1(M)$. The main result of [113] is a calculation of the cohomology of the pure cactus group. Recent papers by Genevois [128, genevoisnote] establishes properties of the cactus group which are relevant in geometric group theory. In particular, he shows that cactus groups are acylindrically hyperbolic and deduces consequences of this.

Part III
Coxeter Groups, Artin Groups, Buildings

Chapter 4
Coxeter Groups, Artin Groups, Buildings

The basic notion in this chapter is that of a "Coxeter system," (W, S). There are three related types of groups that depend on this notion: Coxeter groups, Artin groups, and chamber-transitive automorphism groups of buildings. In each case the group acts on an associated polyhedron. In the case of a Coxeter system the polyhedron is the "Davis–Moussong complex;" in case of an Artin group it is the "Deligne complex;" in the case of a building it is the "standard realization" of the building. The action of any one of these groups has a strict fundamental domain called the "standard fundamental chamber." The fundamental chamber depends only on the Coxeter system. In each case we get a system of groups with the structure of a complex of groups. The fact that the fundamental domain is strict means that in each case we have a simple complex of groups. The Davis–Moussong complex and the standard realization of a building are both CAT(0), hence, contractible. For a general Artin group, it is an open question if its Deligne complex is contractible. For an Artin group, there is a related polyhedron called the "Salvetti complex." The Artin group is the fundamental group of the quotient of the Salvetti complex by the associated Coxeter group. The conjecture that the Deligne complex is contractible is equivalent to the conjecture that the quotient of the Salvetti complex is the classifying space for the Artin group. (This is $K(\pi, 1)$-Conjecture for the Artin group.)

4.1 Some Simple Complexes of Groups

In this Sect. 4.1.1, we summarize the theory of "simple complexes of groups" as developed in Bridson-Haefliger in [35, II.12] or in Appendix A of this book. In

---

The original version of the chapter has been revised. A correction to this chapter can be found at https://doi.org/10.1007/978-3-031-48443-8_8

M. W. Davis, *Infinite Group Actions on Polyhedra*, Ergebnisse der Mathematik
und ihrer Grenzgebiete. 3. Folge / A Series of Modern Surveys in Mathematics 77,
https://doi.org/10.1007/978-3-031-48443-8_4

Sect. 4.1.2, we define the three basic types of examples discussed in this chapter: Coxeter groups, Artin groups, and buildings. In Sect. 4.1.3 we discuss the $K(\pi, 1)$-Question for each type of example.

4.1.1 Quick Summary

Roughly, a *simple complex of groups* $G\mathcal{Q}$ is a poset of groups, i.e., it consists of a poset $\mathcal{Q}$, a collection of groups $\{G_\sigma\}_{\sigma\in\mathcal{Q}}$ and monomorphisms $\varphi_{\tau\sigma} : G_\sigma \hookrightarrow G_\tau$ defined whenever $\sigma > \tau$. (See Definition A.1 in Appendix A.) N.B. If $\mathcal{Q}$ is the poset of faces of a polyhedron, then the monomorphisms $G_\sigma \hookrightarrow G_\tau$ are in the opposite directions from the inclusions $\tau \hookrightarrow \sigma$. The precise definitions and the details of this theory will be postponed until Appendix A. Here we shall be content with giving a few explanatory comments.

Simple complexes of groups come from group actions on spaces which have a strict fundamental domain (see Definition 3.5). The local groups G_σ are the isotropy subgroups at points in the fundamental domain. Two points in the fundamental domain belong to the same (open) stratum if the isotropy subgroups at those points are equal.

Since a simple complex of groups is a system of groups as in Sect. 3.1.2, it has a "direct limit," $\lim G\mathcal{Q}$, which we will often denote simply by G. The simple complex of groups $G\mathcal{Q}$ is *developable* if it is induced from a group action with a strict fundamental domain. Equivalently, it is developable if the natural homomorphism from each local group G_σ to the direct limit G is injective.

We shall use two forms of the basic construction of Sect. 3.1.1. First, there is the poset $D(G, \mathcal{Q})$ defined as the disjoint union of cosets:

$$D(G, \mathcal{Q}) = \bigsqcup_{\sigma\in\mathcal{Q}} G/G_\sigma, \tag{4.1}$$

where $gG_\tau < hG_\sigma$ if and only if $\sigma > \tau$ and we have reverse inclusion of cosets (i.e., hG_σ is a subset of gG_τ).

Recall that the *geometric realization* of the poset $\mathcal{Q}$ is the simplicial complex $|\mathcal{Q}|$ whose simplices are the finite chains in $\mathcal{Q}$. For each point x in $|\mathcal{Q}|$, let $\sigma(x) \in \mathcal{Q}$ be the maximum vertex of the simplex containing x in its relative interior. The *stratum* of $|\mathcal{Q}|$ indexed by σ is defined by

$$|\mathcal{Q}|_\sigma := \{x \in |\mathcal{Q}| \mid \sigma(x) \le \sigma\}. \tag{4.2}$$

In other words, $\sigma(x)$ indexes the smallest stratum such that $x \in |\mathcal{Q}|_\sigma$.

The second incarnation of the basic construction is the polyhedron,

$$D(G, |\mathcal{Q}|) = (G \times |\mathcal{Q}|)/\sim; \tag{4.3}$$

defined as in Sect. 3.1.1, where given $x \in |\mathcal{Q}|$, an equivalence relation $\sim$ is then defined the same way as in (3.9):

$$(g,x) \sim (g',x') \iff x = x' \text{ and } g^{-1}g' \in G_{\sigma(x)}\,, \tag{4.4}$$

Hence, $(g,x) \sim (g',x')$ if and only if $x = x'$ and $gG_{\sigma(x)}$ and $g'G_{\sigma(x)}$ are the same coset. So, $D(G,|\mathcal{Q}|)$ is just the geometric realization of the poset $D(G,\mathcal{Q})$. The poset $D(G,\mathcal{Q})$ and the polyhedron $D(G,|\mathcal{Q}|)$ are both called the *development* of $G\mathcal{Q}$ with respect to G.

4.1.2 The Basic Examples

We start with some relevant definitions from Sects. 4.2–4.4.

Suppose L^1 is a simplicial graph with edges labeled by integers ≥ 2. Put $S = \mathrm{Vert}(L^1)$. The edge labeling is a function $m : \mathrm{Edge}(L^1) \to \{2,3,\dots\}$ written as $\{s,t\} \mapsto m(s,t)$. The labeled graph (L^1, m) is the *defining graph* of a Coxeter system. For $s = t$, put $m(s,s) = 1$ and if distinct elements s and t do not span an edge, put $m(s,t) = \infty$. The symmetric $(S \times S)$-matrix $(m(s,t))$ is called the *Coxeter matrix* associated to (L^1, m). This data can be used to define presentations for two groups, a Coxeter group W and an Artin group A. The set of generators S for W coincides with the vertex set of L^1. The relations are:

$$s^2, \text{ for all } s \in S, \quad \text{and} \quad (st)^{m(s,t)}, \text{ for all } \{s,t\} \in \mathrm{Edge}(L^1). \tag{4.5}$$

That is to say, each generator in S is an involution and each relation in the presentation for W is a word involving only two letters that correspond to the vertices of an edge of L^1. The group W is the *Coxeter group* and the pair (W,S) is the *Coxeter system* associated to the labeled graph (L^1, m). For any subset T of S, the subgroup W_T generated by T is a *special subgroup*. It turns out that (W_T, T) is itself a Coxeter system (see [28, pp. 12–13] or [82, Thm.4.1.6]). If W_T is finite, then the subset T and the subgroup W_T that it generates are said to be *spherical*. Let $\mathcal{S}$ (or $\mathcal{S}(W,S)$) denote the poset of spherical subsets of S. Let $\mathcal{S}^{\mathrm{op}}$ denote the opposite poset to $\mathcal{S}$. The usual partial order on $\mathcal{S}$ corresponding to inclusion of subsets is denoted by $<$ so that $T < T'$ means that T is a proper subset of T'. In $\mathcal{S}^{\mathrm{op}}$ the partial order $\prec$ is reverse inclusion.

To define the presentation for the associated Artin group A, introduce new symbols $\{a_s\}_{s\in S}$ for the generators. The relations are given by omitting the relations corresponding to the s^2 in (4.5) and by rewriting the relations corresponding to $(st)^{m(s,t)}$ as

$$\underbrace{a_s a_t \cdots}_{m(s,t)\text{ terms}} = \underbrace{a_t a_s \cdots}_{m(s,t)\text{ terms}} \tag{4.6}$$

In other words, an alternating word of length $m(s,t)$ in a_s and a_t is equal to the alternating word of the same length beginning with the other generator. The relation in (4.6) is an *Artin relation* (or a *braid relation* when $m(s,t) = 3$). Since each s has order two, it follows that in (W, S) the Artin relation is simply a rewriting of $(st)^{m(s,t)}$. Hence, $a_s \mapsto s$ defines an epimorphism $A \to W$.

A *spherical coset* in W is an element of W/W_T, where W_T is a spherical special subgroup. Let

$$\mathrm{Coset}(W) := \bigsqcup_{T \in \mathcal{S}} W/W_T \tag{4.7}$$

denote the set of spherical cosets. Then $\mathrm{Coset}(W)$ has the structure of a poset, where the order relation is reverse inclusion of one coset in another. Similarly, a *spherical coset* in A is a left coset of A_T, where $T \in \mathcal{S}$ and where A_T denotes the special subgroup of A generated by $\{a_s\}_{s\in T}$. Let

$$\mathrm{Coset}(A) := \bigsqcup_{T \in \mathcal{S}} A/A_T \tag{4.8}$$

denote the *poset of spherical cosets* in A. (The geometric realization of the order complex of $\mathrm{Coset}(W)$ is the barycentric subdivision of the Davis–Moussong complex $\Sigma(W, S)$. This zonotopal cell complex will be described again in Sect. 4.2.3 below. Similarly, the geometric realization of the order complex of $\mathrm{Coset}(A)$ is the barycentric subdivision of the Deligne complex, $\Lambda(W, S)$, that will be described in Sect. 4.3.1.)

Definition 4.1 There is an associated abstract simplicial complex $L(W, S)$ (or L for short) called the *nerve* of (W, S). Its vertex set is S and a nonempty subset of S spans a simplex if and only if it is spherical. In other words, the poset of simplices in L is $\mathcal{S}_{>\emptyset}$.

Note that a subset $\{s, t\}$ consisting of two distinct vertices is spherical if and only if it is an edge in the defining graph. So, the 1-skeleton of L is equal to the defining graph L^1. In the case (W, S) is right-angled, L is the flag completion of the defining graph. The general case is analogous: a complete subgraph of L^1 determines a simplex in L if and only if its vertex set is spherical.

Example 4.2 (The Fundamental Chamber) The geometric realization of $\mathcal{S}^{\mathrm{op}}$ (or of $\mathcal{S}$) is denoted by K and called the *fundamental chamber* for (W, S). For any $T \in \mathcal{S}^{\mathrm{op}}$, the geometric realization of the poset $\mathcal{S}^{\mathrm{op}}_{\leq T}$ is denoted by K_T and called the *T-stratum*. For any $s \in S$, put $K_s := K_{\{s\}}$. If W is the Coxeter group generated by reflections across the facets (= "codimension-one faces") of a simple convex polytope P in $\mathbb{E}^n$ or $\mathbb{H}^n$, then K is identified with P so that $\mathcal{S}^{\mathrm{op}}$ is isomorphic to the poset of faces of P. In other words, a stratum of K is a face of P. The poset $\mathcal{S}$ is the face poset of the dual simplicial complex $L(W, S)$.

Example 4.3 (The Simple Complex of Groups Associated to a Coxeter Group) There is a simple complex of groups $W\mathcal{S}^{\mathrm{op}} = \{W_T\}_{T\in\mathcal{S}^{\mathrm{op}}}$ over $\mathcal{S}^{\mathrm{op}}$ called the *complex of spherical subgroups of* W. Its direct limit is equal to W. On the level of posets, $D(W, \mathcal{S}^{\mathrm{op}})$ is equal to the poset of spherical cosets, Coset(W), defined in (4.7). As a polyhedron, the basic construction $D(W, K)$ is the geometric realization of Coset(W). Thus, $D(W, K)$ is a union of its set of chambers $\{wK\}_{w\in W}$. Alternatively, it is dual to the zonotopal cell complex $\Sigma(W, S)$ of Sect. 4.2.3. (The strata of $D(W, K)$ are dual cones to the cells in $\Sigma(W, S)$.) Two chambers wK and $w'K$ of $D(W, K)$ are *s-adjacent* if and only if $w' = ws$ for some $s \in S$ and if this is the case, their intersection $wK \cap wsK$ is isomorphic to the "mirror" K_s.

Example 4.4 (The Simple Complex of Groups Associated to an Artin Group A) This is the simple complex of groups $A\mathcal{S}^{\mathrm{op}}$ over $\mathcal{S}^{\mathrm{op}}$ defined as $\{A_T\}_{T\in\mathcal{S}^{\mathrm{op}}}$. If $T' \prec T$, then A_T is a subgroup of $A_{T'}$. We also have $\lim A\mathcal{S}^{\mathrm{op}} = A$. On the level of posets, the development $D(A, A\mathcal{S}^{\mathrm{op}})$ is equal to Coset(A), the set of spherical cosets defined in (4.8). On the level of spaces, the development $D(A, K)$ is the Deligne complex, $\Lambda(W, S)$, defined in Sect. 4.3.1 below.

Example 4.5 (The Complex of Spherical Residues in a Building) As a combinatorial object a building consists of a set $\mathcal{C}$ of "chambers" together with some additional structure. Part of this structure is a Coxeter system (W, S). A building has a geometric realization $D(\mathcal{C}, K)$ in which each $C \in \mathcal{C}$ is isomorphic to the fundamental chamber K of Example 4.2. As with the Davis–Moussong complex, there is a notion of two chambers being "s-adjacent"; the difference from the Davis–Moussong complex being that in a building two or more chambers can be s-adjacent along a mirror of the form K_s. Given a chamber $C \in \mathcal{C}$ and a spherical subset $T \in \mathcal{S}^{\mathrm{op}}$, there is also the notion of the "T-residue" containing C: it consists of all chambers C' such that $C \cap C'$ contains a stratum of type K_T. The poset for $\mathcal{C}$ is $R(\mathcal{C})$, the poset of spherical residues. An "apartment" in the building $D(\mathcal{C}, K)$ is an embedded copy of the Davis–Moussong complex $D(W, K)$.

The building $\mathcal{C}$ has a *chamber-transitive* automorphism group if there is a group G that acts transitively on $\mathcal{C}$ and preserves the s-adjacency relations. If G acts chamber transitively on $\mathcal{C}$, then there is an associated simple complex of groups $G\mathcal{S}^{\mathrm{op}} := \{G_T\}_{T\in\mathcal{S}^{\mathrm{op}}}$, where G_T is the stabilizer of the T-residue containing some given chamber C. The direct limit of $G\mathcal{S}^{\mathrm{op}}$ is G; the poset $D(G, \mathcal{S}^{\mathrm{op}})$ is $R(\mathcal{C})^{\mathrm{op}}$; and $D(G, K) = D(\mathcal{C}, K)$.

4.1.3 The $K(\pi, 1)$-Question for Simple Complexes of Groups

Here is a short summary of some material about the $K(\pi, 1)$-Question that is given in more detail in Sects. 4.2.3, 4.3.4, 4.4.2 (in particular, in Theorems 4.12, 4.76, 4.107), and in Appendix A.4.

Suppose $G\mathcal{Q} = \{G_\sigma\}_{\sigma\in\mathcal{Q}}$ is a simple complex of groups over the poset $\mathcal{Q}$. Let $\{BG_\sigma\}_{\sigma\in\mathcal{Q}}$ be the corresponding collection of classifying spaces. Since the monomorphisms $\varphi_{\sigma\tau} : G_\tau \to G_\sigma$ can be realized as inclusions of spaces $\overline{\varphi}_{\sigma\tau} : BG_\tau \to BG_\sigma$, we get a poset of aspherical spaces $\{BG_\sigma\}_{\sigma \in \mathcal{Q}}$. Using iterated mapping cylinders we can glue together the spaces BG_σ to form a new space $BG\mathcal{Q}$ called the *aspherical realization* of the simple complex of groups. The CW complex $BG\mathcal{Q}$ is equipped with a projection $p : BG\mathcal{Q} \to |\mathcal{Q}|$ so that for each vertex $\sigma \in \mathcal{Q}$, $p^{-1}(\sigma)$ is homotopy equivalent to BG_σ. The details of this construction are given in Appendix A.4. When $|\mathcal{Q}|$ is simply connected, it follows from van Kampen's Theorem that the fundamental group of $BG\mathcal{Q}$ is the direct limit G. By a theorem of Haefliger [138], the space $BG\mathcal{Q}$ is well-defined up to homotopy equivalence (cf. Theorem A.18). The $K(\pi, 1)$-Question for the simple complex of groups $\{BG_\sigma\}_{\sigma \in \mathcal{Q}}$ is the following.

Question 4.6 Is the aspherical realization of a simple complex of groups aspherical? In other words, is $BG\mathcal{Q}$ a $K(\pi, 1)$? (When $|\mathcal{Q}|$ is simply connected, $\pi_1(BG\mathcal{Q}) = G$.)

The next theorem provides an answer. A more detailed proof and discussion will be given in Appendix A.4. Here we just sketch the proof.

Theorem 4.7 (cf. Theorem A.19 in Appendix A.4) *Suppose $G\mathcal{Q}$ is a developable simple complex of groups and that $|\mathcal{Q}|$ is simply connected. Then $BG\mathcal{Q}$ is homotopy equivalent to BG if and only if the development $D(G, |\mathcal{Q}|)$ is contractible.*

Sketch of Proof Consider the Borel construction

$$D(G, |\mathcal{Q}|) \times_G EG.$$

There is a projection map $\pi : D(G, |\mathcal{Q}|) \times_G EG \to |Q|$ induced by projection onto the first factor, where $|\mathcal{Q}|$ is identified with $D(G, |\mathcal{Q}|)/G$. The fiber over the vertex $\sigma \in \mathcal{Q}$ is identified with EG/G_σ, which is homotopy equivalent to BG_σ. So, $D(G, |\mathcal{Q}|) \times_G EG$ is a model for $BG\mathcal{Q}$. Since $D(G, |\mathcal{Q}|) \times EG$ is G-homotopy equivalent to EG if and only if $D(G, |\mathcal{Q}|)$ is contractible, the result follows.

This theorem shows that if $D(G, |\mathcal{Q}|)$ is CAT(0), then the $K(\pi, 1)$-Question for $G\mathcal{Q}$ has a positive answer. Indeed, this is a reason why we know the answer is positive for $W\mathcal{S}^{\mathrm{op}}$ and for a chamber-transitive building $\mathcal{C}$. It is also the reason we conjecture that the answer is positive for the Artin complex $A\mathcal{S}^{\mathrm{op}}$.

Remark 4.8 Of course, the answer to the $K(\pi, 1)$-Question depends on the choice of simple complex of groups whose direct limit is G. Indeed, the development $D(G, |\mathcal{Q}|)$, as well as the aspherical realization $BG\mathcal{Q}$, both depend on $G\mathcal{Q}$. For example, consider the simple complex of groups $G\mathcal{Q} = W\mathcal{S}^{\mathrm{op}}$ of Example 4.3 where we assume that the Coxeter group W is finite. If $\mathrm{Card}(S) = n$, then the fundamental chamber $|\mathcal{Q}| = |\mathcal{S}^{\mathrm{op}}|$ is the cone on an $(n-1)$-simplex Δ while the development $D(W, \mathrm{Cone}\,\Delta)$ is the disk D^n (cf. Sect. 4.2.2). If we replace the poset $\mathcal{Q}$ by the poset of proper spherical subsets $\mathcal{Q}' := \mathcal{S}^{\mathrm{op}} - \{S\}$, then the direct limit W is unchanged, the fundamental chamber $|\mathcal{Q}'|$ is Δ^{n-1} and the development is S^{n-1}. The aspherical realization $EG\mathcal{Q}$ is $S^{n-1} \times_W EW$, which is a sphere bundle over BW, and hence, is not homotopy equivalent to BW. it is not even aspherical if $n \geq 3$. So, the $K(\pi, 1)$-Question has a negative answer for $G\mathcal{Q}'$.

4.2 Coxeter Groups

There are at least two methods for constructing a geometric object with an action of a given Coxeter group. First there is the piecewise euclidean CAT(0) cell complex $\Sigma(W, S)$. Basic examples of Coxeter groups occur as a groups W generated by reflections across the facets of a polytope in a space of constant curvature. The second method for realizing (W, S) as a complex Σ generalizes such examples of reflection groups. We will describe several different interrelated perspectives from which to view geometric actions of Coxeter groups on polyhedra. We organize these viewpoints into the areas (A), (B), (C), (D), (E) listed below. A basic reference for this material is [28]. Much of the exposition in this section is taken from the fuller account given in my book [82].

Outline of This Section

(A) The Davis Moussong complex, $\Sigma(W, S)$ (Sect. 4.2.3)

- (A.1) Spherical Coxeter groups and Coxeter zonotopes (Sects. 4.2.1 and 4.2.2).
- (A.2) Moussong's Theorem: $\Sigma(W, S)$ is CAT(0) (Sect. 4.2.4).
- (A.3) The $K(\pi, 1)$-Question for Coxeter groups (Sect. 4.2.4).

(B) Fundamental polytopes and convex cocompact actions on $\mathbb{H}^n$ (Sect. 4.2.8).

(C) Word hyperbolic and relatively hyperbolic Coxeter groups (Sects. 4.2.4, 4.2.5, and 4.2.8).

- (C.1) Moussong's Conditions for word hyperbolicity (Lemma 4.21).
- (C.2) Caprace's conditions for relative hyperbolicity (Theorem 4.26).

(D) Linear and projective representations of Tits and Vinberg (Sects. 4.2.7 and 4.2.9).

- (D.1) The fundamental chamber in the Tits representation is a simplicial cone.
- (D.2) In Vinberg's representations, the fundamental chambers are polyhedral cones.

(E) Convex, cocompact actions on proper convex domains in $\mathbb{R}P^N$ (Sect. 4.2.9).

The next three paragraphs contain comments on various parts of the above outline.

(A) and (C). The cells of the Davis–Moussong complex are "Coxeter zonotopes." Associated to a spherical Coxeter subsystem (W_T, T) there is a Coxeter zonotope $Z(W_T, T)$. It is a convex cell in a euclidean space of dimension equal to Card(T) (cf., Example 3.73 in Sect. 3.3.1). Copies of these cells are glued together to form $\Sigma(W, S)$. The main result is Moussong's Theorem: the natural piecewise euclidean metic on $\Sigma(W, S)$ is CAT(0). The cells $Z(W_T, T)$ can also be realized as convex cells in $\mathbb{H}^n$; so we get a piecewise hyperbolic metric on $\Sigma(W, S)$. This leads to the question of when is this metric (with suitably scaled edge lengths) CAT(−1)? Moussong also answered this: the piecewise hyperbolic metric on $\Sigma(W, S)$ is CAT(−1) if and only if (i) no special subgroup of W is a euclidean Coxeter group of rank at least 3 and (ii) W has no special subgroups of the form $W_{T_1} \times W_{T_2}$, where W_{T_1} and W_{T_2} are both infinite (see Lemma 4.21). Moreover, W is word hyperbolic if and only if $\Sigma(W, S)$ can be given a CAT(−1) structure. (See [81, 82, 84] for general discussions of reflection groups and nonpositive curvature.)

(C). There is a closely related notion of a "relatively hyperbolic" Coxeter group. The basic example is a group generated by reflections across the facets of a polytope of finite volume in $\mathbb{H}^n$. (Such a polytope is allowed to have "ideal vertices" lying on the sphere at infinity. The stabilizer of such an ideal vertex is a euclidean Coxeter group.) There is a general notion of what it means for a group to be relatively hyperbolic (relative to a family of subgroups). In the case of Coxeter groups the family of subgroups is the family of maximal free abelian subgroups. It turns out that W is relatively hyperbolic if and only if $\Sigma(W, S)$ does not have "isolated flats" of dimension > 1. Caprace proved that this is equivalent to a condition on special subgroups of the form $W_T = W_{T_1} \times W_{T_2}$. His condition is that any such W_T is virtually abelian. In other words, either both factors are euclidean reflection groups or at least one factor is spherical. (See Theorem 4.26.)

(B), (D) and (E). A basic fact is that every Coxeter group W admits faithful linear representations as reflection group on some vector space so that W acts properly on some open convex subset. Given a Coxeter system (W, S) Tits defined a representation of W on $\mathbb{R}^N$, with $N = \text{Card}(S)$, across the facets of a simplicial cone C. (The facets of the simplex are indexed by S.) The *Tits cone* is the union of all translates of C by elements of W. It is a convex cone and it is a proper subset of $\mathbb{R}^N$ whenever W is not spherical. The group W acts properly on the interior of the Tits cone and hence, on the image Ω of the interior of the Tits cone in $\mathbb{R}P^{N-1}$. More generally, Vinberg [227] considered representations of W on $\mathbb{R}^N$ as groups generated by reflections across the facets of a convex polyhedral cone C. As before, a subset of facets of C corresponds to the set of generators S; however, the dimension of the vector space can now be less than Card(S). The image of C in $\mathbb{R}P^{N-1}$ is a convex polytope P (often called the "Poincaré polytope"). The image of W in $PGL(\mathbb{R}^N)$ is discrete if and only if each spherical dihedral subgroup $W_{\{s,t\}}$ acts on $\mathbb{R}^N$ with strict fundamental domain the sector determined by the corresponding facets P_s and P_t. When does such a representation give a proper action of W on some open fundamental domain Ω of $\mathbb{R}P^{N-1}$ with compact quotient, i.e., when is the W-action "convex cocompact"? This is answered in [71]. It turns out that the class of (W, S) which admit such cocompact representations strictly contains the class of all word hyperbolic W and is contained in the class of relatively hyperbolic W. (See Proposition 4.49 below.)

4.2.1 Spherical Coxeter Groups

In this subsection we suppose that the Coxeter group W is finite. The Coxeter system (W, S) has a representation as an orthogonal linear reflection group on $\mathbb{R}^n$, with $n = \text{Card}(S)$, so that each $s \in S$ corresponds to reflection across a supporting hyperplane of a simplicial cone. When (W, S) is irreducible this representation is unique up to homothety (e.g., see [28, p. 70]). It is called the *geometric representation* or the *canonical representation*. The set of all reflections R in W is precisely the set of conjugates of elements of S. The linear hyperplane H_r fixed by a reflection $r \in R$ is called a *wall*. The set of all such hyperplanes, $\mathcal{A} := \{H_r\}_{r\in R}$, is the associated *reflection arrangement*. The arrangement $\mathcal{A}$ cuts $\mathbb{R}^n$ into simplicial cones; this collection of cones together with their faces is called the *fan*, $\text{Fan}(\mathcal{A})$, associated to $\mathcal{A}$. Each top-dimensional simplicial cone in $\text{Fan}(\mathcal{A})$ is a *chamber*. There are two chambers (antipodal to each other) bounded by the walls indexed by S. Let C be one of them. Then C is a strict fundamental domain for the W-action on $\mathbb{R}^n$. A *codimension-one face* C_s of C is the intersection of C with the wall H_s fixed by s. The intersection of $\text{Fan}(\mathcal{A})$ with $\mathbb{S}^{n-1}$ is the *spherical fan*, denoted $\mathbb{S}\text{Fan}(\mathcal{A})$. It is a tessellation of $\mathbb{S}^{n-1}$ by spherical simplices. The underlying simplicial complex is the *Coxeter complex* of (W, S).

The intersection $C \cap \mathbb{S}^{n-1}$ is denoted by $\sigma(W, S)$ (or by σ) and is called the *fundamental spherical simplex* (or the *fundamental chamber*) for W on $\mathbb{S}^{n-1}$. The codimension-one faces (or "facets") of σ $(= \sigma(W, S))$ are the faces $\sigma_s := \sigma \cap H_s$. For each proper subset $T < S$ we have a face σ_T of σ defined by $\sigma_T := \bigcap_{s\in T} \sigma_s$. The isotropy subgroup at a point in the relative interior of σ_T is a special subgroup W_T. Since $W \curvearrowright \mathbb{S}^{n-1}$ with $\sigma(W, S)$ as a strict fundamental domain, the sphere $\mathbb{S}^{n-1}$ can be identified with the basic construction $D(W, \sigma)$ defined in Lemma 3.7 using (3.9).

The angle between bounding lines of a fundamental sector for the dihedral group D_m on $\mathbb{R}^2$ is π/m. For each $s \in S$, let u_s be the outward-pointing unit normal vector to the face C_s of C (or equivalently, the outward-pointing normal vector to σ_s in $\mathbb{S}^{n1}$). Since the angle between u_s and u_t is the exterior dihedral angle along $C_s \cap C_t$, this angle is $\pi - \pi/m(s, t)$. Hence, the matrix of inner products $(u_s \cdot u_t)$ is equal to the *cosine matrix* $c(s, t)$ of (W, S), that is, the $(S \times S)$-symmetric matrix defined by

$$c(s, t) = (u_s \cdot u_t) = -\cos(\pi/m(s, t)). \tag{4.9}$$

When $s = t$, $m(s, s) = 1$ so that $c(s, s) = 1$. (The formula in (4.9) for the cosine matrix $c(s, t)$ makes sense for any Coxeter system (W, S), provided that when $m(s, t) = \infty$, we interpret $-\cos(\pi/\infty)$ to be $-\cos(0)$ (i.e., to be $= -1$), cf. Sect. 4.2.7.) When (W, S) is spherical, $\{u_s\}_{s\in S}$ is a basis for $\mathbb{R}^n$; hence, we see that the cosine matrix is positive definite whenever W is finite. The converse is also true. This gives the following well-known characterization of finite Coxeter groups.

Lemma 4.9 (E.g., See [28, Ch. V §4.8] or [82, Theorem 6.12.9]) *The Coxeter group W is finite if and only its cosine matrix $(c(s,t))$ defined by* (4.9) *is positive definite.*

The *fundamental dual simplex* $\Delta(W,S)$ is the spherical simplex in $\mathbb{S}^{n-1}$ spanned by the unit vectors $\{u_s\}_{s\in S}$. The simplex $\Delta(W,S)$ is dual to $\sigma(W,S)$ in the sense that it consists of all points $y \in \mathbb{S}^{n-1}$ of distance $\geq \pi/2$ from $\sigma(W,S)$, i.e., $\Delta(W,S) = \{y \in \mathbb{S}^{n-1} \mid y \cdot x \leq 0 \text{ for all } x \in \sigma(W,S)\}$. The length of the circular arc connecting the vertices u_s and u_t of $\Delta(W,S)$ is $\pi(1 - \frac{1}{m(s,t)})$.

The information in the labeled graph (L^1, m) is the same as that contained in the *Coxeter diagram*. This is a graph with the same vertex set S but with edges labeled by an integer ≥ 3 or ∞. If $m(s,t) = 2$, the edge is omitted. Otherwise the label is $m(s,t)$ unless $m(s,t) = 3$ in which case the edge is not labeled. The Coxeter diagrams of the irreducible spherical and euclidean Coxeter systems are listed below in Table 4.1.

Table 4.1 Diagrams of spherical and euclidean Coxeter systems

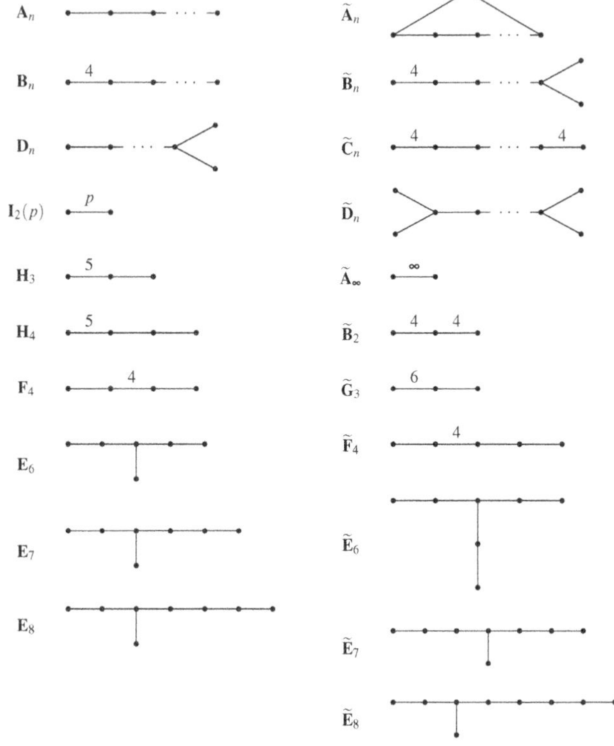

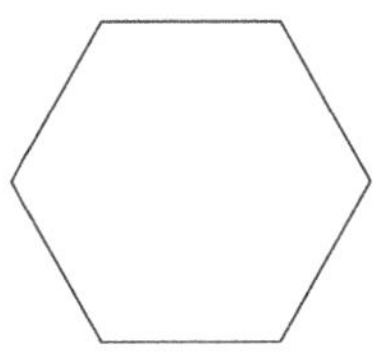

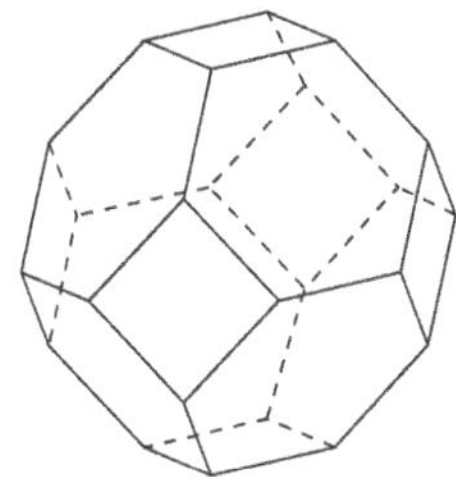

Fig. 4.1 The hexagon and permutohedron

4.2.2 *Coxeter Zonotopes*

Associated to a spherical Coxeter system (W, S) there is a *Coxeter zonotope* Z $(= Z(W, S))$ that is dual to the reflection arrangement $\mathcal{A}$ (cf. Example 3.73 in Sect. 3.3.1 or Example 5.9 in Sect. 5.1.2 below). More precisely, Z is the simple convex polytope such that ∂Z is dual to the spherical fan $\mathbb{S}\mathrm{Fan}(\mathcal{A})$ (a simplicial complex). Here is an explicit description. Let C be a fundamental chamber bounded by the walls indexed by S. Choose a base point x_0 in the interior of C and consider its W-orbit, Wx_0. Then $Z(W, S)$ can be defined as the convex hull of Wx_0.

Lemma 4.10 (See, e.g., [57, Lemma 2.1.3]) *Suppose that (W, S) is a spherical Coxeter system and that $Z(W, S)$ is the corresponding Coxeter zonotope. Then the face poset of $Z(W, S)$ is isomorphic to the poset of spherical cosets* $\mathrm{Coset}(W)$ *defined in* (4.7). *The face corresponding to the coset wW_T has vertex set $(wW_T)x_0$ and this face is isomorphic to the zonotope $Z(W_T, T)$. (In particular, the face corresponding to $T = S$ is the entire zonotope $Z(W, S)$.)*

As examples, if (W, S) is the symmetric group on $n+1$ letters (i.e., if its Coxeter diagram is $\mathbf{A}_n$), then Z is a permutohedron. When $n = 2$, a permutohedron is a hexagon. (See Fig. 4.1). If $W = (\mathbf{C}_2)^n$, then Z is an n-cube. N.B. The action of W on the cell complex Z, "has inversions," e.g., although the maximum cell Z is stabilized by the entire group W, it is not fixed pointwise.

The link of the vertex x_0 in Z, $\mathrm{Lk}(x_0, Z)$, is naturally identified with the dual spherical simplex $\Delta(W, S)$. The edges meeting at the vertex x_0 are parallel to the basis of unit normal vectors, $\{u_s\}_{s \in S}$. The metric on $Z(W, S)$ depends on the position of x_0 in the interior of C. We can normalize this position by choosing x_0 to be the unique point of distance $= 1/2$ from each wall H_s, $s \in S$. This will have the effect of making the length of each edge of $Z(W, S)$ equal to 1. The face corresponding to wW_T is said to be of *type* T; it is a cell whose dimension is $\mathrm{Card}(T)$.

Coxeter zonotopes play the same role in constructing Davis–Moussong complexes for general Coxeter groups as do cubes for RACGs. They are also used in the construction of the Salvetti complex for general Artin groups.

4.2.3 The Davis–Moussong Complex

We return to the situation where the Coxeter system (W, S) is arbitrary, where $\mathcal{S} = \mathcal{S}(W, S)$ is its poset of spherical subsets, and where $W\mathcal{S}^{\text{op}}$ is its poset of spherical cosets (cf. (4.7)). The purpose of this subsection is to describe a cell complex $\Sigma = \Sigma(W, S)$ together with a proper action, $W \curvearrowright \Sigma$. The poset of cells of $\Sigma(W, S)$ is equal to $W\mathcal{S}$; the cell corresponding to the spherical coset wW_T is isomorphic to the zonotope $Z(W_T, T)$ (cf. Lemma 4.10). It follows that the barycentric subdivision of $\Sigma(W, S)$ is the geometric realization $|W\mathcal{S}|$.

The complex $\Sigma(W, S)$ plays a central role in my book [82] and is described in detail there. Here we content ourselves with a fairly brief description of it. In the case where W is right-angled, $\Sigma(W, S)$ is the cubical complex $\tilde{P}_L$ explained at the end of Sect. 3.1.1. Some properties of Σ are listed in the next proposition.

Proposition 4.11 (cf. [179] or [82]) *Let* $\Sigma = \Sigma(W, S)$.

(1) *The poset of cells in the cell complex* Σ *is equal to* $W\mathcal{S}$.
(2) *Each cell of* Σ *is a Coxeter zonotope. The cell corresponding to the spherical coset* wW_T *is isomorphic to the Coxeter zonotope* $Z(W_T, T)$*; it is a convex polytope of dimension equal to* Card T.
(3) *The group* W *acts properly on* Σ. *The stabilizer of the cell* $wZ(W_T, T)$ *is the subgroup* wW_Tw^{-1}.
(4) *The* W*-action on* Σ *has a strict fundamental domain homeomorphic to the geometric realization* $|\mathcal{S}|$ *of the poset* $\mathcal{S}$. *So, when* S *is finite,* Σ/W *is compact.*
(5) *The natural piecewise euclidean metric on* Σ *is* CAT(0). *In particular,* Σ *is contractible by Theorem 2.2.*

As we explained in Sect. 4.1.3, the fact that $\Sigma(W, S)$ is CAT(0) implies that the $K(\pi, 1)$-Question for $W\mathcal{S}^{\text{op}}$ has a positive answer is the following corollary to Proposition 4.11 (5).

Theorem 4.12 (The $K(\pi, 1)$-Question for $W\mathcal{S}^{\text{op}}$ Has a Positive Answer) *The space* $\Sigma(W, S) \times_W EW$ *is homotopy equivalent to* BW.

In order to appreciate Σ, we spell out its construction in more detail. Its 0-skeleton, Σ^0, is identified with W (i.e., with the spherical cosets in $W/W_\emptyset$). Its 1-skeleton is the Cayley graph Cay(W, S); there is an edge connecting w to ws whenever $\{w, ws\}$ is a coset of $W_{\{s\}}$. So, there is an orbit of edges for each generator $s \in S$. Next fill in the 2-cells. There is an orbit of 2-cells of the form $Z(W_{\{s,t\}}, \{s, t\})$, whenever Card$\{s, t\} = 2$ and $m(s, t) < \infty$. Each such 2-cell is a 2-dimensional Coxeter zonotope, that is, a polygon with $2m(s, t)$ saasides. The orbits of these 2-cells correspond precisely to the relations in (4.5). So, Σ^2 is essentially the Cayley 2-complex of (W, S). In particular, Σ^2 is simply connected. We continue by filling in an orbit of 3-dimensional zonotopes of the form $Z(W_T, T)$ for each spherical subset $T \in \mathcal{S}$ with 3 elements. Miraculously, after filling in all such 3-cells, the resulting 3-skeleton, Σ^3 is 2-connected. Continuing, the complex Σ is formed by filling in an orbit of Coxeter zonotopes for each spherical coset in $W\mathcal{S}^{\text{op}}$. It turns out that Σ is contractible. One way to show this is to prove statement (5) in the above proposition, i.e., to show that Σ is CAT(0). This is a theorem of Moussong which will be discussed in the next subsection. We note that when W is finite, $\Sigma(W, S)$ is equal to the single Coxeter zonotope $Z(W, S)$ (cf. Lemma 4.10).

Recall from Sect. 4.1 that the *nerve* of (W, S) is the simplicial complex $L(W, S)$ whose simplices correspond to the nonempty spherical subsets of S. There is natural piecewise spherical metric on $L(W, S)$ where the simplex corresponding to T is identified with the fundamental dual spherical simplex $\Delta(W_T, T)$ spanned by the outward-pointing unit normal vectors to $\sigma(W_T, T)$, $\{u_t\}_{t\in T}$, as in Sect. 4.2.1. In particular, the edge $\{s, t\}$ that corresponds to the spherical subset $T = \{s, t\}$ has length $\pi(1 - 1/m(s, t))$.

A spherical simplex Δ *has size* $\geq \pi/2$ if the distance from any vertex to the opposite face is $\geq \pi/2$.

Definition 4.13 (Metric Flag Complexes) To define this concept suppose L is a simplicial complex with a piecewise spherical metric so that whenever e is an edge of L, its length, $l(e)$, lies in the interval $[\pi/2, \pi)$ (this is implied by the hypothesis that each simplex of L has size $\geq \pi/2$). Then L is a *metric flag complex* if whenever Γ is a subcomplex of L isometric to the 1-skeleton of a spherical k-simplex τ, there actually exists a k-simplex τ in L with edge lengths specified by the $l(e)$, $e \in \tau^{(1)}$.

Note that in the case where the piecewise spherical metric is all right, L is a metric flag complex if and only if it is a flag complex. (See Definition 2.26 for the meaning of "all right".)

The condition in Definition 4.13 also can be expressed as follows. Suppose T is the set of vertices of a subcomplex isomorphic to a complete graph Γ and that for distinct vertices t, t' of Γ, the length $l(t, t')$ of the edge $\{t, t'\}$ lies $[\pi/2, \pi)$. Let $c(t, t')$ be the "cosine matrix" defined by (4.9). The condition that each edge length lies in $[\pi/2, \pi)$ corresponds to the condition that each spherical simplex has size $\geq \pi/2$. Then L is a metric flag complex if and only if the $(T \times T)$-symmetric matrix $c(t, t')$ is positive definite whenever T is the vertex set of a simplex of L. In other words, the simplices of L are determined by the metric on L^1. By Lemma 4.9, for each $T \in \mathcal{S}$, the corresponding cosine matrix is positive definite; so, $L(W, S)$, with its natural piecewise spherical structure is a metric flag complex.

Example 4.14 (Metric Flag Complexes of Dimension 1) When is the graph L^1 equipped with its piecewise spherical metric CAT(1)? The answer was alluded to in Examples 2.12 and 2.13 as well as in Sect. 2.4.5: the length of each circuit in L^1 must be $\geq 2\pi$. If all edge lengths lie in $[\pi/2, \pi]$ and if L^1 is a simplicial graph, then this condition is vacuous except for circuits with 3 edges, say e_1,e_2, e_3, in which case it reads:

$$l(e_1) + l(e_2) + l(e_3) \geq 2\pi \tag{4.10}$$

(cf. Lemma 2.56). This is precisely the condition that the cosine matrix for the 3-circuit fails to be positive definite. Thus, L^1 is a metric flag complex if and only if (4.10) holds for each 3-circuit in L^1 and this is equivalent to the condition that L^1 be CAT(1). Moussong's Lemma, which is stated in the next subsection, is a generalization of this to higher dimensions.

Example 4.15 (Σ Is a 2-Dimensional CAT(0) **Cell Complex, cf. Example 2.13)** Suppose (L^1, m) is a labeled simplicial graph with labeling, m : Edge$(L^1) \to \{2, 3, \dots\}$. Let (W, S) be the associated Coxeter system. As in the previous example, suppose each spherical subset $T \in \mathcal{S}$ has cardinality ≤ 2 so that $L(W, S) = L^1$. Then $\dim \Sigma = 2$. The 2-dimensional zonotope corresponding to an edge e of L^1 is a regular euclidean $2m(e)$-gon. The interior angle at a vertex of such a $2m(e)$-gon is $\pi(1 - 1/m(e))$. If $\{e_1, e_2, e_3\}$ is a 3-circuit of L^1, then the condition in (4.10) is equivalent to

$$\sum_{i=1}^{3} \frac{1}{m(e_i)} \leq 1. \tag{4.11}$$

So, $\dim \Sigma = 2$ exactly when (4.11) holds for all 3-circuits in L^1. Thus, Σ is a 2-dimensional CAT(0) cell complex with an orbit of $2m(e)$-gons for each $e \in L^1$.

Next, consider the question: when does Σ admit a piecewise hyperbolic structure that is CAT(-1)? First, we need to discuss how to make the zonotopes hyperbolic. If W_T is a spherical Coxeter group, then we can represent it as a finite reflection group on $\mathbb{H}^m$, where $m = \operatorname{Card} T$. Let C^h be a fundamental cone for the W_T-action on $\mathbb{H}^m$ and choose a point x_0 in the interior of C^h of distance $\frac{1}{2}\varepsilon$ from each wall. Let $Z^h_\varepsilon(W_T, T)$ $(= Z^h_\varepsilon)$ denote the convex hull of the orbit of x_0. Then $Z^h_\varepsilon(W_T, T)$ is the hyperbolic zonotope with all its edge lengths equal to ε. As before, we can glue together the $Z^h_\varepsilon(W_T, T)$ to obtain a piecewise hyperbolic model $\Sigma^h(W, S)$ of the Davis–Moussong complex. When ε is small, the effect on $\operatorname{Lk}(x_0, Z^h_\varepsilon(W_T, T))$ will be to make a small change in the dual spherical simplex $\Delta(W_T, T)$. Hence, there also will only be a small change in the metric on $\operatorname{Lk}(v, \Sigma^h_\varepsilon)$ (cf. [179, Lemma 5.11] or [82, Lemma I.6.7]).

Example 4.16 (2-Dimensional CAT(-1) **Cell Complexes)** Suppose $\dim \Sigma = 2$. A small deformation of the metric on L^1 is then guaranteed to be CAT(1) if and only if the sum of the edge lengths in each circuit is strictly greater than 2π. As in Example 4.15, we need only consider circuits with ≤ 4 edges. If the number of edges is 3 we must modify (4.11) by requiring the inequality to be strict. If the number of edges is 4, the only case which leads to a closed geodesic of length 2π is when each edge has length exactly $\pi/2$. We exclude this possibility by requiring that for 4-circuits at least one edge has $m(e) \neq 2$. (This is a more general version of the no $\square$-condition of Sect. 2.2.) From the discussion in Sect. 2.1.2 we see that a small deformation of the piecewise euclidean structure on Σ to a piecewise hyperbolic structure Σ^h_ε is CAT(-1) if and only if the above two conditions hold for L^1, i.e., if L^1 is extra large in the sense of Definition 2.16. When this is the case, each of the deformed vertex links will be CAT(1). Hence, for small enough ε, Σ_ε^h will be CAT(-1) if and only if L^1 is extra large.

Example 4.17 (Gromov Polyhedra) Suppose L^1 is a complete graph on p vertices with edges labeled by a constant integer $m \geq 3$. Let $W(p, m)$ be the Coxeter group associated to (L^1, m). Then there is a simply connected, 2-dimensional polyhedron $X_{p,2m}$ so that each 2-cell is a $2m$-gon and so that the link of each vertex is a complete graph on p-vertices. To see this, apply the construction in the previous example to the case where L^1 is a complete graph, each edge of which is labeled by the same integer m, and $X_{p,2m} = \Sigma(W(p, m))$ (or its 2-skeleton when $m = 2$.) If $m \geq 3$, then $X_{p,2m}$ is CAT(0). When $m \geq 4$, it can be given a CAT(-1) metric where each polygon is hyperbolic.

4.2.4 Moussong's Lemma

Lemma 4.18 (Moussong's Lemma, cf. [179, Appendix I.7] or [82]) *A piecewise spherical simplicial complex L with all simplices of size $\geq \pi/2$ is* CAT(1) *if and only if it is a metric flag complex.*

Moussong's Theorem, stated below, is a corollary of the above lemma.

Theorem 4.19 (Moussong [179, Lemma I.7.4], also cf. [103] or [82]) *The piecewise euclidean cell complex $\Sigma(W, S)$ is* CAT(0).

Comments on the Proof As explained in Sect. 4.2.2, the link of a vertex in the zonotope $Z(W_T, T)$ is isometric to the dual spherical simplex $\Delta(W_T, T)$. It follows that the link of each vertex in $\Sigma(W, S)$ is isometric to $L(W, S)$. By Moussong's Lemma, this link is CAT(1); hence, $\Sigma(W, S)$ is NPC. Since the 2-skeleton of Σ is the Cayley 2-complex for (W, S), we see that Σ is simply connected; hence, by Theorem 2.3, Σ is CAT(0).

When (W, S) is right-angled, its nerve $L(W, S)$ is a flag complex and it has an all right, piecewise spherical structure. By Lemma 2.33, $L(W, S)$ is extra large if and only if it satisfies the no $\square$-condition. So, the cubical metric on $\Sigma(W, S)$ can be deformed to a piecewise hyperbolic metric that is CAT(-1) if and only if $L(W, S)$ satisfies the no $\square$-condition. This establishes the following proposition. We will give some details of its proof later in Theorem 4.23 below.

Proposition 4.20 *Suppose (W, S) is a* RACS . *Then (W, S) is word hyperbolic if and only if $L(W, S)$ satisfies the no $\square$-condition.*

Moussong continued along the line he began in Lemma 4.18 by determining when a piecewise spherical, metric flag complex L of size $\geq \pi/2$ is extra large (cf. Definition 2.16). As usual $S = \operatorname{Vert} L$. For $T < S$, let L_T denote the full subcomplex spanned by T. Then L_T is a totally geodesic subspace of L. Two types of subcomplexes L_T obviously can lead to a closed geodesic of length 2π in L. They are:

(i) The complex L_T is isometric to the polar dual of a euclidean simplex, i.e., L_T is isometric to the round sphere of dimension $\text{Card}(T) - 2$ and L_T is combinatorially isomorphic to the boundary complex of a simplex of dimension $\text{Card}(T) - 1$. (The polar dual of a convex polytope is the piecewise spherical complex spanned by the outward-pointing unit normals to its facets.)
(ii) The complex L_T decomposes as a spherical join $L_{T_1} * L_{T_2}$, where neither L_{T_1} nor L_{T_2} is a spherical simplex.

To see that (ii) yields a closed geodesic of length 2π, for $i = 1, 2$, choose points x_i, y_i in L_{T_i} of distance at least $\pi/2$. Then the join $\{x_1, y_1\} * \{x_2, y_2\}$ is a 4-circuit where each edge has length $\pi/2$. In [179, Lemma 10.3] Moussong proved that the metric flag complex L has no closed geodesics of length 2π if and only if it contains no subcomplexes L_T of type (i) or (ii). Applying this in the special case $L = L(W, S)$ we get the following.

Lemma 4.21 *The piecewise spherical complex $L(W, S)$ is extra large if for each subset T of S neither of the following conditions hold for W_T.*

(i) *The special subgroup W_T is an irreducible euclidean reflection group (generated by the reflections across the faces of a euclidean simplex of dimension ≥ 2).*
(ii) *(W_T, T) decomposes as $(W_{T_1} \times W_{T_2}, T_1 \sqcup T_2)$, where both W_{T_1} and W_{T_2} are infinite.*

If conditions (i) and (ii) of this lemma both fail to hold, we say that (W, S) satisfies *Moussong's Conditions*.

Remark 4.22 (Andreev's Conditions, [8]) When the simplicial complex L is homeomorphic to S^2, Moussong's Conditions are equivalent to Andreev's Conditions (e.g., see [82, §6.10]). Andreev gave conditions are necessary and sufficient for L to be the polar dual to the boundary complex of a compact simple convex polytope in $\mathbb{H}^3$ so that the dihedral angle along an edge dual to an edge e in L is $\pi/m(e)$. The Coxeter group W is then a cocompact reflection group on $\mathbb{H}^3$.

Theorem 4.23 (Moussong's Characterization of Word Hyperbolic Coxeter Groups) *The following conditions on a finitely generated Coxeter system (W, S) are equivalent.*

(a) *The group W is word hyperbolic.*
(b) *The group W does not contain $\mathbb{Z} \times \mathbb{Z}$ (the product of two infinite cyclic groups).*
(c) *The Coxeter system (W, S) satisfies Moussong's Conditions.*
(d) *The* CAT(0) *metric on $\Sigma(W, S)$ can be deformed to a piecewise hyperbolic,* CAT(−1) *metric.*

Proof Since a word hyperbolic group cannot contain $\mathbb{Z} \times \mathbb{Z}$, (a) $\Longrightarrow$ (b). If a special subgroup W_T is a euclidean reflection group, then its translation subgroup is free abelian of rank Card$(T) - 1$ and this rank is ≥ 2 whenever (b) holds. Similarly, if $W_T = W_{T_1} \times W_{T_2}$, where both factors are infinite, then $\mathbb{Z} \times \mathbb{Z} < W_T$. So, (b) $\implies$ (c). By Lemma 4.21, Moussong's Conditions imply that $L(W, S)$ is extra large. As in Example 4.16, Proposition 2.18 implies that $\Sigma(W, S)$ can be given a piecewise hyperbolic CAT(-1) metric. So, (c) $\implies$ (d). By Lemma 2.21, (d) $\implies$ (a).

Abelian Subgroups in Coxeter Groups An isometrically embedded copy of euclidean n-space, $\mathbb{E}^n$, in a CAT(0) space X is called an *n-flat*. If A is a free abelian group of rank n acting properly on a CAT(0) space X, then by the Flat Torus Theorem 2.37 of Sect. 2.3, its "minset" contains an n-flat.

By Moussong's Theorem 4.19, $\Sigma(W, S)$ is CAT(0). By the Bruhat–Tits Fixed Point Theorem 2.34, every finite subgroup G of W is conjugate into a subgroup of an isotropy subgroup of W on Σ, i.e., to a spherical special subgroup W_T. In other words, the fixed set of G coincides with the fixed set of a spherical parabolic subgroup. (Recall that a subgroup of W is a *parabolic subgroup* if it is conjugate to some special subgroup W_T. Similarly, a *parabolic subgroup of an Artin group* is a conjugate of a special subgroup A_T.) As in the proof of Theorem 4.23, there are two obvious ways in which a W_T can contain a virtually free abelian subgroup G of rank at least two:

(i) W_T is a euclidean Coxeter group of rank ≥ 2, or
(ii) W_T decomposes as $W_{T_1} \times W_{T_2}$ where both factors are infinite.

The *algebraic rank* (W, S) is the maximum integer n such that W contains a free abelian subgroup of rank n. The *geometric rank* is the maximum n such that $\Sigma(W, S)$ contains an n-flat. By the Flat Torus Theorem 2.37, the geometric rank is at least the algebraic rank. By Theorem 4.23, (W, S) has algebraic rank equal to 1 if and only if W is word hyperbolic and is not spherical.

Next, suppose that a special subgroup W_T has an irreducible decomposition,

$$W_T = W_{T_1} \times \cdots \times W_{T_k}, \tag{4.12}$$

where each irreducible factor is infinite and either is a word hyperbolic or a euclidean Coxeter group. (Only the infinite dihedral group is both word hyperbolic and euclidean.) The algebraic rank of W_T is then the number of hyperbolic factors plus the sum of the ranks of the euclidean factors.

Theorem 4.24 (Caprace-Haglund [51]) *The geometric rank of* (W, S) *equals its algebraic rank. Moreover, the* algebraic rank *is the maximum of the algebraic ranks of any special subgroup* W_T *which has an irreducible decomposition of the form given in* (4.12).

In fact, any free abelian subgroup G of W is virtually a subgroup of a parabolic subgroup that is conjugate to some W_T in the form given in (4.12).

4.2.5 Relatively Hyperbolic Coxeter Groups

The notion of a relatively hyperbolic group was introduced in [136]. Other versions were given by Farb [114] and by Bowditch [31]. More recently, there have been further generalizations (e.g., "acylindrically hyperbolic groups"), a description of these generalizations can be found in [70]. When P is a subgroup of a finitely generated group G, then the *coned off Cayley graph* $\Omega(G, P)$ is formed by attaching to the Cayley graph $\mathrm{Cay}(G)$ a cone on each coset of P in G. (Such a coset is a subset of the vertex set of $\mathrm{Cay}(G)$.) If $\{P_1, \dots, P_k\}$ is a finite collection of subgroups, define $\Omega(G, \{P_1, \dots, P_k\})$ similarly. At first approximation, one says that *G is hyperbolic relative to* $\{P_1, \dots, P_k\}$ if $\Omega(G, \{P_1, \dots, P_k\})$ is a hyperbolic metric space. For the actual definition of a relatively hyperbolic group, Farb adds the requirement the coned off Cayley graph has the "Bounded Coset Penetration Property." (See [114, §3.3].) Bowditch's definition of relative hyperbolicity is along a different line, which we shall not explain here. The prototypical example for a relatively hyperbolic pair (G, P) is when G is the fundamental group of a hyperbolic manifold or orbifold of finite volume and the subgroup P is the fundamental group of a cusp (so that P is virtually abelian). For example, G could be the Coxeter group generated by reflections across the facets of a hyperbolic polytope of finite volume and $P_1, \dots, P_k$ could be the stabilizers of its ideal vertices.

In the most important cases, the P_i are maximal virtually free abelian subgroups of rank ≥ 2. When G acts geometrically on a $\mathrm{CAT}(0)$ space X, such P_i can occur as stabilizers of maximal flat subspaces of X. There is the following important result of Hruska and Kleiner.

Theorem 4.25 (Hruska–Kleiner [150]) *Suppose G acts properly and cocompactly via isometries on a* CAT(0) *space X. Let $\mathcal{F}$ be the family of maximal free abelian subgroups of G. Then G is relatively hyperbolic with respect to $\mathcal{F}$ if and only if X has isolated flats (i.e., the maximal flats of X are isolated).*

In the case of a Coxeter group W acting on the Davis–Moussong complex $\Sigma(W, S)$, Theorem 4.24 determines the virtually abelian subgroups of W of rank ≥ 2. In [47] Caprace determined which of these subgroups correspond to isolated flats. Using the Hruska-Kleiner Theorem he gets the following characterization of when W is hyperbolic relative to its virtually abelian subgroups.

Theorem 4.26 (Isolated Flats in Coxeter Groups, cf. Caprace [47, Cor. D], [48]) *The following conditions are equivalent.*

(i) *For any two disjoint nonspherical subsets T_1, T_2 in S such that T_1 and T_2 commute, $W_{T_1} \times W_{T_2}$ is virtually abelian.*
(ii) *The Coxeter group W is relatively hyperbolic with respect to the collection of virtually abelian subgroups of rank at least 2.*
(iii) *The* CAT(0) *space $\Sigma(W, S)$ has isolated flats.*

4.2.6 A Different Piecewise Euclidean Metric on the Fundamental Chamber

Let $K = |\mathcal{S}^{\mathrm{op}}|$ $(= |\mathcal{S}|)$ be the fundamental chamber defined in Example 4.2 and put $K_T = |\mathcal{S}^{\mathrm{op}}_{\leq T}|$. In other words, K_T is the union of simplices in the order complex of $\mathcal{S}^{\mathrm{op}}$ with maximum vertex T. Let $\Sigma = \Sigma(W, S)$. Since $W \curvearrowright \Sigma$ with strict fundamental domain K (cf. Definition 3.5), we get a different description of Σ in terms of the basic construction of (4.4) (or (3.9)), $\Sigma = D(W, K)$ where $D(W, K) = (W \times K)/\sim$ and where $\sim$ is defined by:

$$(u, x) \sim (v, x') \iff x = x' \text{ and } uW_{S(x)} = vW_{S(x)}.$$

Here $S(x)$ is the index of the smallest stratum containing x. This stratification is dual to the cellulation of Σ by Coxeter zonotopes described in Proposition 4.11.

The fundamental domain K defined in Example 4.2 is a cube complex. We describe its cubical structure in more detail. For any $T \in \mathcal{S}$, the poset $\mathcal{S}_{\leq T}$ $(= \mathcal{S}^{\mathrm{op}}_{\geq T})$ is the power set of T. This means that the order complex of $\mathcal{S}^{\mathrm{op}}_{\geq T}$ can be identified with a standard subdivision of a cube $\square^T$ of dimension $\mathrm{Card}(T)$. More generally, for any subset T' of T consider the interval $[T, T']$ in $\mathcal{S}^{\mathrm{op}}$ defined by $[T, T'] = \{J \in \mathcal{S} \mid T' \leq J \leq T\}$. The geometric realization of $[T, T']$ is a standard subdivision of a cube $\square^{T-T'}$ We denote this subdivision of the cube by $\square^{[T,T']}$. Explicitly, $\square^{[T,T']}$ is the subcomplex of the geometric realization of $\mathcal{S}^{\mathrm{op}}$ consisting of all simplices in $|\mathcal{S}^{\mathrm{op}}|$ with minimum vertex T and maximum vertex T'. In other words, the simplices in $|\mathcal{S}^{\mathrm{op}}|$ $(= K)$ can be amalgamated into cubes, giving K the structure of a cube complex. By taking the union of all translates of such cubes in $|W\mathcal{S}|$ (i.e., in Σ), we see that Σ is also a cube complex. We state this as the following.

Lemma 4.27 *The fundamental chamber K has the structure of a combinatorial cube complex:*

$$K = \bigcup_{T \in \mathcal{S}^{\mathrm{op}}} \square^{[T,\emptyset]}$$

and similarly, for the zonotopal complex $\Sigma(W, S)$.

The subcomplexes $\square^{[T,T']}$ are only combinatorially equivalent to cubes. What is the relationship between these combinatorial cubes and the Coxeter zonotopes in Σ? Suppose $Z(W_T, T)$ is the Coxeter zonotope in $\mathbb{R}^T$ corresponding to W_T. Let $\mathrm{Fan}(W_T)$ be the fan cut out by the reflecting hyperplanes. A top-dimensional simplicial cone in $\mathrm{Fan}(W_T)$ is called a *sector*.

Definition 4.28 The intersection $B_T := Z(W_T, T) \cap \mathrm{Fan}(W_T)$ is called a *Coxeter block*; it is combinatorially isomorphic to the cube $\square^{[T,\emptyset]}$.

However, B_T need not be isometric to a unit cube. (The link in the Coxeter block of the vertex corresponding to $T \in \mathcal{S}_{\leq T}$ (the cone point of $\mathrm{Fan}(W_T)$) is isometric to the spherical simplex $\sigma(W_T, T)$ while the link at the opposite vertex corresponding to $\emptyset \in \mathbb{S}_{\leq T}$ is the dual spherical simplex $\Delta(W_T, T)$ defined near the end of Sect. 4.2.1.) The *natural piecewise euclidean metric on* Σ is the path metric defined by giving each zonotopal cell in Σ its natural metric as a Coxeter zonotope (say, with each edge of length 2). Similarly, the *natural piecewise euclidean metric on* K is the path metric defined by giving each Coxeter block its natural metric as a convex subset of the Coxeter zonotope. We just showed in Theorem 4.19, that, with its natural piecewise euclidean metric, Σ is CAT(0). On the other hand, we could give K the piecewise euclidean metric where each subcomplex $\square^{[T,T']}$ is isometric to the unit cube. Denote K with this cubical metric by $K^\square$. Similarly, the induced cubical metric on Σ is denoted by $\Sigma^\square$. The cube complexes $K^\square$ and $\Sigma^\square$ may fail to be CAT(0) since links need not be flag complexes. The condition that we need to insure that all links are flag complexes is given in the following definition.

Definition 4.29 A Coxeter system (W, S) is *type FC* if its nerve $L(W, S)$ is a flag complex. Similarly, an Artin group is *type FC* if its associated Coxeter system is FC.

The cubical structures on these polyhedra induce the piecewise euclidean cubical metrics, denoted $K^\square$ and $\Sigma^\square$, respectively. N.B. In general, $K^\square$ and $\Sigma^\square$ will not be CAT(0). On the other hand, by Theorem 4.19, Σ is CAT(0) when it is given the metric induced from its zonotopal cells.

Proposition 4.30 *The cube complex* $\Sigma^\square$ *is* CAT(0) *if and only if the Coxeter system* (W, S) *is type FC.*

Proof Of course, the proof of this should follow from the version of Gromov's Lemma stated as Corollary 2.31; however, it does not immediately follow from the fact that $L(W, S)$ $(= L)$ is a flag complex. Although it is not true that the link of any cube in Σ is isomorphic to L or to a sublink of the form $\mathrm{Lk}(\sigma, L)$, it is true that the link of every 0-cube in Σ is isomorphic to a join $\mathbb{S}\mathrm{Fan}(W_T, T) * \mathrm{Lk}(\sigma_T, L)$, where $\mathbb{S}\mathrm{Fan}(W_T, T)$ is the spherical Coxeter complex with its all right piecewise spherical structure and σ_T is the simplex corresponding to T. Since any spherical Coxeter complex is a triangulation of a sphere as a flag complex (cf. [57, Lemma 4.3.2]), both factors of the join are flag complexes and hence, by Lemma 2.25 (ii), so is the join.

This shows that when (W, S) is type FC one can prove that $\Sigma(W, S)$ has a CAT(0) structure without using Moussong's Lemma. In Sect. 4.3.3 essentially the same observation will be used to show that if an Artin group is FC, then its Deligne complex has a CAT(0) cubical structure.

4.2.7 The Tits Representation

When W is infinite, it also has a representation on $\mathbb{R}^N$, with $N = \mathrm{Card}\, S$, as a group generated by linear reflections (which need not be orthogonal linear transformations). More generally, Vinberg [227] studied representations of W as reflections across the facets of a polyhedral cone in $\mathbb{R}^N$, where N can be less than Card(S).

Let $c(s,t)$ be the cosine matrix defined as in (4.9) with the proviso that $c(s,t) = -1$ when $m(s,t) = \infty$. There is a symmetric bilinear form $B : \mathbb{R}^S \times \mathbb{R}^S \to \mathbb{R}$ defined by $B(e_s, e_t) = c(s,t)$. For each $s \in S$, let ρ_s be the linear reflection on $\mathbb{R}^S$ defined by

$$\rho_s(x) := x - 2B(e_s, x)e_s. \tag{4.13}$$

(The eigenvalue -1 of ρ_s has eigenvector e_s.) The map $s \mapsto \rho_s$ extends to a homomorphism $\rho : W \to GL(\mathbb{R}^S)$, called the *canonical representation* in [28]. More interesting to us is the dual of the canonical representation, $\rho^* : W \to GL((\mathbb{R}^S)^*)$, which we call the *geometric representation* or the *Tits representation* of W. The main fact about the geometric representation is that it defines a W-action on a certain open convex set Ω with a strict fundamental domain. This allows us to conclude that the representation ρ^* is discrete and faithful. To see this, let C be the simplicial cone in $(\mathbb{R}^S)^*$ defined by the inequalities: $e_s(v) \leq 0$, for $s \in S$. (Each basis vector e_s gives a linear form, $v \mapsto e_s(v)$ on $(\mathbb{R}^S)^*$.) For each subset $T \leq S$, let C_T be the face of C defined by $e_s(v) = 0$, for $s \in T$. Let C^f denote the complement in C of the union of all faces C_T with $T \notin \mathcal{S}$. That is to say, $C^f = \{x \in C \mid S(x) \in \mathcal{S}\}$, where $S(x) = \{s \in S \mid x \in C_s\}$. The *Tits cone* $\overline{\Omega}$ is defined by

$$\overline{\Omega} := \bigcup_{w \in W} wC.$$

Let Ω denote the interior of $\overline{\Omega}$. Tits established the following facts (e.g., see [28, Prop. 6, p. 102]).

(a) For any nontrivial $w \in W$, $w(\operatorname{int} C) \cap (\operatorname{int} C) = \emptyset$.
(b) $\overline{\Omega}$ is a convex cone. So, its interior Ω is an open W-stable subset in $\mathbb{R}^S$.
(c) $\Omega = \bigcup_{w \in W} wC^f$.
(d) The isotropy subgroup at a point $x \in C^f$ is W_T where C_T is the smallest face such that $x \in C_T$.

It follows from (a) that $\rho^* : W \to GL((\mathbb{R}^S)^*)$ is injective and that its image is a discrete subgroup of $GL((\mathbb{R}^S)^*)$. By (d) each isotropy subgroup is finite and the orbit space C^f is Hausdorff. Hence, W acts properly on Ω. The next result implies that $W \curvearrowright \Omega$ as a reflection group.

Theorem 4.31 (Tits, cf. [28], [227]) *The group W acts properly on the open set Ω with strict fundamental domain C^f. Thus, Ω is equivariantly homeomorphic to the basic construction $D(W, C^f)$ defined by formula (3.9) in Sect. 3.1. It follows that there is a W-equivariant, piecewise linear embedding $\Sigma(W,S) \hookrightarrow \Omega$ so that Ω equivariantly deformation retracts on to $\Sigma(W,S)$. In other words, $\Sigma(W,S)$ is a "spine" for Ω.*

4.2.8 Groups of Isometries of $\mathbb{H}^n$ Versus Word Hyperbolic Groups

We return to the line that we began in Theorem 4.23 at the end of Sect. 4.2.4. Suppose P is a (not necessarily compact) polytope in $\mathbb{H}^n$ and that $\{P_s\}_{s\in S}$ is its set of facets. The polytope P is a *Poincaré polytope* if for each pair of distinct facets P_s, P_t either $P_s \cap P_t$ is empty or is equal to a codimension-two face with the dihedral angle along $P_s \cap P_t$ of the form $\pi/m(s,t)$, where $m(s,t)$ is an integer ≥ 2. Let s denote isometric reflection across P_s and let S be the collection of all such s. Poincaré showed that the group W generated by S is a discrete subgroup of $\mathrm{Isom}(\mathbb{H}^n)$ acting properly on $\mathbb{H}^n$ with strict fundamental domain P. (Since the dihedral angle is an integral submultiple of π, the sector bounded by the walls corresponding to s and t is a strict fundamental domain for $W_{\{s,t\}}$, and this implies that P is a strict fundamental domain for W.) Moreover, (W, S) is a Coxeter system. The subgroup $W < \mathrm{Isom}(\mathbb{H}^n)$ is called a *hyperbolic reflection group*. The polytope P, *does not have asymptotic faces* if there is no pair of distinct faces F, F' in P such that F has codimension one, F' has arbitrary codimension, $F \cap F' = \emptyset$, and $d(F, F') = 0$ (in other words, F, F' are "asymptotic"). This is equivalent to the condition that W does not contain a parabolic element in $\mathrm{Isom}(\mathbb{H}^n)$.

The classical picture of a reflection group on $\mathbb{H}^n$ is where the Poincaré polytope P is compact and W is generated by the reflections across its facets. Then $\mathbb{H}^n/W \cong P$. If $\Gamma < W$ is a torsion-free subgroup of finite index, then $\mathbb{H}^n/\Gamma$ is a closed hyperbolic manifold tiled by finitely many copies of P. If P is allowed to have some ideal vertices but still is required to have finite volume, then the isotropy subgroup at an ideal vertex is a euclidean reflection group. (This is a case where W has asymptotic faces.) Moreover, $\mathbb{H}^n/\Gamma$ is a hyperbolic manifold of finite volume. Examples where P is compact occur primarily in dimensions 2 and 3. When $n = 2$, P is a convex polygon in $\mathbb{H}^2$; so, L^1 is a circle, $L(W, S) = L^1$ and any edge labeling, $m : \mathrm{Edge}(L^1) \to \{2, 3, \dots\}$, can occur subject to the constraints given by the Gauss-Bonnet Theorem. When $n = 3$, $L = L(W, S)$ is a triangulation of S^2 and any edge labeling can occur subject only to the conditions that the vertices of each 2-simplex define a spherical subgroup and that Moussong's Conditions hold. In dimension 3 these conditions are called *Andreev's Conditions* (see Remark 4.22). (Compare Vinberg [229]) Andreev [8] proved that such an edge labeling of the 1-skeleton of a triangulation of S^2 corresponds to a compact Poincaré polytope $P^3 < \mathbb{H}^3$, with dihedral angles prescribed by the edge labels; moreover, P^3 is unique up to isometry. So, when K is a compact simple polytope of dimension $n \leq 3$, the Coxeter group is word hyperbolic if and only if it can be represented as a reflection group on $\mathbb{H}^n$. When $n = 4$, there are exactly five examples where P is a 4-simplex. (See [82, Table 6.2, p. 105].) There is also a well-known example of a 4-dimensional RACG where P is a right-angled 120-cell (cf. [75], [82, Ex. 6.11.1] or Example 4.34 below). In this case W is a subgroup of index $(120)^2$ in one of the simplicial examples. There are further sporadic examples in dimensions ≤ 8.

In contrast to the situation in dimensions 2 and 3, Vinberg [228] proved that for $n > 28$, there do not exist compact hyperbolic polytopes with dihedral angles integral submultiples of π. In fact, if the fundamental polytope is required to be compact and to be right-angled, then such hyperbolic polytopes do not exist in dimensions > 4. (See [82, Cor. 6.11.7].) The proof only uses the fact that (W, S) satisfies Moussong's Conditions (cf. Lemma 4.21) and that (W, S) is type HM as in Definition 4.32 below. Moussong observed that this gives a generalization which can be stated as Theorem 4.33 below.

In Sect. 3.2 we explained a completely combinatorial condition which is necessary and sufficient (at least in the right-angled case) for a Coxeter group W to act as a reflection group on an open contractible n-manifold with compact fundamental chamber (see [82, Def. 10.6.2]). Roughly, this condition says that the fundamental chamber K homologically resembles a compact convex simple polytope or dually, that the nerve $L(W, S)$ homologically resembles a triangulation of an $(n-1)$-sphere (or, in other words, that $L(W, S)$ is a GHS^{n-1} as in Sect. 3.2.1). When this holds we say that (W, S) is *type* HM^n. Here is the precise definition.

Definition 4.32 (Coxeter Groups of Type HM) A Coxeter group is *type* HM^n if $L(W, S)$ is a GHS^{n-1} (i.e., a generalized homology sphere as defined in Sect. 3.2).

For example, if $L(W, S)$ is combinatorially isomorphic to the boundary complex of a simplicial n-polytope, then the fundamental chamber K is isomorphic to the dual simple polytope and $D(W, K)$ is diffeomorphic to $\mathbb{R}^n$. More generally, if $L(W, S)$ is a GHS^{n-1}, then the fundamental chamber can be resolved to be a simple homotopy n-cell K' so that $D(W, K')$ is a contractible n-manifold (cf. Proposition 3.35). Vinberg's argument then yields the following.

Theorem 4.33 (cf. [228], [179], [159], [82, Section 6.11]) *There do not exist word hyperbolic Coxeter groups of type* HM^n *for* $n \geq 29$. *Moreover, there do not exist word hyperbolic* RACGs *of type* HM^n *for* $n > 4$.

Example 4.34 (The 120-Cell) The regular 120-cell can be realized as a compact simple polytope P in $\mathbb{H}^4$ so that whenever two facets have nonempty intersection they make a dihedral angle of $\pi/2$. This gives a right-angled reflection group acting cocompactly on $\mathbb{H}^4$. So, the dimension bound in the last sentence of Theorem 4.33 is the best possible. The symmetry group of the regular 120-cell is the spherical Coxeter group of order $(120)^2$ whose diagram $\mathbf{H}_4$ is a straight line with edges labeled $(5, 3, 3)$. One of the four Coxeter groups with fundamental chamber a 4-simplex has diagram $(5, 3, 3, 4)$. It follows that $W(P)$ is a subgroup of index $(120)^2$ in $(5, 3, 3, 4)$. If we double the 120-cell P along a facet, we obtain an index-two subgroup of $W(P)$ whose fundamental chamber is a doubled 120-cell. Continuing in this fashion we obtain infinitely many subgroups of finite index in $W(P)$ and with fundamental chamber a right-angled polytope in $\mathbb{H}^4$. (Analogously, one can produce infinitely many right-angled polygons in $\mathbb{H}^2$ by pasting together regular right-angled hyperbolic pentagons in $\mathbb{H}^2$ along edges.)

Example 4.35 (Word Hyperbolic RACGs of Type HM^4) In [192] Przytycki and Świątkowski prove that any 3-dimensional simplicial complex can be subdivided into a flag complex satisfying the no-□ condition. The construction uses the 600-cell (the simplicial sphere dual to the boundary complex of the 120-cell). It follows that any simplicial sphere of dimension ≤ 3 can be subdivided into a flag complex, L. As in Sect. 3.2 the resulting NPC cube complex P_L is covered by a CAT(-1) piecewise hyperbolic 4-manifold $\Sigma(W_L, S)$ that is diffeomorphic to $\mathbb{R}^4$ (and hence also to $\mathbb{H}^4$). The word hyperbolic RACG, W_L, is the group of lifts of the $(\mathbf{C}_2)^S$-action. It seems likely that some of these word hyperbolic RACGs are not commensurable with the $(5, 3, 3, 4)$ Coxeter group of the previous example; however, it is not clear which can be realized as cocompact isometric reflection groups on $\mathbb{H}^4$. If, instead, we start with a triangulation of a homology 3-sphere that is not simply connected and apply the subdivision of [192], then $\Sigma(W_L, S)$ is a homology manifold but not a 4-manifold. The word hyperbolic RACG, W_L, is still type HM^4 (cf. Theorem 3.40). Since $\Sigma(W, S)$ is not simply connected at infinity, it follows from Theorem 3.40 that $\Sigma(W_L, S)$ does not have the same proper homotopy type as $\mathbb{R}^4$.

Example 4.36 (More Word Hyperbolic Coxeter Groups of Type HM^4) We resurrect an old example from [74, pp. 219–220]. Suppose J is a flag triangulation of S^2 with vertex set T. Suspend J to get a triangulation L of S^3. Call the suspension vertices s_+ and s_- and let $S = T \sqcup \{s_+, s_-\}$. Label the edges of J by 2 and the edges connecting to a suspension vertex by 3. Then $K(L)$ is combinatorially equivalent to a convex polytope, namely, the prism on the dual polytope J^* to J. Thus, J^* is a 3-dimensional simple polytope. If F, F' are two facets of J^* that intersect, then the vertical facets $F \times I$ and $F' \times I$ of $J^* \times I$ make a dihedral angle of $\pi/2$. The top and bottom facets of $J^* \times I$ intersect each vertical facet at a dihedral angle of $\pi/3$. Let (W', S') be the spherical Coxeter system of rank 4 with Coxeter diagram $\mathbf{D}_4$, as in Table 4.1 of Sect. 4.2.1. (This diagram is a "Y", with extreme nodes $\{s_0, s_1, s_2\}$, with central node s_3 and with each edge labeled 3.) It follows that each tetrahedron in L has diagram $\mathbf{D}_4$. So, L is a metric flag complex and the corresponding Coxeter system is type HM^4. If J satisfies the no □-condition (for example, if J is the boundary complex of an icosahedron), then it follows from Andreev's Theorem (cf. Remark 4.22) that J^* can be realized as a convex polytope in $\mathbb{H}^3$ and that the reflection group corresponding to J^* is hyperbolic. Since J has no empty 4-cycles, L has no empty 4-cycles with each edge labeled 2. So, the corresponding Coxeter system (W, S) is word hyperbolic. (Is it ever possible to realize $K(L)$ as a convex polytope in $\mathbb{H}^4$? The answer is probably no.)

Remark 4.37 (Word Hyperbolic RACGs with Arbitrarily High vcd) Theorem 4.33 shows that there does not exist a word hyperbolic Coxeter group which is virtually the fundamental group of a closed n-manifold for $n > 28$. This fact led Moussong [179] to ask if there is an upper bound on the virtual cohomological dimension of word hyperbolic Coxeter groups. In [159] Januszkiewicz–

Świątkowski showed that there is no such upper bound. In fact, for every $k \geq 1$, they show that there is a k-dimensional, finite simplicial complex L which

(a) is a connected k-dimensional pseudomanifold,
(b) is a flag complex, and
(c) satisfies the no $\square$-condition of Sect. 2.2.

Let (W, S) be the associated RACS with $L(W, S) = L$. Then $\operatorname{vcd} \Gamma = k + 1$. To see this, observe that W has a finite index, torsion-free subgroup Γ such that Σ/Γ is an orientable $(k+1)$-dimensional pseudomanifold and hence, has a nonzero fundamental homology class in dimension $k + 1$.

Suppose W is a reflection group on $\mathbb{H}^n$ with Poincaré polytope P. When it is compact, P can be identified with the fundamental chamber $K(W, S)$ of Sect. 4.2.6. When it is not compact, P can be regarded as a thickening of $K(W, S)$. Similarly, $\Sigma(W, S)$ $(= D(W, K))$ equivariantly embeds as the cocompact core (or the "spine") of $D(W, P)$. When P is not compact, the reflection subgroup $W(P) < \operatorname{Isom}(\mathbb{H}^n)$ need not be word hyperbolic. The condition that is needed is that $W(P)$ is "convex cocompact" as defined in the next paragraph.

Convex Cocompact Subgroups of Isom($\mathbb{H}^n$) A discrete subgroup $\Gamma < \operatorname{Isom}(\mathbb{H}^n)$ without parabolic elements is *convex cocompact* if the action on the convex hull of its limit set is cocompact. Equivalently, Γ is convex cocompact if it is cocompact on some convex subset $C < \mathbb{H}^n$. A third equivalent condition is that the orbit map $\Gamma \to \mathbb{H}^n$, defined by $\gamma \mapsto \gamma x_0$ for some $x_0 \in \mathbb{H}^n$, is a quasi-isometric embedding. It follows from the second condition that Γ is quasi-isometric to the convex subset $C < \mathbb{H}^n$. Since the metric on $\mathbb{H}^n$ restricts to a CAT(-1) metric on the convex set C, it follows that Γ is word hyperbolic (cf. Sect. 4.2.9).

Note that the reflection group $W(P)$ contains no parabolic elements if and only if its Poincaré polytope contains no pair of "asymptotic faces." (Two faces are *asymptotic* if they have nonempty intersection with the sphere at infinity of $\mathbb{H}^n$.) The following result is well-known.

Theorem 4.38 (Desgroseilliers–Haglund [108, Theorem 4.7]) *Suppose $P < \mathbb{H}^n$ is a Poincaré polytope and $W(P) < \operatorname{Isom}(\mathbb{H}^n)$ is the hyperbolic reflection group generated by the reflections across the facets of P. Then $W(P)$ is convex cocompact on $\mathbb{H}^n$ if and only if P contains no pair of asymptotic faces. Consequently, $W(P)$ is word hyperbolic whenever this condition holds.*

Definition 4.39 Suppose that $\{P_s\}_{s \in S}$ is the set of facets of a hyperbolic convex polytope $P < \mathbb{H}^n$ and that $u_s \in \mathbb{R}^{n,1}$ is the outward-pointing unit normal to P_s. The *Gram matrix* of P is the $(S \times S)$-matrix $(u_s \cdot u_t)_{(s,t) \in S \times S}$.

The Gram matrix of P is closely related to the cosine matrix of $W(P)$. Indeed, if $P_s \cap P_t \neq \emptyset$, then $u_s \cdot u_t = -\cos(\pi/m(s, t))$; while if $P_s \cap P_t = \emptyset$, then $m(s, t) = \infty$ and $u_s \cdot u_t \in (-\infty, -1]$. Moreover, $u_s \cdot u_t = -1$ if and only if P_s and P_t are asymptotic. So, when $S \times S$ contains no pair (s, t) with $m(s, t) = \infty$, Theorem 4.38

means that $W(P)$ is convex cocompact on $\text{int}(\bigcup wP) < \mathbb{H}^n$ if and only if it contains no special subgroup W_T that is an irreducible euclidean reflection group of rank ≥ 3. If this is the case, then $W(P)$ is word hyperbolic since it acts cocompactly on a convex domain in $\mathbb{H}^n$ and hence, has a CAT(-1) metric. Alternatively, the fact that $W(P)$ is word hyperbolic can be proved directly from Moussong's Theorem 4.23. Indeed, since W is a discrete subgroup of $\text{Isom}(\mathbb{H}^n)$ with no parabolic elements, it satisfies Moussong's Conditions, i.e., it does not have any obvious $\mathbb{Z} \times \mathbb{Z}$ subgroups. Conversely, as we shall see in Corollary 4.50 below, if W is word hyperbolic, then it admits a projective representation $W < PGL(\mathbb{R}^{n+1})$ so that W acts cocompactly on a convex subset of some strictly convex, open domain $\Omega < \mathbb{R}P^n$. So, when P does not contain a pair of asymptotic faces, the group $W = W(P)$ is word hyperbolic.

Remark 4.40 In the Tits representation we require that $u_s \cdot u_t = -1$ whenever $m(s,t) = \infty$. On the other hand, if $m(s,t) = \infty$, then one can make a small change in the cosine matrix so that $c_{st} = c_{ts} < -1$ and all other entries are unchanged. In other words, if (W, S) contains no euclidean special subgroups of rank ≥ 3, one can deform the fundamental simplex in the Tits representation so that it contains no pair of asymptotic facets.

2-Spherical Coxeter Groups In general, a representation of (W, S) as a hyperbolic reflection group need not be identical to its Tits representation; however, there is one case when the two representations coincide—when (W, S) is 2-spherical as defined below.

Definition 4.41 (2-Spherical Coxeter Groups) A Coxeter system (W, S) is *2-spherical* if $m(s,t) < \infty$ for all $(s,t) \in S \times S$. Equivalently, (W, S) is 2-spherical if its defining graph L^1 is the complete graph on S. (A 2-spherical Coxeter system is sometimes said to be "∞ free," for example, in [156].)

So, if (W, S) is 2-spherical, then it has a representation as a hyperbolic reflection group on $\mathbb{H}^n$ with $n \leq \text{Card}(S) - 1$, if and only if its cosine matrix has signature $(n, 1)$. One of the main results in Desgroseilliers–Haglund [108] is that when (W, S) is 2-spherical, the cosine matrix often has signature $(n, 1)$. For example, they prove the following.

Lemma 4.42 ([108, Cor. 4.22]) *Suppose L^1 is a complete graph on $n+1$ vertices with $n > 1$ and each edge is labeled by an integer $m(e) \geq 4$. Then (W, S) has a convex cocompact representation on $\mathbb{H}^n$.*

Proof The point is that when each $m(e) \geq 4$, the signature of the cosine matrix is $(n, 1)$. It also has signature $(n, 1)$ when each edge is labeled 3 and $n \geq 3$; however, the representation is not convex cocompact in this case since each 3-cycle in L^1 gives a special subgroup which is a euclidean triangle group.

Example 4.43 (*Gromov Polyhedra*, Continued From Example 4.17) Suppose L^1 is a complete graph with each edge labeled by a constant integer m. By a previous lemma, if $m \geq 4$, then the Gromov polyhedron $X_{n,2m}$ embeds as the cocompact core of $W(n+1, m)$ on $\mathbb{H}^n$. If $m = 3$ and $n \geq 3$, then the cosine matrix also has signature $(n, 1)$ and so, the Coxeter group $W(n+1, 3)$ again has a representation as a hyperbolic reflection group (with parabolic elements).

Neighborly Polytopes and Their Duals We will use the notation $J(d+m, d)$ to mean a d-dimensional simplicial polytope with $d+m$ vertices. Similarly, $J^*(d+m, d)$ will stand for a d-dimensional simple polytope with $d+m$ facets. The polytope $J(d+m, d)$ is *k-neighborly* if its k-skeleton is equal to the k-skeleton of Δ^{d+m-1}. (In other words, a simplicial polytope is k-neighborly if and only if any collection of $(k+1)$ vertices spans a k-simplex.) For example, Δ^d has type $J(d+1, d)$ and is $(d+1)$-neighborly. For $m > 1$, 1-neighborly polytopes of type $J(d+m, d)$ exist if and only if $d \geq 4$. If $J(d+2, d)$ is the dual of the product of p-simplex and a $(d-p)$-simplex, then its boundary complex is the join $\partial\Delta^p * \partial\Delta^{d-p}$, which is k-neighborly for $k = \min\{p, d-p\} - 1$. (Note that the boundary complex of a simplicial d-polytope is a triangulation of S^{d-1}.) So, for each $d > 3$, there exists a compact, d-dimensional, simplicial polytope whose 1-skeleton is the complete graph on $d+m$ vertices with m an arbitrary integer ≥ 1. An example of such a simplicial polytope is the *cyclic polytope* $C(d+m, d)$, defined below.

Definition 4.44 The d-dimensional *cyclic polytope* with $d+m$ vertices is the simplicial polytope $C(d+m, d)$ defined as the convex hull in $\mathbb{R}^d$ of the image of $d+m$ points on the moment curve $t \mapsto (t, t^2, \dots t^d)$. Its dual $C^*(d+m, d)$ is a *dual cyclic polytope*.

The cyclic polytope $C(d+m, d)$ is $\lfloor d/2 \rfloor$-neighborly. In particular, if $d \geq 4$, its 1-skeleton is the complete graph on its vertex set.

The Four Color Theorem asserts that when $d = 3$, four colors suffice to color the vertices of any 3-dimensional simplicial polytope. The existence of cyclic polytopes shows that the situation for $d > 3$ stands in stark contrast to the situation for $d \leq 3$. For $d > 3$, the number of colors needed to color the vertices of $C(d+m, d)$ is $d+m$ and this can be arbitrarily large. If the 1-skeleton L^1 of the nerve $L(W, S)$ of a Coxeter system is a complete graph, then, by definition, (W, S) is 2-spherical. So, if $L(W, S)$ is the boundary complex of a 2-neighborly simplicial polytope $J(d+m, d)$, then (W, S) is type HM^d. If, in addition, (W, S) satisfies Moussong's Conditions, then (W, S) is word hyperbolic. The fundamental chamber $K(W, S)$ is combinatorially isomorphic to the simple convex polytope $J^*(d+m, d)$ that is dual to $J(d+m, d)$. If $J^*(d+m, d)$ cannot be realized as a convex polytope in $\mathbb{H}^d$ with prescribed Gram matrix (= the cosine matrix), then (W, S) cannot be a cocompact reflection group on $\mathbb{H}^d$. What's more, if the cosine matrix does not have signature $(d+m-1, 1)$, then no thickening of $J^*(d+m, d)$ can be a Poincaré polytope for W on $\mathbb{H}^{d+m}$; so, W cannot be realized as a convex cocompact hyperbolic reflection group on any hyperbolic space. This raises the question of finding a word hyperbolic Coxeter systems (W, S) with $L(W, S) = \partial J(d+m, d)$, where $J(d+m, d)$ is a 2-neighborly simplicial d-polytope and the cosine matrix does not have signature $(d+m-1, 1)$.

Remark 4.45 (The Fundamental Chamber Is a Simplex) When $m = 1$, the simple polytope $J^*(d+1, d)$ must be a d-simplex σ^d. Every irreducible Coxeter group with fundamental chamber a simplex is a reflection group of either euclidean or hyperbolic type. There are complete lists of the diagrams of such Coxeter groups (e.g., see Table 4.1 in Sect. 4.2.1 or [82, Table 6.1, p.104] for the euclidean case and [82, Table 6.2, p.105] for the hyperbolic case). The list where $J^*(d+1,d)$ is a hyperbolic simplex can be summarized as follows. For $d=2$, there are infinitely many hyperbolic triangle groups; for $d=3$ there are 9 examples with fundamental polytope a hyperbolic tetrahedron; for $d=4$, there are 5 examples with fundamental polytope a hyperbolic 4-simplex; and in higher dimensions there are no further hyperbolic examples.

Consider a product of simplices $\sigma_1 \times \cdots \times \sigma_p$. If each σ_i is of dimension ≥ 2, then the product is the dual of a 1-neighborly polytope $J(d+p,d)$, where $\partial J(d+p,d)$ is the join $\partial\Delta_1 * \cdots * \partial\Delta_p$. Here d is the dimension of the product and Δ_i denotes the simplex dual to σ_i (i.e., $\sigma_i = \Delta_i^*$). In other words, if $W(\sigma_i)$ is a euclidean or hyperbolic Coxeter group with fundamental chamber σ_i and nerve $\partial\Delta_i$, then the product $W = W(\sigma_1) \times \cdots \times W(\sigma_p)$ has fundamental chamber $\sigma_1 \times \cdots \times \sigma_p$. When $p > 1$, the product is never word hyperbolic, since it contains the product of two or more infinite special subgroups. Note that if each σ_i is a hyperbolic simplex (say of dimension d_i), then the cosine matrix has an orthogonal decomposition into blocks of signature $(d_i, 1)$; hence, the Tits representation of the product has signature (d, p), with $d = \sum d_i$.

Moussong's Examples Consider the product of two hyperbolic triangle group $W(\sigma_1) \times W(\sigma_2)$. Its Coxeter diagram has two components, each with 3 vertices. The cosine matrix has signature $(4, 2)$. Its fundamental domain K is $\sigma_1 \times \sigma_2$ which is combinatorially equivalent to the dual cyclic polytope $C^*(6,4)$. If we introduce a new edge labeled 3 connecting the components, then the resulting Coxeter system (W, S) will be word hyperbolic. If we are careful where we put this edge, then the resulting nerve $L(W, S)$ often still will be isomorphic to the join $\partial\Delta_1 * \partial\Delta_2$, i.e., (W, S) will be type HM^4 and the fundamental chamber again will be $C^*(6,4)$. Moussong considered such examples in [179, p. 47]. We list these diagrams in Table 4.2. The signature of the cosine matrix of $W(\sigma_1) \times W(\sigma_2)$ is $(4, 2)$. The cosine matrix of a diagram in Table 4.2 is almost the same as the cosine matrix of $W(\sigma_1) \times W(\sigma_2)$, the only difference being that the pair of entries corresponding to the edge connecting the two components has been changed from 0 to -1/2. A direct computation shows that the signature remains $(4, 2)$. Hence, the Coxeter group corresponding to the diagram on the left in Table 4.2 does not have representations

Table 4.2 Diagrams with 4-dimensional dual cyclic polytopes as fundamental chambers

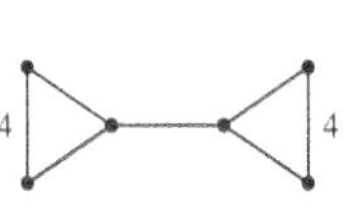

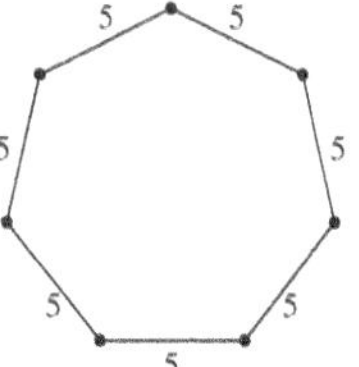

Table 4.3 Diagrams with 5-dimensional dual cyclic polytopes as fundamental chambers

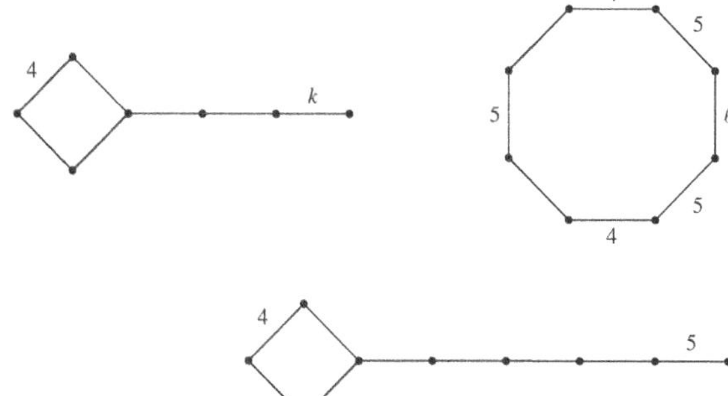

Table 4.4 Diagram with 7-dimensional dual cyclic polytope as fundamental chamber

as convex cocompact reflection groups on $\mathbb{H}^n$. Similar remarks apply to the Coxeter groups with diagrams in Tables 4.3 and 4.4.

Proposition 4.46 (Moussong [179, p. 47], [180, pp. 542–544]) *The diagrams in Tables 4.2, 4.3, 4.4 give examples of word hyperbolic, 2-spherical Coxeter groups of type* HM. *In each case the fundamental chamber is combinatorially isomorphic to a dual cyclic polytope. However, none of these can be realized as a convex cocompact reflection group on* $\mathbb{H}^n$.

4.2.9 *Convex Cocompact Projective Representations*

As was mentioned previously, Vinberg [227] extended the work of Tits that was described in Sect. 4.2.7 to study representations of a Coxeter group W into $GL(\mathbb{R}^{n+1})$ as a discrete group generated by reflections across the facets of a polyhedral cone C. In [227] the reflections are not required to preserve a symmetric bilinear form. However, groups generated by reflections across the facets of the cone on a hyperbolic polytope do fit into Vinberg's framework. For example, if the hyperbolic polytope is not a simplex, then Vinberg's reflection group does not come from a Tits representation. In [227] this gives a representation of W into $PGL(\mathbb{R}^{n+1}) = GL(\mathbb{R}^{n+1})/\mathbb{R}^*$ which often defines an action of W as a reflection group on some open convex set $\Omega < \mathbb{R}P^n$. When this is the case, the quotient orbifold Ω/W will have a projective structure. The projective representation is *convex cocompact* if there is a smaller convex subset $\Omega' < \Omega$ on which the action is cocompact. (In [71] this is called "naively convex cocompact.") It is proved in [71] that any word hyperbolic W admits a convex cocompact representation on some $\Omega < \mathbb{R}P^n$. (See Proposition 4.49 below.) Certain relatively hyperbolic Coxeter groups which are not word hyperbolic can still admit convex cocompact projective representations. The point is that certain euclidean special subgroups of rank ≥ 3 are allowed to occur, namely, euclidean special subgroups of type $\tilde{\mathbf{A}}_n$ with $n \geq 3$. Relatively hyperbolic Coxeter systems for which this property holds are said to satisfy the "weak version of Moussong's Conditions.") Proposition 4.49 below states that these are precisely the Coxeter groups which admit convex cocompact projective representations.

A subset Ω in $\mathbb{R}P^n$ is *convex* if it is a convex subset in some affine chart; it is *properly convex* if it is a bounded subset; it is *strictly convex* if $\partial\Omega$ contains no nontrivial line segment. Any properly convex domain Ω has a "Hilbert metric," defined as follows. Suppose A, B are distinct points in Ω and that the line they determine intersects $\partial\Omega$ in points X, Y, where the order of the points is X, A, B, Y. In the *Hilbert metric* the distance from A to B is defined using cross ratios as in [71] by:

$$d(A,B) := \log \frac{|Y-A|\,|X-B|}{|Y-B|\,|X-A|}. \tag{4.14}$$

It is a Finsler metric. If Ω is isomorphic to the round disk (or the interior of an ellipsoid), then the Hilbert metric is the hyperbolic metric on the Klein model of $\mathbb{H}^n$. So, it makes sense to ask if the Hilbert metric is Gromov hyperbolic. Benoist [16] showed that the Hilbert metric is Gromov hyperbolic if and only if Ω is strictly convex. A discrete subgroup $\Gamma < PGL(\mathbb{R}^{n+1})$ that is cocompact on an open convex domain Ω is said to be *divisible*. If Γ is divisible, then Ω/Γ is a compact orbifold with a projective structure. When Γ is divisible, the orbit map, $\Gamma \to \Omega$, is a quasi-isometry. Hence, if Ω is Gromov hyperbolic, then so is Γ, that is to say, Γ is word hyperbolic.

A *reflection* on $\mathbb{R}^{n+1}$ (or on $\mathbb{R}P^n$) is a linear involution $r : \mathbb{R}^{n+1} \to \mathbb{R}^{n+1}$ such that the eigenvalue (-1) has multiplicity 1. Thus, r is determined by a (-1)-eigenvector $e \in \mathbb{R}^{n+1}$ and a linear form $\alpha \in (\mathbb{R}^{n+1})^*$ such that $\ker(\alpha)$ is the $(+1)$-eignenspace. If we normalize by requiring $\alpha(e) = 2$, then the corresponding reflection is the linear transformation, $v \mapsto v - \alpha(v)e$. (In the presence of a nonsingular symmetric bilinear form, $\langle\ ,\ \rangle : \mathbb{R}^{n+1} \times \mathbb{R}^{n+1} \to \mathbb{R}$, given by $(v, v') \mapsto \langle v, v'\rangle \in \mathbb{R}$, by using the natural identification of $\mathbb{R}^{n+1}$ and its dual space, we see that the reflection is defined by $v \mapsto v - \langle v, e\rangle e$, when e is normalized so that $\langle e, e\rangle = 2$.)

Following [227], suppose W is a discrete subgroup in $GL(\mathbb{R}^{n+1})$ generated by reflections across the facets of a polyhedral cone C so that the set of reflections across these facets form a fundamental set of generators, $S < W$. (We do not require that each facet be assigned a reflection.) The image of C in $\mathbb{R}P^n$ is the *Poincaré polytope*. Since a reflection $s \in S$ is determined, as above, by a pair $(\alpha_s, e_s) \in (\mathbb{R}^{n+1})^* \times \mathbb{R}^{n+1}$, we get an $(S \times S)$-matrix $(A(s,t))$ called the *Cartan matrix* of (W, S) defined by $A(s,t) = \alpha_s(e_t)$. Each diagonal element of $A(s,t)$ is equal to 2, while the off-diagonal elements lie in $(-\infty, 0]$ and satisfy

$$A(s,t)A(t,s) \begin{cases} = 4\cos^2(\pi/m(s,t)), & \text{if } m(s,t) \neq \infty, \text{ and} \\ \geq 4, & \text{if } m(s,t) = \infty. \end{cases} \tag{4.15}$$

The matrix $(A(s,t))$ is not necessarily symmetric. If it is symmetric, then for $m(s,t) \neq \infty$, $A(s,t)$ is twice the corresponding entry $c(s,t)$ of the cosine matrix defined in (4.9). The matrix is said to be *symmetrizable* if, after a change of basis, it is equivalent to an $(n+1) \times (n+1)$ symmetric matrix. If this is the case, then there a W-invariant symmetric bilinear on $\mathbb{R}^{n+1}$ such that $(A(s,t))$ is essentially the Gram matrix defined by the outward-pointing unit normals to the facets of C. A result of [227, Prop. 20] asserts that $(A(s,t))$ is always symmetrizable if and only if the Coxeter diagram of (W, S) does not contain a circuit. (Such a circuit could have some edges labeled ∞.)

Example 4.47 (Triangle Groups of Kac–Vinberg [230]) In 1967 Kac and Vinberg [230] showed that if W is isomorphic to a hyperbolic triangle group whose diagram is a 3-cycle, then W admits a representation as a divisible subgroup of $PGL(\mathbb{R}^3)$ which cannot be deformed to a hyperbolic representation (i.e., into $PO(2,1)$). Suppose L is a triangle and m is a labeling of its edge set with each $m(e) \geq 3$. Define the Cartan matrix A_{ij} for $W(L)$ with $A_{11} = A_{22} = A_{33} = 2$, $A_{12} = A_{21} = -\cos(\pi/m(1,2))$, $A_{23} = A_{32} = -\cos(\pi/m(2,3))$ and $A_{13} \neq A_{31}$ (but with $A_{13}A_{31} = \cos^2(\pi/m(1,3))$ so that (4.15) still holds). Then A_{ij} is not symmetrizable and the representation of $W(L)$ in $GL(\mathbb{R}^3)$ is not conjugate to a hyperbolic triangle group in $O(2,1)$.

Example 4.48 (*Type* $\tilde{\mathbf{A}}_n$) Suppose (W, S) is an irreducible euclidean Coxeter system of rank $n+1$, with symmetric Cartan matrix and with $n+1 \geq 3$. Then its cosine matrix is positive semidefinite. It has open convex domain in $\mathbb{R}P^n$ defined by (4.17) below that is isomorphic to an affine chart $\mathbb{R}^n$ (and hence, is not properly convex). The Coxeter diagrams of irreducible euclidean Coxeter systems are trees except for those of type $\tilde{\mathbf{A}}_n$, the diagram of which is a cycle of length $n+1$, with $n \geq 2$. Hence, if the diagram of a euclidean Coxeter group is not type $\tilde{\mathbf{A}}_n$, its Cartan matrix is symmetrizable. On the other hand, for type $\tilde{\mathbf{A}}_n$, the Cartan matrix can be nonsymmetrizable (cf. (4.16) below). Similarly, if W is a Coxeter group of type $\tilde{\mathbf{A}}_n$ then, as in Example 4.47, W can have a projective representation as a divisible subgroup of $PGL(\mathbb{R}^{n+1})$; the domain Ω is the interior of an affine n-simplex. For example, for $\tilde{\mathbf{A}}_2$ we have:

$$(A(s,t)) = \begin{pmatrix} 2 & -1 & -\frac{1}{a} \\ -1 & 2 & -1 \\ -a & -1 & 2 \end{pmatrix}, \quad \text{with } a > 1. \tag{4.16}$$

This Cartan matrix can be used to define a representation of the Coxeter group of type $\tilde{\mathbf{A}}_2$ (that is, the $(3,3,3)$ triangle group) into $PGL(\mathbb{R}^3)$. The domain Ω, defined by (4.17) below, is bounded and is isomorphic to the interior of a triangle in the affine plane. Hence, Ω is properly convex but not strictly convex. There are similar Cartan matrices and projective representations for $\tilde{\mathbf{A}}_n$. Thus, although a special subgroup of type $\tilde{\mathbf{A}}_n$ obstructs the word hyperbolicity of W, it does not obstruct the existence of a convex cocompact projective representation.

Let P denote the image of the polyhedral cone C in $\mathbb{R}P^n$ and, as in Sect. 4.2.7, let P^f denote the complement in P of the nonspherical faces. Put

$$\Omega = \bigcup_{w \in W} wP^f. \tag{4.17}$$

Vinberg [227] establishes that properties (a), (b), (c), (d) of Sect. 4.2.7 hold in this generality, as does Theorem 4.31.

We say that (W, S) satisfies the *weak version of Moussong's Conditions* if for each $T < S$, neither of the following conditions holds for any special subgroup W_T (cf. conditions (i) and (ii) in Lemma 4.21):

(i) W_T is an irreducible euclidean reflection group of rank ≥ 3 whose diagram is different from $\tilde{\mathbf{A}}_n$.
(ii) W_T decomposes as $(W_{T_1} \times W_{T_2}, T_1 \sqcup T_2)$, where both W_{T_1} and W_{T_2} are infinite.

In other words, in contrast with the strong version, the weak version of Moussong's Conditions allows the possibility of special subgroups of type $\tilde{\mathbf{A}}_n$.

Proposition 4.49 (Danciger-Guéritaud-Kassel-Lee-Marquis [71, Thm. 1.3]) *A Coxeter system (W, S) satisfies the weak version of Moussong's Conditions if and only if it has a representation as a projective reflection group acting convex cocompactly on a properly convex domain $\Omega < \mathbb{R}P^n$.*

The strong version of Moussong's Conditions together with the above proposition implies that when W is word hyperbolic (cf. Moussong's Theorem 4.23) the following holds.

Corollary 4.50 ([71, Cor. 1.11]) *An infinite Coxeter group W is word hyperbolic if and only if it has a representation as a projective reflection group acting convex cocompactly on a strictly convex domain $\Omega < \mathbb{R}P^n$. When this is the case the Hilbert metric on Ω is Gromov hyperbolic.*

So, when W is word hyperbolic, it admits a convex cocompact projective representation, independent of any consideration of the signature of Tits representation.

Example 4.51 (cf. Lemma 4.42) Suppose L^1 is the complete graph on n+1 vertices, with $n \geq 2$. and with each edge labeled by an integer $m(e) \geq 3$. Then (W, S) admits a convex cocompact representation on a properly convex subset of $\mathbb{R}P^n$.

Example 4.52 The following example was explained to me by Ryan Greene. Suppose the Coxeter diagram of (W, S) consists of m triangles arranged along the integer points in line segment $[0, 2m]$ so that there is an edge connecting the ith triangle to the $(i + 1)^{th}$ as in Table 4.5. Each edge is labeled 3 (meaning that it is not labeled). Then L^1 is a complete graph on $3m$ vertices and the fundamental chamber K is combinatorially equivalent to product of m triangles. So, (W, S) is a Coxeter system of type HM^{2m}; moreover, it satisfies the weak version of Moussong's Conditions. Although it is not word hyperbolic (since it has special

Table 4.5 A diagram with fundamental chamber a product of 2-simplices

subgroups of type $\tilde{A}_2$), it does admit a convex cocompact representation on $\mathbb{R}P^{3m-1}$ by Proposition 4.49.

Example 4.47 raises the possibility that there are word hyperbolic Coxeter groups of type HM^n that, although they cannot be represented as subgroups of $\mathrm{Isom}(\mathbb{H}^n)$, have divisible projective representations. Some possible candidates are provided in Tables 4.2, 4.3, 4.4 where it might be possible that fundamental polytope is an n-dimensional dual cyclic polytope in $\mathbb{R}P^n$. Benoist [16] showed that among Moussong's examples there was one family that had such a projective representation with this property.

Proposition 4.53 (Benoist [16, Prop. 3.1]) *Let (W, S) be the Coxeter system whose diagram is the one on the left in Table 4.2. Then W is 2-spherical, type HM^4 and word hyperbolic. Although it does not act as a reflection group on $\mathbb{H}^n$ for any n, there is a representation $W \hookrightarrow PGL(\mathbb{R}^5)$ so that W is a reflection group on a open convex set D in $\mathbb{R}P^4$), with compact fundamental polytope P that is combinatorially isomorphic to a product of two 2-simplices.*

The remarkable aspect of this proposition is that the polytope P is compact; hence, that D is quasi-isometric to W.

Since Davis–Moussong complexes of word hyperbolic Coxeter groups all have piecewise hyperbolic CAT(-1) metrics, the examples in this section of fundamental chambers of word hyperbolic Coxeter groups of type HM all have the structure of compact orbifolds with NPC polyhedral metrics. Benoist's example from the previous proposition has a negatively curved Finsler metric. The leads to the question if there is such an example which is not actually a compact polytope in hyperbolic space, yet which admits a smooth riemannian metric (as an orbifold) with negative sectional curvature. The answer is probably no.

4.3 Artin Groups

There are two different cell complexes associated to an Artin group A_W. First, there is the Deligne complex, $\Lambda(W, S)$ defined as a development $D(A_W, |\mathcal{S}^{\mathrm{op}}|)$ of the simple complex of groups $A\mathcal{S}^{\mathrm{op}}$. The corresponding poset $D((A_W, \mathcal{S}^{\mathrm{op}})$ can be identified with the poset of spherical Artin cosets defined in (4.8). A major drawback of $\Lambda(W, S)$ is that it is not locally finite. The second cell complex is the pure Salvetti complex, $X(W, S)$ (also called the "oriented Davis complex"). It is homotopy equivalent to the complement of the union of the reflection hyperplanes in a certain open convex subset of the complexification of the Tits cone. The fundamental group of this hyperplane arrangement complement is the pure Artin group, PA_W. The group A_W is the fundamental group of the quotient of this complement by the free W-action. The $K(\pi, 1)$-Conjecture for Artin groups states that the *Salvetti complex* $X(W, S)/W$ is the classifying space for A_W. Of course, the $K(\pi, 1)$-Conjecture is implied by the conjecture that the universal cover of $X(W, S)$ is contractible. In Sect. 4.3.4 we show that this conjecture is equivalent to the conjecture that the Deligne complex $\Lambda(W, S)$ is contractible. When (W, S) is right-angled, A_W is a RAAG. In this case, the Salvetti complex is the standard toral complex $\mathbb{T}_L$ of Sect. 3.1.1 (see also Sect. 3.1.2).

When W is spherical, there is third simplicial complex $B(W, S)$, called "Bestvina's normal form complex," which will be explained in Sect. 4.3.5. The group G that acts on $B(W, S)$ is $A_W/\langle\Delta^2\rangle$, where Δ^2 is an element of infinite order in the center of the spherical Artin group A_W. The complex $B(W, S)$ is locally finite and the G-action is cocompact. Moreover, it has a "distance function" which has much in common with distance in a CAT(0) metric. In particular, $B(W, S)$ is contractible. The proof of this is closely related to the proof by Deligne [107] that $\Lambda(W, S)$ is contractible.

Outline of This Section

(A) The Deligne complex $\Lambda(W, S)$ (Sect. 4.3.1).

- (A.1) The spherical Deligne complex $\Lambda'(W, S)$ of a spherical Artin group.
- (A.2) Apartments in a Deligne complex.
- (A.3) Conjecture 4.54: $\Lambda(W, S)$ is CAT(0).

(B) The Salvetti complex (Sect. 4.3.3).

- (B.1) Reflection arrangement complements (Sect. 4.3.2).
- (B.2) Deligne's Theorem on asphericity of arrangement complements (Theorem 4.62).

(C) The $K(\pi, 1)$-Conjecture for Artin groups (Sect. 4.3.4).

(D) Bestvina's normal form complex for spherical Artin groups (Sect. 4.3.5).

4.3.1 The Deligne Complex

The material in this section is taken from [107] and [56].

Given an Artin group A $(= A_W)$, let $\mathcal{G} = A\mathcal{S}^{\mathrm{op}}$ be the complex of spherical Artin subgroups defined in Example 4.4 of Sect. 4.1. The basic construction of Theorem A.6 in Appendix A.2 defines a poset $D(A, \mathcal{S}^{\mathrm{op}})$, as well as, a space $D(A, K)$, with fundamental chamber $K = |\mathcal{S}^{\mathrm{op}}|$ (cf. (4.3)). The poset $D(A, \mathcal{S}^{\mathrm{op}})$ is equal to Coset(A), the poset of spherical cosets in A from (4.8). The space

$D(A, K)$ is called the *Deligne complex* for A. It will be denoted by Λ $(= \Lambda(W, S))$. So, $\Lambda(W, S)$ is the analog of the Davis–Moussong complex $\Sigma(W, S)$ of Sect. 4.2.3 defined using the basic construction and fundamental chamber K.

If A is a RAAG, then it is a graph product over L^1 of infinite cyclic groups; $D(A, \mathcal{S}^{op})$ is the associated right-angled building described in Sect. 3.1.2 and in Sect. 4.4.3 below. Moreover, $D(A, K)$ is the standard realization of this building.

As in Definition 4.28 of Sect. 4.2.6, each copy of the fundamental chamber K is cellulated by Coxeter blocks, where each block is the intersection of a Coxeter zonotope and a sector in a fan for a reflection arrangement. This induces a piecewise euclidean metric on $\Lambda(W, S)$ that is analogous to the natural piecewise euclidean metric on the zonotopal complex $\Sigma(W, S)$. In view of Proposition 4.11 (5), this leads to the following conjecture.

Conjecture 4.54 (cf. [56, Conjecture 4.4.4, p. 622]) With its natural piecewise euclidean metric, $\Lambda(W, S)$ is CAT(0).

By Theorem 2.2 this conjecture implies that $\Lambda(W, S)$ is contractible. The issue is to show that the link of each cell of Λ satisfies the Link Condition of Definition 2.10.

When (W, S) is spherical, define its *spherical Deligne complex* $\Lambda'(W, S)$ by using $|\mathcal{S}_{<S}|$ as fundamental chamber instead of $|\mathcal{S}|$. Instead of a Coxeter block, the fundamental chamber for $A \curvearrowright \Lambda'(W, S)$ then becomes the fundamental spherical simplex $\sigma(W, S)$ for the W-action on the unit sphere $\mathbb{S}^{n-1}$ in the reflection representation, $W \curvearrowright \mathbb{R}^n$. This induces the *natural piecewise spherical metric* on $\Lambda'(W, S)$. As we will explain at the end of this subsection when (W, S) is spherical, $\Lambda'(W, S)$ is a union of subcomplexes, called *apartments*, each of which is isometric to the round sphere $\mathbb{S}^{n-1}$. When (W, S) is spherical, Deligne [107] proved that $\Lambda'(W, S)$ has the homotopy type of a wedge of $(n-1)$-spheres. For example, when (W, S) has diagram $\mathbf{A}_2$ so that A_W is the braid group on three strands, then $\Lambda'(W, S)$ is a graph of infinite valence at each vertex; each apartment is a 6-cycle and the girth of $\Lambda'(W, S)$ is 6 (cf. Proposition 4.56 below). When (W, S) is spherical, Deligne [107] also define a complex $\Lambda''(W, S)$ obtained by filling in each apartment of Λ' with a round ball and he proved that Λ is contractible.

To prove Conjecture 4.54 there are three types of links of cells in Λ which must be shown to be CAT(1).

(a) The link of a Coxeter block B_T. Such a link corresponds to the order complex of $\mathcal{S}_{>T}$, i.e., it is the link of a simplex corresponding to T in $L(W, S)$.
(b) A link isomorphic to $\Lambda'(W_T, T)$. (This is the link of the cone point of B_T in the subcomplex $\Lambda(W_T, T)$.)
(c) A join of a link of type (a) with one of type (b).

Links of type (a) are metric flag complexes; hence, they are CAT(1) by Moussong's Lemma 4.18. The spherical join of two CAT(1) piecewise spherical complexes is CAT(1) (see [35, Prop. 5.15, p. 64] or [55, Appendix, Theorem A9]). So, it suffices to consider links of type (b). This leads to the following conjecture of [57].

Conjecture 4.55 (cf. [57, Conjecture 3, p. 6]) If (W, S) is spherical, then the natural piecewise spherical metric on the spherical Deligne complex $\Lambda'(W, S)$ is CAT(1).

So, Conjecture 4.55 is equivalent to Conjecture 4.54. (When proving the natural metric on Σ is CAT(0), the links of type (b) are replaced by Coxeter complexes of the form $D(W_T, \sigma(W_T, T))$ which is a round sphere and hence, is CAT(1). So, to prove Moussong's Theorem that Σ is CAT(0), it was not necessary to consider links of type (b).)

Proposition 4.56 (cf. [56, Proposition 4.4.5]) *Suppose* (W, S) *is spherical with* $\mathrm{Card}(S) = 2$ *so that* W *is a finite dihedral group. Then the spherical Deligne complex* $\Lambda'(W, S)$ *is* CAT(1).

It is observed in [56] that Proposition 4.56 follows from a lemma of Appel-Schupp [10, Lemma 6] which asserts that, for $S = \{s, t\}$, the shortest loop in $\Lambda'(W, S)$ has length $2m(s, t)$ (corresponding to the Artin relation for $\{a_s, a_t\}$). This proposition has the following corollary.

Corollary 4.57 *Suppose* $\dim \Lambda \leq 2$. *Then* Λ *is* CAT(0) *and hence, is contractible.*

So, 2-dimensional Deligne complexes provide examples of CAT(0) polygonal complexes that are not locally finite. The 2-cells are Coxeter blocks with their natural euclidean metrics.

The cubical structure on K defined in Sect. 4.2.6 gives Λ the structure of a combinatorial cube complex. When equipped with the piecewise euclidean metric in which each cell is a regular euclidean cube, this complex is denoted by $\Lambda^{\square}$. When is this cube complex CAT(0)? The answer is essentially the same as the answer provided for $\Sigma^{\square}$ in Proposition 4.30. In the following proposition it is shown that $\Lambda^{\square}$ is CAT(0) if and only if (W, S) is type FC (see Definition 4.29). Once one knows that links of type (b) are flag complexes (a fact proved in [56, Lemma 4.3.4]), the proof of Proposition 4.30 shows $\Lambda^{\square}$ is CAT(0) if and only if $L(W, S)$ is a flag complex. This gives the following analog of Proposition 4.30.

Proposition 4.58 (cf. [56, Theorem 4.3.5]) *Let* $\Lambda^{\square}$ *denote the Deligne complex with its cubical metric. This metric is* CAT(0) *if and only if* (W, S) *is type FC.*

This yields another case where we know that the Deligne complex $\Lambda(W, S)$ is contractible.

Corollary 4.59 (cf. [56, Theorem A]) *If* (W, S) *is type FC, then* $\Lambda(W, S)$ *is contractible.*

Definition 4.60 Let $p : A_W \to W$ be the natural epimorphism. The kernel of p is the *pure Artin group* denoted by PA_W.

Apartments in $\Lambda(W, S)$ If $s_1 \dots s_k$ is a reduced decomposition of an element $w \in W$, then $a_w = a_{s_1} \dots a_{s_k}$ is a well-defined element of A that depends only on w and not on the choice of reduced decomposition. (This follows from the solution to the word problem for Coxeter groups by Tits; see [82, §3.4].) Hence, $q : w \mapsto a_w$ is a section of the canonical epimorphism $p : A \to W$. (N.B. The function q is only a map of sets; it is not a homomorphism.) The epimorphism $A \to W$ induces a projection $\Lambda \to b\Sigma$ and q induces a section $q : b\Sigma \to b\Lambda$, where $b\Sigma$ means the barycentric subdivision of the zonotopal cell complex Σ and $b\Lambda$ is the analogous subdivision of Λ. Moreover, for any $a \in A$, after composing with translation by a, we get another section $aq : b\Sigma \to \Lambda$. The image any such section is called an *apartment* of Λ (cf. [107]). As is the case with buildings, Λ retracts onto any one of its apartments. However, Λ is usually *not* a building. (It is a building only in the right-angled case; this comes down to the fact that for each edge $\sigma \in L^{(1)}$, the corresponding group A_σ is $\mathbb{Z} \times \mathbb{Z}$, which is abelian.)

We have that $PA_W \curvearrowright \Lambda$, the orbit space is Σ, and the quotient map is $p : \Lambda \to \Sigma$; moreover, $q(\Sigma)$ is a strict fundamental domain for the PA_W-action. This implies that $\Lambda(W, S)$ is the development of a simple complex of groups whose direct limit is PA_W. Its underlying poset is $\mathrm{Coset}(W)^{\mathrm{op}}$, the poset of spherical cosets in W, cf. (4.7). There is an associated simple complex of groups:

$$\mathcal{G} := \{PA_{w\sigma}\}_{w\sigma \in \mathrm{Coset}(W)^{\mathrm{op}}}, \tag{4.18}$$

where $\sigma \in \mathcal{S}$, where $w\sigma$ means the coset $wW_\sigma \in (W/W_\sigma)$, and where $PA_{w\sigma} = (a_w)A_\sigma(a_w)^{-1}$. The fact that $\Sigma(W, S)$ $(= |\mathrm{Coset}(W)^{\mathrm{op}}|)$ is a strict fundamental domain for PA_W on $\Lambda(W, S)$ gives the following.

Proposition 4.61 (cf. Appendix A.2) *The system $\mathcal{G}$ is a simple complex of groups over* $\mathrm{Coset}(W)^{\mathrm{op}}$*, its direct limit is PA_W, and its development is the Deligne complex, $\Lambda(W, S)$.*

4.3.2 The Complement of a Reflection Arrangement

First suppose that (W, S) is a spherical Coxeter system so that W acts orthogonally on $\mathbb{R}^n$ $(= \mathbb{R}^S)$ via the geometric representation of Sect. 4.2.7. Let $\mathcal{A}$ be the corresponding reflection arrangement of hyperplanes in $\mathbb{R}^n$. Complexification gives $W \curvearrowright \mathbb{C}^n$ and we get an arrangement $\mathcal{A} \otimes \mathbb{C}$ of complexified hyperplanes in $\mathbb{C}^n$. The complement of this arrangement is denoted by

$$M(\mathcal{A} \otimes \mathbb{C}) := \mathbb{C}^n - \bigcup_{H \in \mathcal{A} \otimes \mathbb{C}} H. \tag{4.19}$$

It is well-known that $\pi_1(M(\mathcal{A} \otimes \mathbb{C})) = PA_W$ (cf. Definition 4.60). The group W acts freely on M $(= M(\mathcal{A} \otimes \mathbb{C}))$. So, $\pi_1(M/W)$ is the Artin group A $(= A_W)$. For example, if W is the symmetric group S_{n+1} acting on $\mathbb{R}^{n+1}$ by permuting the coordinates, then M is identified with the configuration space of $n + 1$ distinct ordered points in $\mathbb{C}$ and M/S_{n+1} is the unordered configuration space. Define B_{n+1} to be $\pi_1(M/S_{n+1})$ and call it the *braid group* on $n + 1$ strands. Define PB_{n+1} to be $\pi_1(M)$, the *pure braid group*. Braid groups were introduced in 1947 by Emil Artin

[11]. He showed that the fundamental group of the unordered configuration space M/S_{n+1} has a presentation in the form of (4.6). In other words, its presentation is the one associated to the Coxeter diagram $\mathbf{A}_n$. Thus, Artin groups are "generalized braid groups." The main result of [107] is the following.

Theorem 4.62 (Deligne's Theorem in [107]) *Suppose (W, S) is spherical. Then the arrangement complement, $M = M(\mathcal{A} \otimes \mathbb{C})$, is aspherical. Hence, M is a model for BPA_W and M/W is a model for BA_W. In other words, the $K(\pi, 1)$-Conjecture holds for spherical Artin groups.*

In fact, Deligne proved that whenever $\mathcal{A}$ is a real simplicial arrangement in $\mathbb{R}^n$, the complement $M(\mathcal{A} \otimes \mathbb{C})$ of the complexified arrangement is aspherical.

Remark 4.63 Spherical Artin groups have faithful linear representations (see [64]) and hence, are residually finite.

Almost all cases of Theorem 4.62 had been proved slightly earlier by Brieskorn [36] and Brieskorn-Saito [37, 38]. In 2015 this result was extended to finite groups generated by complex reflections in Bessis [20].

Remark 4.64 To prove Theorem 4.62 Deligne considered the complex $\Lambda''(W, S)$ obtained by filling in each apartment of the spherical Deligne complex, $\Lambda'(W, S)$, with a round ball. He identified $\Lambda''(W, S)$ with the nerve of a certain open cover of the universal cover of the arrangement complement $M(\mathcal{A} \otimes \mathbb{C})$. He then proved that $\Lambda''(W, S)$ was contractible (cf. Sect. 4.3.5) and hence, that $\Lambda'(W, S)$ has the homotopy type of a wedge of spheres. Using the Nerve Lemma (see Lemma 4.67 below) he concluded that the universal cover of $M(\mathcal{A} \otimes \mathbb{C})$ was also contractible.

Even when W is not required to be finite, it can still be realized as a group generated by reflections on a open convex subset $\Omega < \mathbb{R}^n$ by using the geometric representation $\rho^* : W \to GL(\mathbb{R}^n)$ of Sect. 4.2.7. (Here Ω denotes the interior of the Tits cone.) The group W acts properly on the open subset $\mathbb{R}^n + i\Omega$ of $\mathbb{C}^n$ and the action is free on the arrangement complement, M $(= M(\mathcal{A} \otimes \mathbb{C}))$, defined by

$$M = (\mathbb{R}^n + i\Omega) - \bigcup_{H \in \mathcal{A} \otimes \mathbb{C}} H\,, \tag{4.20}$$

where $\mathcal{A} \otimes \mathbb{C}$ is the collection of all reflecting hyperplanes in $\mathbb{C}^n$.

The following conjecture, due independently to Arnold, Pham, and Thom, asserts that a generalization of Theorem 4.62 holds for all Artin group. It is the most important open question about Artin groups. The conjecture is explained in detail in [56, 57] and [171].

Conjecture 4.65 (The $K(\pi, 1)$-Conjecture for Artin Groups) The space M/W is a model for BA_W.

4.3.3 The Salvetti Complex and Its Universal Cover

Given an arrangement $\mathcal{A}$ consisting of finitely many affine hyperplanes in $\mathbb{R}^n$, Salvetti [203] constructed a finite cell complex $X(\mathcal{A})$ that is homotopy equivalent to the complement in $\mathbb{C}^n$ of the union of the complexified hyperplanes of the arrangement. If the arrangement is essential, then $\dim X(\mathcal{A}) = n$. The cells of $X(\mathcal{A})$ are zonotopes—each zonotope is the dual of a central arrangement normal to a subspace that is given as an intersection of hyperplanes in $\mathcal{A}$. When W is infinite and $\mathcal{A}$ is the arrangement associated to the W-action on Ω, it is shown in [57] that the definition of X $(= X(W, S))$ still makes sense and that it is homotopy equivalent to the complement M as defined in (4.20). The definition of $X(W, S)$ depends only on (W, S); there is a partial order on $W \times \mathcal{S}$ and the barycentric subdivision of X is the order complex of $W \times \mathcal{S}$.

The partial order on $W \times \mathcal{S}$ is closely related to the partial order on Coset(W). Its definition is that $(u, T) < (v, T')$ if and only if the following two conditions hold:

(1) $uW_T < vW_{T'}$ as in (4.7), i.e., $v^{-1}uW_T < W_{T'}$,
(2) $v^{-1}u$ is the shortest element in the coset $v^{-1}uW_T$.

The projection $p : W \times \mathcal{S} \to W\mathcal{S}$ defined by $(w, T) \mapsto wW_T$ is clearly order-preserving. With regard to item (2), an element u of W is said to be *$(\emptyset, T)$-reduced* if it is the shortest element in the left coset uW_T. (This is equivalent to the condition that $l(ut) > l(u)$ for all $t \in T$, see [82, Section 4.3].) Moreover, each $w \in W$ has a unique decomposition as $w = uw_0$, where u is $(\emptyset, T)$-reduced and $w_0 \in W_T$.

Theorem 4.66 (cf. Salvetti [203] and Charney-Davis [57]) *Given a Coxeter system (W, S), the arrangement complement M of* (4.20) *is homotopy equivalent to $|W \times \mathcal{S}|$, the geometric realization of the order complex of $W \times \mathcal{S}$.*

We note that $(W \times \mathcal{S})_{\leq(w,T)} \cong W\mathcal{S}_{\leq wW_T} \cong W_T(\mathcal{S}_{\leq T})$. So, $(W \times \mathcal{S})_{\leq(w,T)}$ is isomorphic to the poset of faces of the Coxeter zonotope $Z(W_T, T)$. Hence, there is a cell structure on $|W \times \mathcal{S}|$ defined by identifying the union of all in simplices in the order complex that have maximum vertex (w, T) with the zonotope $wZ(W_T, T)$.

The *pure Salvetti complex* $X(W, S)$ is a CW complex whose poset of cells is $W \times \mathcal{S}$.[1] The cell corresponding to (w, T) is the Coxeter zonotope $wZ(W_T, T)$ defined in Sect. 4.2.2. So, the barycentric subdivision of $X(W, S)$ is the geometric realization, $|W \times \mathcal{S}|$. (N.B. The cell complex $X(W, S)$ does not satisfy the classical conditions of Sect. 2.1—distinct faces of a zonotope can be identified with one another.)

The proof of Theorem 4.66 is based on the following basic result, called the "Nerve Lemma" by combinatorialists.

[1] More often, the term "Salvetti complex" has been used for the CW complex formed as the quotient space of the free W-action on $X(W, S)$. The space $X(W, S)/W$ is often a model for the classifying space BA_W.

Lemma 4.67 (The Nerve Lemma) *Suppose $\mathcal{U}$ is a locally finite, open cover of a paracompact space Y such that each $U \in \mathcal{U}$, as well as each finite, nonempty intersection of such U, is contractible. Then Y is homotopy equivalent to the nerve of the open cover,* $\text{Nerve}(\mathcal{U})$.

Theorem 4.66 is proved by applying the Nerve Lemma to a certain open cover $\mathcal{U}$ of M, the nerve of which is the order complex of $W \times \mathcal{S}$. Although M is defined as a subset of $\mathbb{R}^n + i\Omega$, we can replace it by the homotopy equivalent space $M \cap (b\Sigma \times b\Sigma)$, where the barycentric subdivision $b\Sigma$ of Σ is identified with the order complex of $\text{Coset}(W)$. For each $(w, T) \in W \times \mathcal{S}$, define $\text{Star}(w, T)$ to be the open star of the vertex corresponding to wW_T in $b\Sigma$; let $\text{Sec}(w, T)$ be the open "sector" in Σ that is bounded by the walls that are indexed by wTw^{-1} and that contains the open chamber $\text{Star}(w, \emptyset)$. Put

$$U(w, T) := \text{Star}(w, T) \times \text{Sec}(w, T).$$

It is clear that both $\text{Star}(w, T)$ and $\text{Sec}(w, T)$ are contractible; hence, so is $U(w, T)$ If $x \in \text{Star}(w, T)$ lies in a wall, then that wall is indexed by a reflection in wW_Tw^{-1}. Since the open sector $\text{Sec}(w, T)$ intersects no such wall, we see that $U(w, T) < M$ and hence, that $\mathcal{U} = \{U(w, T))\}_{(w,T)\in W\times\mathcal{S}}$ is an open cover of M. It is proved in [57, Lemma 1.5.2 (ii)] that $U(w_0, T_0) \cap \cdots \cap U(w_k, T_k)$ is nonempty precisely when $\{(w_0, T_0), \ldots, (w_k, T_k)\}$ is a chain in $W \times \mathcal{S}$. Thus, $\text{Nerve}(\mathcal{U}) = |W \times \mathcal{S}|$. Theorem 4.66 follows.

Combining Salvetti's Theorem 4.66 with Deligne's Theorem 4.62, we see that when W is spherical, the finite CW complex $X = |W \times \mathcal{S}|$ is a model for BPA_W (and hence, that X/W is a model for BA_W). A corollary is that any spherical Artin group has a finite CW complex as its classifying space. In other words, spherical Artin groups are type F. (Recall that a group is *type F* if it has a classifying space that is a finite CW complex.) Using Theorem 4.66, we get the following analog of Proposition 4.11 which was about $\Sigma(W, S)$.

Proposition 4.68 (cf. [203], [57]) *Let $X(W, S)$ be the pure Salvetti complex defined above.*

(1) *The poset of cells in $X(W, S)$ is equal to $W \times \mathcal{S}$. The cell corresponding to (w, T) is the Coxeter zonotope $wZ(W_T, T)$; its dimension is equal to* $\text{Card}\, T$.
(2) *The fundamental group of $X(W, S)$ is the pure Artin group PA_W.*
(3) *There is a free action $W \curvearrowright X(W, S)$. The quotient space $X(W, S)/W$ is a finite CW complex (provided S is finite).*

Remark 4.69 Here is another way to understand the cell structure on X ($= X(W, S)$). Each cell of X can be represented as a pair (v, F), where F is a zonotopal cell in Σ ($= \Sigma(W, S)$) and $v \in \text{Vert}\, F$. The partial order on the set of such (v, F) is defined by $(v, F) < (v', F')$ if and only if $F < F'$ and the shortest edge path from v' to F terminates at v. The geometric realization of the projection $p : W\times\mathcal{S} \to W\mathcal{S}$ is also denoted $p : X \to \Sigma$. If F is a cell of Σ of type (W_T, T), then the set of connected components of $p^{-1}(\text{int}\, F)$ is naturally bijective with $\text{Vert}\, F$. The closure of such a component is isomorphic to F. To understand the W-action on X it is better to consider an equivalence relation $\sim$ on the set of all pairs (v, F), where v is a vertex of Σ and F is a cell in Σ and $\sim$ is defined by

$$(v, F) \sim (v', F') \iff F = F' \text{ and } \mathrm{Proj}_F(v) = \mathrm{Proj}_F(v'),$$

where $\mathrm{Proj}_F(v)$ is the terminal point of the shortest edge path from v to F. Each equivalence class $[v', F]$ then has a unique representative of the form (v, F) where $v \in \mathrm{Vert}(F)$. When described in this fashion the cell complex X is sometimes called the "oriented Davis complex" (see [156, §3.4]) instead of the Salvetti complex. (This is done so that term "Salvetti complex" can be reserved for the quotient X/W.)

Just as we did for $\Sigma(W, S)$ in Sect. 4.2.3, in the following proposition we spell out the cell structure on $X(W, S)$ in more detail.

Proposition 4.70 (Some Properties of $X(W, S)$)

(a) *A vertex of X has the form (v, v) where $v \in \mathrm{Vert}\,\Sigma$. Hence, the 0-skeleton of X is naturally isomorphic to the 0-skeleton of Σ. Both correspond bijectively to W.*
(b) *If E is an edge of Σ with end points u and v, then there are two edges in X lying above it: (u, E) and (v, E). The edge (u, E) is directed so that its initial vertex is (u, u) and its terminal vertex is (v, v). If E is labeled by the element s, then label the two directed edges lying above it by the Artin generator a_s. Thus, each edge of Σ is doubled to get a circuit labeled $(a_s)^2$ consisting of two directed edges in X.*
(c) *Similarly, if F is a $2m$-gon in Σ corresponding to an edge $\{s, t\} \in L^{(1)}$ with $m(s, t) = m$, then $p^{-1}(F)$ consists of $2m$ copies of F. The face (v, F) is glued to the 1-skeleton as follows. The two edge paths from v to the antipodal vertex $-v$ in F are labeled by alternating words of length m, namely, $st \cdots$ and $ts \cdots$. If we direct these edge paths as traveling from v to $-v$, then corresponding to an edge labeled by s (or t) there is a directed doubled edge labeled by a_s (or a_t). Then (v, F) is glued onto X^1 so that the two edge paths of F correspond to the positively oriented edge paths $a_s a_t \cdots$ and $a_t a_s \cdots$ in X^1.*
(d) *The projection $\pi : W \times \mathcal{S} \to \mathrm{Coset}(W)$ has a section $f : \mathrm{Coset}(W) \to W \times \mathcal{S}$ defined by $wW_T \mapsto (u, T)$ where u is the shortest element in the coset wW_T. This induces a continuous map $f : \Sigma \to X$ that is a section of $\pi : X \to \Sigma$.*

The 1-skeleton of X is similar to the Cayley graph of (W, S) except that each edge is doubled and each of the new edges is then assigned a direction. Each doubled edge corresponds to a conjugate of the square of an Artin generator by an element of A_W and these conjugates give a set of generators for PA_W. By property (c) of the above proposition, the 2-cells in X correspond to conjugates of the Artin relations in (4.5). It follows that $\pi_1(X^2) = PA_W$. (This also follows from the fact that X is homotopy equivalent to the arrangement complement, $M(\mathcal{A} \otimes \mathbb{C})$.) Let $\tilde{X}$ denote the universal cover of X. Some additional properties of X and $\tilde{X}$ are listed in the following proposition.

Proposition 4.71 (More Properties of $X(W, S)$)

(e) *The group W acts freely on X and $\pi_1(X/W) = A_W$ (cf. Theorem 4.66).*
(f) *There is an equivalence relation $\sim$ on Σ so that $X/W = \Sigma/\sim$. (By* (d) *there is a injection $f : \Sigma \hookrightarrow X$ so that the composition $\Sigma \to X \to X/W$ induces the equivalence relation on Σ.)*
(g) *The space $Y = X/W$ is a finite CW complex (provided S is finite) with one vertex and with one cell of dimension k for each spherical subset $T \in \mathcal{S}$ with* Card $T = k$.
(h) *The projection $X \to |\mathcal{S}^{\mathrm{op}}|$ descends to a projection $Y \to |\mathcal{S}^{\mathrm{op}}|$. This gives Y the structure of a complex of spaces over $|\mathcal{S}^{\mathrm{op}}|$ $(= K)$. This means that Y can be written as a union of subcomplexes*

$$Y = \bigcup_{T \in \mathcal{S}} Y_T,$$

where $Y_T = X(W_T, T)/W_T$ denotes the quotient of the Salvetti complex for W_T by W_T.
(i) *Let $\tilde{X}$ denote the universal cover of X/W. The 1-skeleton $\tilde{X}^1$ is the Cayley graph of A_W with respect to the standard generating set $\{a_s\}_{s \in S}$. Similarly, $\tilde{X}^2$ is the Cayley 2-complex of A_W.*
(j) *The cell complex $\tilde{X}$ satisfies the classical condition for a convex cell complex in Sect. 2.1: the intersection of two cells is either empty or is a common face of both.*

For example, if W is the cyclic 2-group $\mathbf{C}_2$, then X is a circle cellulated by two 1-cells and two 0-cells, and Y is a circle with a single edge and a single vertex. Similarly, when $W = (\mathbf{C}_2)^S$, with Card$(S) = n$, then X is an n-torus cellulated by 2^n i-cells for each i with $0 \leq i \leq n$, while Y is the n-torus with single i-cell for each i. In the general situation, each i-cell of Y is a Coxeter zonotope $Z(W_T, T)$, Card$(T) = i$ (cf. Sect. 4.2.2) with some identifications on its boundary. If (W, S) has diagram $\mathbf{A}_2$ (i.e., if W is the symmetric group on 3 letters), then $\Sigma(W, S)$ is a hexagon while the pure Salvetti complex $X(W, S)$ is a 2-complex with 6 vertices, 12 edges, and 6 hexagons. The Salvetti complex $Y = X/W$ is the presentation complex for the braid group on 3 strands. It has one 0-cell, two 1-cells, and one 2-cell.

The following result of Goodelle and Paris states that the Salvetti complex for a parabolic subgroup A_T of an Artin group A is a retract of $X(W, S)$. (Also see [189].)

Proposition 4.72 (Goodelle-Paris [133, Theorem 2.2]) *Let T be a subset of S. Then $X(W_T, T)$ is a retract of $X(W, S)$.*

The $K(\pi, 1)$-Conjecture can be rewritten as follows.

Conjecture 4.73 (The $K(\pi, 1)$-Conjecture for Artin Groups, Second Version) The finite CW complex X/W is a model for BA_W.

A corollary to Proposition 4.72 is the following.

Corollary 4.74 ([133, Corollary 2.4]) *If the $K(\pi, 1)$-Conjecture is true for an Artin group A_W, then it is also true for any parabolic subgroup A_T, for $T < S$.*

The universal cover of $X(W, S)$ has a natural piecewise euclidean structure where each cell is a Coxeter zonotope. In analogy with the Davis–Moussong complex $\Sigma(W, S)$, one might be tempted to speculate that the universal cover of $X(W, S)$ is CAT(0). Of course, when A_W is a RAAG, then, as we showed in Sect. 3.1, the Salvetti complex X/W is $\mathbb{T}_L$ and its universal cover is CAT(0); however, this is definitely not the case when A_W is not a RAAG. For example, as was pointed out in [57, Remark 2.3.2], when W is the symmetric group on 3 letters (so that A_W is the braid group on 3 strands), then the hexagonal complex $X(W, S)$ cannot be given any reasonable NPC structure.

Nevertheless, many people have speculated on the possibility of finding nonpositively curved classifying spaces for Artin groups. When A_W is spherical a likely candidate for an NPC classifying space has been proposed by T. Brady and J. McCammond. Brady used the "dual Garside structure" to define a BA_W, as a finite piecewise euclidean cell complex with an NPC metric whenever W is spherical of rank ≤ 4. The definition of BA_W works for any W; however, proving that the Link Condition holds becomes increasingly difficult as the rank increases. When W is irreducible and spherical, let G_W denote the quotient of A_W by its center (which is isomorphic to $\mathbb{Z}$). Bestvina [22] described a model for BG_W which has many features of an NPC space. We will explain Bestvina's complex in Sect. 4.3.5 below.

Remark 4.75 Another approach to Conjecture 4.73 using nonpositive curvature has been proposed by Allcock [5]. Consider the finite arrangement of hyperplanes $\mathcal{A} \otimes \mathbb{C}$ in $\mathbb{C}^d$ associated to a finite Coxeter group W. The arrangement gives $\mathbb{C}^d$ the structure of an orbihedron; the strata are intersections of subspaces and the local groups are the isotropy subgroups in W. Put the standard flat metric on $\mathbb{C}^d$ and form the "universal branched cover" $\hat{\mathbb{C}}^d$ of $\mathbb{C}^d$. This means that we take the metric completion of the universal cover of the hyperplane complement $M(\mathcal{A} \otimes \mathbb{C})$. The conjecture is that $\hat{\mathbb{C}}^d$ is CAT(0). Since removing the cone on a contractible subspace does not change a space's homotopy type, this conjecture implies that the universal cover of $M(\mathcal{A} \otimes \mathbb{C})$ is contractible by induction on dimension. Given this conjecture on can deduce that the universal branched cover of $\Sigma \times \Sigma$ along the union of the diagonal walls is also CAT(0) where Σ is the Davis–Moussong complex of an arbitrary Coxeter system. So, by taking complements we would get Conjecture 4.73 in full generality. In other words, the statement that $\hat{\mathbb{C}}^d$ is CAT(0) for all spherical reflection groups implies Conjecture 4.73 for all Artin groups.

4.3.4 The $K(\pi, 1)$-Conjecture for Artin Groups

The development of $A\mathcal{S}^{\mathrm{op}}$ is the Deligne complex $D(A, K)$ (also denoted by $\Lambda(W, S)$ in Sect. 4.3.1). The Salvetti complex X/W is the aspherical realization of $A\mathcal{S}^{\mathrm{op}}$. So, as we explained in Theorem 4.7 of Sect. 4.1.3, the $K(\pi, 1)$-Question for $A\mathcal{S}^{\mathrm{op}}$ has a positive answer if and only if the Deligne complex is contractible.

Theorem 4.76 (cf. Charney-Davis [56, Thm. 1.5.1] and Appendix A.4) *The universal cover $\tilde{X}$ of the pure Salvetti complex is contractible if and only if the Deligne complex $\Lambda(W, S)$ is contractible. So, $\Lambda(W, S) \times_A \tilde{X}$ is homotopy equivalent to BA.*

In other words, if $\Lambda(W, S)$ is contractible, then the $K(\pi, 1)$-Question for $A\mathcal{S}^{\mathrm{op}}$ has a positive answer.

Remark 4.77 By [56], $\Lambda(W, S)$ is contractible whenever (a) A_W is 2-dimensional (cf. [147]). or (b) (W, S) has type FC (cf. Corollaries 4.57 and 4.59)). Hence, $K(\pi, 1)$-Question for $A\mathcal{S}^{\mathrm{op}}$ has a positive answer in both of these cases. Huang–Osajda [157] proved that Artin groups of type FC were "Helly groups" meaning that their Salvetti complexes had certain NPC type properties

Remark 4.78 As explained in the remark following Lemma 4.3.7 in [56], when (W, S) is type FC, one can prove that $\Lambda^{\square}(W, S)$ is contractible by a simple inductive argument on the number of vertices. First, observe that it holds when (W, S) is spherical. Second, if $L(W, S)$ is a flag complex, then there are vertices s and t which are not connected by an edge so that A_S is an amalgam of $A_{S-\{s\}}$ and $A_{S-\{t\}}$ along $A_{S-\{s,t\}}$. By induction, Deligne complexes of the three smaller groups are contractible, so the same is true for $\Lambda(W, S)$. This same argument was used by Godelle–Paris [133] and by Ellis–Sköldberg [112] to show that the $K(\pi, 1)$-Question has a positive answer whenever there is a family of special subgroups $\{W_T, T\}$, closed under passing to smaller special subgroups and containing the spherical subgroups, such that the Deligne complex is contractible for each of the (W_T, T) in the family.

A positive answer to the $K(\pi, 1)$-Question for $A\mathcal{S}^{\mathrm{op}}$ implies that the classifying space of an Artin group A is a finite complex i.e., it is type F. (See Sect. 7.2 for the definition a discussion of "type F.") The conjecture that this be true is slightly weaker than having a positive answer to the $K(\pi, 1)$-Question. It is often called the "$K(\pi, 1)$-Conjecture for Artin groups." It is the most important open question in the area. For example without it, we do not even known how to prove that A is torsion-free. (For a survey of the results on this question as of 2014 see [190].)

Fairly recently there has been a good deal of work greatly extending the class of Artin groups for which the $K(\pi, 1)$-Conjecture was known, notably:

- Paolini–Salvetti [188] have proved it for A_W whenever (W, S) is a euclidean Coxeter system.

- By using the Goodelle–Paris method, Huang [156] has proved the conjecture for a large class of A_W including all those whose Dynkin diagram is a tree. Huang's results hold promise that the general conjecture will be proved soon.

Other recent results include Goldman [134] and several papers by Haettel, e.g., [139].

4.3.5 Bestvina's Normal Form Complex

Let A be the Artin group associated to a Coxeter system (W, S) and let A^+ be the *positive monoid* consisting of those $a \in A$ that are represented by words in positive powers of the Artin generators a_s. Let $p : A \to W$ denote the canonical epimorphism. Recall from the paragraph on apartments in the Deligne complex at the end of Sect. 4.3.1, the epimorphism p has a set theoretic section $W \to A^+$ denoted by $w \mapsto a_w$ and defined as follows: if w has reduced decomposition $w = s_1 \cdots s_k$, then

$$a_w = a_{s_1} \cdots a_{s_k}.$$

The element a_w is independent of the choice of a reduced word $s_1 \cdots s_k$ for w; a_w is called an *atom*. It is characterized by the fact that the word length of a_w in A^+ is equal to $l(w)$, the word length of w in W.

The Left Greedy Normal Form For the remainder of this subsection W is a finite Coxeter group and A $(= A_W)$ is the associated spherical Artin group. There is a unique element $w_0 \in W$ such that $l(w_0) \geq l(w)$ for all $w \in W$. It is called the *element of longest length*. Here are some of its properties (cf. [82, §4.6]):

- w_0 is an involution.
- In the geometric representation of W on $\mathbb{R}^S$, w_0 takes the fundamental chamber C whose walls are labeled by S to the antipodal chamber $-C$ (whose walls are also labeled by S).
- w_0 lies in the center of W if and only if it acts on $\mathbb{R}^n$ as multiplication by -1.

Define a partial order on A^+ by $x \leq y$ if there is $z \in A^+$ such that $y = xz$. The homomorphism p induces a bijection between the set of atoms in A^+ and W. The atom corresponding to w_0 is denoted by Δ. Inner automorphism by Δ takes an atom a to another atom $\overline{a}$. Since inner automorphism by Δ is an involution, the element Δ^2 is in the center of A (in fact, Δ is central whenever w_0 is the antipodal map on $\mathbb{R}^S$). This partial order gives A^+ the structure of a lattice. This means that any two elements $x, y \in A^+$ have a greatest lower bound $x \wedge y$, as well as a least upper bound $x \vee y$. A key fact is that for every $x \in A$, there is an integer $k \geq 0$ such that $\Delta^k x \in A^+$.

The normal form in which we are interested is essentially due to Deligne [107]. It is based on earlier work of Garside on a normal form for the braid group. Its definition is based on the following proposition.

Proposition 4.79 (cf. [22, Prop. 1.2] and [107]) *For any $g \in A^+$ there is a unique atom $\alpha(g)$ such that*

(1) $\alpha(g) \leq g$,
(2) *if $a \in A^+$ is any atom with $a \leq g$, then $a \leq \alpha(g)$.*

In other words, $\alpha(g)$ is the longest atom with which g can begin. In [61] $\alpha(g)$ is called the *left front* of g. It follows that any $g \in A^+ - \{1\}$ can be written as a nontrivial product of atoms: $g = a_1 \cdots a_k$ such that $a_i = \alpha(a_i \cdots a_k)$. This is called the *left greedy normal form* for g. Note that if $a_1 \cdots a_k$ is such a normal form, then all Δ s occur at the beginning.

In [22] Bestvina works with the group $G = A/\langle \Delta^2 \rangle$. Note that the homomorphism $A \to G \times \mathbb{Z}$, which in the first coordinate is the natural projection and in the second coordinate sends each Artin generator to $1 \in \mathbb{Z}$, is an injection onto a subgroup of finite index.

Bestvina's Complex Following [22] one can define a simplicial flag complex $X(G)$ with G-action. Its vertex set V is the set of cosets $A/\langle \Delta \rangle$. Each vertex has a representative $g\langle \Delta \rangle$ with $g \in A^+$. (If x is an arbitrary representative for a vertex $x\langle \Delta \rangle$, then we can multiply x by a high power of Δ^2 to obtain a representative $g = (\Delta^2)^k x \in A^+$.) Let $g = \Delta^m \cdot a_1 \cdot a_2 \cdots a_k$ be a representative with each a_i an atom and with no $a_i = \Delta$. We can shift all the Δ s to the end to obtain a *special representative* in the same coset as g that does not contain any Δ s. (If m is even, this special representative is $a_1 \cdots a_k$, while if m is odd, it is $\overline{a}_1 \cdots \overline{a}_k$.) The *atomnorm* of a vertex v, denoted $|v|$, is the number of atoms in this special representative. After some work (cf. Lemma 2.1 and Proposition 2.2 in [22]) one can show that the atomnorm extends to a distance function $d_{at} : V \times V \to \mathbb{Z}_+$, with $d_{at}(*, v) = d_{at}(v, *) = |v|$, where $*$ denotes the base vertex corresponding to $1 \in A^+$.

Define a simplicial graph $X^1(G)$ which has vertex set V and which has an edge between vertices v_0 and v_1 if and only if $d_{at}(v_0, v_1) = 1$. Let $X(G)$ be the flag complex determined by this graph. The group G is not torsion-free. Indeed it has a subgroup $\langle \Delta \rangle / \langle \Delta^2 \rangle$ which is isomorphic to $\mathbb{Z}/2$. The dimension of $X(G)$ is the word length of Δ, i.e., it is $l(w_0)$, where $w_0 \in W$ is the element of longest length and $l(w_0)$ denotes its word length in W.

Remark 4.80 The picture of $X(G)$ is not what one might expect after studying the Deligne and Salvetti complexes. For example, if A is the braid group on three strands, then G is virtually a free group and the complex $X(G)$ is a union of 2-simplices so that the intersection of any two 2-simplices is either empty or a common vertex. In other words, $X(G)$ looks like a regular trivalent tree where each vertex has been expanded to a 2-simplex and each edge has been shrunk to a 0-simplex. (See [21, Figure 1, p. 280].)

Define a metric on $X(G)$ in the usual fashion by declaring each edge to have length 1 and then taking the induced path metric. Using the normal form, there is a canonical edge path from $*$ to any other vertex v in $X(G)$. Translating by an element of G we obtain a canonical edge path between any two vertices. Such an edge path is a *geodesic*.

Bestvina defines the *wordnorm* $\|v\|$ of a vertex v to be the word length (with respect to the standard set of Artin generators) of a special representative for v. This extends to a left invariant function $d_{wd} : V \times V \to \mathbb{Z}^+$ with $d_{wd}(*, v) = \|v\|$. (N.B. While similar to a metric, the function d_{wd} is not symmetric. For example, if b is an atom, then $d_{wd}(*, b)$ is the word length of b, while $d_{wd}(b, *)$ is the word length of the complementary atom c defined by $cb = \Delta$. The wordnorm is used in establishing various nonpositive curvature properties for $X(G)$ which resemble those in a CAT(0) space.

Let $d + 1$ $(= \mathrm{Card}(S))$ be the dimension of the geometric representation of W.

Theorem 4.81 (Bestvina [22, Theorem 3.6]) *The group G acts properly and cocompactly on $X(G)$. Moreover, $X(G)$ is contractible. It is properly homotopy equivalent to a d-dimensional cell complex which is $(d-2)$-connected at infinity.*

The universal cover of the Salvetti complex, denoted $\tilde{X}(W, S)$, is a $(d + 1)$-dimensional contractible cell complex on which the group A acts freely and cocompactly. Since A and $G \times \mathbb{Z}$ are commensurable, $\tilde{X}(W, S)$ serves many of the same purposes as does $X(G)$. Thus, Theorem 4.81 gives another proof of Deligne's Theorem 4.62. The disadvantage of $\tilde{X}(W, S)$ is that it is not CAT(0) and so, does not enjoy the same properties of nonpositive curvature as does $X(G)$. On the other hand, the Deligne complex $\Lambda(W, S)$ is conjectured to be CAT(0). Its disadvantages are that it is not locally finite and that the A-action is not proper.

Two other features of nonpositive curvature which were discussed in Sect. 2.3 persist to $X(G)$: one can define a *circumcenter* of any bounded subset of $X(G)$ and the *minset* $\mathrm{Min}(g)$ for any $g \in G$ of infinite order. In both cases something close to the standard definition works. Although circumcenters are not necessarily unique, the set of circumcenters spans a simplex of $X(G)$. It follows that the Bruhat–Tits Fixed Point Theorem holds for $X(G)$. As in Sect. 2.3 this gives the following theorem (cf. Corollary 2.35).

Theorem 4.82 (cf. [22]) *For any finite subgroup $F < G$, the fixed point set, $X(G)^F$, is a nonempty convex subset of $X(G)$. Hence, $X(G)$ is a cocompact model for $\underline{E}G$.*

Also recall from Sect. 2.3 the *translation length* $\tau(g)$ of an element $g \in G$ is defined by:

$$\tau(g) := \lim_{n\to\infty} \frac{\|g^n(*)\|}{n}.$$

Lemma 4.83 (Bestvina [22, Thm. 4.1]) *Let δ be the length of w_0, the element of longest length in W. Then the set of translation lengths of infinite order elements in G is bounded below by $\frac{1}{2\delta}$.*

By using minsets and translation lengths, Bestvina then establishes an algebraic version of the Flat Torus Theorem and some of its consequences (cf. Theorems 2.37 and 2.38 in Sect. 2.3).

Theorem 4.84 (Bestvina [22, Corollaries 4.2, 4.3, 4.4])

(i) *Every abelian subgroup of G (or of A) is finitely generated.*
(ii) *The group G does not contain any infinitely divisible elements of infinite order.*
(iii) *Every solvable subgroup of G (or of A) is virtually free abelian.*

Remark 4.85 It was proved by Bigelow and Krammer that spherical Artin groups have faithful linear representations. It follows that they satisfy the Tits Alternative: every finitely generated subgroup of A is either free abelian or contains a nonabelian free subgroup.

4.4 Buildings

In the 1950s and 1960s, J. Tits [216, 218] (and see [219]) developed the theory of buildings in connection with his work on representing algebraic groups as automorphism groups of incidence geometries. Buildings provide examples of highly symmetric polyhedra. Each building has an associated Coxeter system (W, S). For buildings that arise from algebraic groups, only spherical and euclidean Coxeter groups play a role. Nevertheless, Tits was careful to develop the theory for arbitrary Coxeter systems; in fact, his motivation for introducing the notion of a Coxeter system was to use it in developing the theory of buildings. A remarkable aspect of Tits' work is that the group of symmetries of a building is not involved its definition. So, a building need not be defined as the development of a simple complex of groups.

In Sect. 4.4.1 we explain the combinatorial approach to buildings. This is similar to the description of a simple complex of groups as a poset of groups given in Appendix A.1. In the next subsection, Sect. 4.4.2, we discuss various geometric realizations of a building based on a choice of "model chamber." The standard choice of model chamber is the fundamental chamber $K(W, S)$ of the associated Coxeter system and this leads to the "standard realization" of the building. Another possible choice for model chamber is a simplex Δ with its codimension one faces indexed by S. We call this choice is the "classical realization" of a building. It is the appropriate choice for spherical buildings and for irreducible euclidean buildings. In the spherical case, Δ is a spherical simplex and the classical realization has a piecewise spherical, CAT(1) metric. In the (irreducible) euclidean case, Δ is a euclidean simplex and the classical realization is given a piecewise euclidean, CAT(0) metric.

The most celebrated result in the theory of buildings is Tits' classification of spherical buildings of rank ≥ 3 in [216]. "Spherical" means that the associated Coxeter group is spherical. Roughly speaking, Tits' theorem states that a thick, irreducible spherical building of rank ≥ 3 is the building of an algebraic group over a field $\mathbb{F}$. (See [1, Chapter 9] and [198, Chapter 6, 8] for discussions of this result.) The reason that this is only a "rough" statement of the Classification Theorem is that there are classical groups which are not algebraic groups but yet still give rise to spherical buildings (for example, the field can be replaced by a noncommutative division ring). The theorem also holds for rank two spherical buildings (also known as "generalized polygons") provided we add the requirement that the building has the "Moufang Property." This property always holds for irreducible spherical building of rank at least 3. The classification of such Moufang polygons was accomplished by Tits–Weiss in [225]. We shall not define the Moufang Property in this book. Suffice it to say that it implies a strong transitivity property for the group generated by the "root groups" and it implies that the automorphism groups of such spherical buildings are associated to "BN pairs" (also not defined in this book). Tits also succeeded in classifying thick, irreducible euclidean buildings of rank ≥ 4 in [222] (cf. [1, §11.9]). The result extends to euclidean buildings of rank 3 if the spherical building at infinity is assumed to be Moufang. The result is that such euclidean buildings are in one-to-one correspondence with absolutely simple algebraic groups or classical groups or mixed groups defined over a field that is complete with respect to a discrete valuation. Finally, there are similar results relating certain "twin-buildings" whose Coxeter systems are 2-spherical and Kac–Moody groups over finite fields.

In Sect. 4.4.5 we explain how to "pullback" a combinatorial building $\mathcal{C}'$ over a Coxeter system (W', S') via a homomorphism of Coxeter systems $(W, S) \to (W', S')$. This gives us a method of constructing new examples in Sect. 4.4.6 by starting from classical examples. The main source of examples are the Bruhat–Tits buildings of spherical type. Section 4.4.8 contains a discussion of classical examples of spherical buildings, euclidean buildings and Kac–Moody buildings.

Lattices in automorphism groups of buildings give a beautiful class of examples for geometric group theory. Frequently, such lattices turn out to be infinite simple groups. For example, a product of two trees is a building and the Burger–Mozes examples of Sect. 2.4.4 are cocompact lattices in the automorphism group of the product of two regular trees.

Outline of This Section

(A) The combinatorial theory of buildings (Sect. 4.4.1).

- (A.1) Chamber systems, galleries, residues, Weyl distance function (Sect. 4.4.1).
- (A.2) Apartments (Definition 4.90).

(A.3) Rank two buildings (Example 4.93).
(A.4) Projective spaces (Example 4.99).

(B) Geometric realizations.

(B.1) The standard geometric realization is CAT(0) (Theorem 4.101).
(B.2) The $K(\pi, 1)$-Question for chamber-transitive buildings (Theorem 4.107).

(C) Regular right-angled buildings (Sect. 4.4.3).
(D) A local approach to buildings

References for the material in this section include [1], [198], [142], [122], [216, 218, 220–222], [77, 83], [82, Ch. 18], as well as the survey article [115].

4.4.1 The Combinatorial Theory of Buildings

A *chamber system* is a set $\mathcal{C}$ (of "chambers") together with a family of equivalence relations on $\mathcal{C}$ indexed by another set S. For $s \in S$, distinct s-equivalent chambers $C, D \in \mathcal{C}$ are said to be *s-adjacent*; they are *adjacent* if s is not specified. A *gallery* in $\mathcal{C}$ is a finite sequence of chambers $(C_0, \dots, C_k)$ such that C_{j-1} is adjacent to C_j, for $1 \le j \le k$. The *type* of this gallery is the word $\mathbf{s} = (s_1, \dots, s_k)$ where C_{j-1} is s_j-adjacent to C_j. If each s_j belongs to a given subset T of S, then $(C_0, \dots, C_k)$ is a *T-gallery*. Two chambers are in the same *T-connected component* if they can be connected by a T-gallery. The T-connected components of a chamber system $\mathcal{C}$ are its *residues* of *type T*. An s-equivalence class is the same thing as a residue of type $\{s\}$. If $C \in \mathcal{C}$ and $T \le S$, then $\operatorname{Res}_T(C)$ denotes the residue of type T that contains C. If for some chamber C, $\operatorname{Res}_S(C) = \mathcal{C}$, then $\mathcal{C}$ is *gallery connected.*

Example 4.86 (The Chamber System Associated to a Family of Subgroups) Suppose that G is group, B is a subgroup of G, and $(G_s)_{s\in S}$ a family of subgroups of G such that each G_s contains B. Define a chamber system $\mathcal{C} = \mathcal{C}(G, B, (G_s)_{s\in S})$ as follows: $\mathcal{C} = G/B$; chambers gB and $g'B$ are *s-adjacent* if they have the same image in G/G_s. For each subset T of S, let G_T be the subgroup generated by $\{G_s\}_{s\in T}$. The chamber system is connected if $G = G_S$. A T-residue in $\mathcal{C}(G, B, (G_s)_{s\in S})$ is a coset in G/G_T. Put $G_\emptyset = B$. As we shall see in Appendix A, if $\mathcal{P}$ denotes the power set of S (or the poset of its proper subsets or spherical subsets), then $\{G_T\}_{T\in\mathcal{P}^{\mathrm{op}}}$ is a simple complex of groups over the opposite poset to $\mathcal{P}$. The examples from Sect. 4.1 where G is a Coxeter group or Artin group and B is the trivial group are examples of such chamber systems.

In the following, a "building" is defined as a chamber system equipped with the extra structure that depends on a Coxeter system (W, S) and a "Weyl distance" as defined below.

Definition 4.87 (cf. [1, Section 4.1]) Suppose (W, S) is a Coxeter system. A *building of type* (W, S) is a pair $(\mathcal{C}, \delta)$ consisting of a nonempty set $\mathcal{C}$ (the elements of which are called *chambers*), and a function $\delta : \mathcal{C} \times \mathcal{C} \to W$ (called the *Weyl distance*) so that the following three conditions hold for all $C, D \in \mathcal{C}$.

(WD1) $\delta(C, D) = 1$ if and only if $C = D$.
(WD2) If $\delta(C, D) = w$ and $C' \in \mathcal{C}$ satisfies $\delta(C', C) = s \in S$, then $\delta(C', D)$ is either sw or w. If, in addition, $l(sw) = l(w) + 1$, then $\delta(C', D) = sw$.
(WD3) If $\delta(C, D) = w$, then for any $s \in S$ there is a chamber $C' \in \mathcal{C}$ such that $\delta(C', C) = s$ and $\delta(C', D) = sw$.

Given a word $\mathbf{s} = (s_1, \ldots, s_k)$ in the set S of fundamental generators for W, its *value* is the corresponding group element $w(\mathbf{s}) = s_1 \cdots s_k$. The word $\mathbf{s}$ is a *reduced decomposition* for $w(\mathbf{s})$ if $l(w(\mathbf{s})) = k$.

The Weyl distance can be used to give $\mathcal{C}$ the structure of a chamber system over S: given $s \in S$, chambers C and D are defined to be *s-adjacent* if and only if $\delta(C, D) = s$. There is a close connection between galleries in $\mathcal{C}$, words in S and the Weyl distance: if $(C_0, C_1, \ldots C_k)$ is a gallery in $\mathcal{C}$ of type $\mathbf{s} = (s_1, \ldots, s_k)$, it follows from the axioms that there is the following relationship between the Weyl distance and galleries.

(G) Suppose $\mathbf{s}$ is a reduced decomposition and that C, D are chambers of $\mathcal{C}$. Then there is a gallery of type $\mathbf{s}$ from C to D if and only if $\delta(C, D) = w(\mathbf{s})$ (see [1, Section 5.2]).

It follows from **(G)** that the Weyl distance is a skew symmetric, i.e., $\delta(D, C) = \delta(C, D)^{-1}$. (See [1, Cor. 5.17 (2)].)

By **(WD3)** every s-equivalence class has at least two elements. A building $\mathcal{C}$ is *thick* if each s-equivalence class contains at least 3 elements.

Example 4.88 The Coxeter group W itself has the structure of a building: the Weyl distance $\delta : W \times W \to W$ is defined by $\delta(v, w) = v^{-1}w$. When considered as a building, the Coxeter group is the *thin building* of type (W, S).

A building is *spherical* if its type is a spherical Coxeter system (see Sect. 4.2.1). The building is a *right-angled building* (abbreviated as RAB) if its type is a right-angled Coxeter system (see Sect. 3.1.1). In practice the set of chambers $\mathcal{C}$ in a building will usually be associated to a family of subgroups $\mathcal{C}(G, B, (G_s)_{s\in S})$ as in Example 4.86. When this is the case, $\mathcal{C} = G/B$ and a residue of type T is the image of a left coset gG_T in G/B.

A residue of type T in a building is itself a building; its type is (W_T, T). Let $R_T(\mathcal{C})$ denote the set of residues in $\mathcal{C}$ of type (W_T, T). Let $R^{\text{all}}(\mathcal{C})$ denote the disjoint union of all $R_T(\mathcal{C})$ with $T \leq S$. We have subsets $R^{\text{prop}}(\mathcal{C})$ (resp., $R(\mathcal{C})$) of $R^{\text{all}}(\mathcal{C})$ defined as the union of all $R_T(\mathcal{C})$, where T is a proper subset of S (resp., a spherical subset of S). There is a partial order on $R^{\text{all}}(\mathcal{C})$: $\text{Res}_T(C) < \text{Res}_{T'}(C')$ if $T < T'$ and the first residue is a subset of the second. The subposet $R(\mathcal{C})$ is the poset of *spherical residues*.

Remark 4.89 Let X be a topological space equipped with a collection of closed subspaces $\{X_s\}_{s \in S}$ indexed by S. The X_s are the *mirrors* Of X. We can use X to serve as a "model chamber" for a building of type (W, S). We shall show in the next subsection how to define a geometric realization of the building $\mathcal{C}$ by pasting together models chambers, one for each element of $\mathcal{C}$. When the model chamber is a simplex of dimension $\mathrm{Card}(S) - 1$, this realization is called the *classical realization* of $\mathcal{C}$; its barycentric subdivision can be identified with the order complex of $R^{\mathrm{prop}}(\mathcal{C})$. When the model chamber is the fundamental domain $K(W, S)$ $(= |\mathcal{S}^{\mathrm{op}}|)$ for the Davis–Moussong complex, the corresponding realization is denoted $|\mathcal{C}|$ and called the *standard realization* of $\mathcal{C}$; its barycentric subdivision can be identified with the order complex of $R(\mathcal{C})$ (which, of course, is the same as the order complex of $R(\mathcal{C})^{\mathrm{op}}$).

Definition 4.90 An *apartment* in $\mathcal{C}$ is a subset which is W-isometric to the thin building W. In other words, if $\rho_C : \mathcal{C} \to W$ is the function $D \mapsto \delta(C, D)$ for a fixed chamber C, then a subset $\mathcal{A}$ of $\mathcal{C}$ is an apartment if $\rho_C|_{\mathcal{A}} : \mathcal{A} \to W$ is an isomorphism. The function $\mathcal{C} \to \mathcal{A}$, defined as the composition of ρ_C with the inverse of $\rho_C|_{\mathcal{A}}$, is called the *retraction onto $\mathcal{A}$ centered at C*.

In the standard realization the barycentric subdivision of each apartment is isomorphic to $b\Sigma(W, S)$, the barycentric subdivision of the Davis–Moussong complex from Sect. 4.2.3.

Definition 4.91 Given a chamber C, for each $s \in S$, let $q_s(C)$ be the number of chambers s-adjacent to C (i.e., $q_s(C)$ is one less than the number of elements in the $\{s\}$-residue containing C). The building is *regular* if $q_s(C)$ is independent of the choice of chamber C (for example, this is the case whenever the automorphism group of $\mathcal{C}$ is chamber-transitive). For a locally finite, regular building $\mathcal{C}$ of type (W, S), $(q_s)_{s \in S}$ is called the *S-tuple of parameters* of $\mathcal{C}$.

There is a simple criterion on the Coxeter diagram for determining if elements $s, t \in S$ are conjugate to one another: they are if and only if there is a sequence $s_1, \dots, s_k$ of elements of S with $s_1 = s$, $s_k = t$ and $m(s_j, s_j + 1)$ is an odd integer for $1 \le j < k$ (see [28, Prop. 3, p. 5]). If s is conjugate to t in W, then $q_s = q_t$. If the parameters of a spherical building are all finite, then the building is finite. Hence, when all parameters of $\mathcal{C}$ are finite, its geometric realization $|\mathcal{C}|$ is a locally finite complex (since each spherical residue is a finite set). If $\mathcal{C}$ is the building for an algebraic group over a finite field $\mathbb{F}_q$ of order q, then it is often the case each $\{s\}$-residue can be identified with the projective line over $\mathbb{F}_q$ and hence, has $q + 1$ elements and so, $q_s = q$.

The *rank* of a building of type (W, S) is $\mathrm{Card}(S)$.

Example 4.92 (Rank One Buildings) The only Coxeter diagram with one vertex is $\mathbf{A}_1$, i.e., the only Coxeter system of rank one is $(\langle s \rangle, \{s\})$, where $\langle s \rangle$ means the cyclic group of order two. So, a rank one building is any set with at least two elements (the "chambers"); the Weyl distance $\delta : \mathcal{C} \times \mathcal{C} \to \langle s \rangle$ is defined by

$$\delta(C, D) = \begin{cases} 1, & \text{if } C = D, \\ s, & \text{if } C \neq D. \end{cases}$$

Example 4.93 (Rank Two Buildings) If $(W, \{s, t\})$ is a Coxeter system of rank two and $m = m(s, t)$, then W is a dihedral group D_m of order $2m$ when $m < \infty$, and is D_∞, the infinite dihedral group when $m = \infty$. Either the classical realization or the standard realization of a building of type $(D_\infty, \{s, t\})$ is the same thing as a tree without terminal vertices, that is to say, $\mathcal{C}$ is the set of edges in such a tree. An apartment in such a building is a subtree isomorphic to the real line, that is, it is isomorphic to the Coxeter complex of $(D_\infty, \{s, t\})$. When $m < \infty$, a building $\mathcal{C}$ of type $(D_m, \{s, t\})$ is a *generalized m-gon*. The classical realization of a generalized m-gon $\mathcal{C}$ is a bipartite graph of girth $2m$ and diameter m: the edges of the graph are the chambers of $\mathcal{C}$. Each vertex of an edge determines a residue of type $\{s\}$ or type $\{t\}$. An apartment in $\mathcal{C}$ is a subgraph that is a circuit of length $2m$. The classical realization of a generalized 2-gon is the same thing as a complete bipartite graph. A generalized 3-gon is called a *projective plane*. A generalized 4-gon is a generalized *quadrangle*. A theorem of Feit–Higman [117] asserts that there are generalized m-gons that are finite and thick only for $m \in \{2, 3, 4, 6, 8\}$ (cf. Theorem 4.130).

Example 4.94 (Products) For $i = 1, 2$, suppose $(\mathcal{C}_i, \delta_i)$ is a building of type (W_i, S_i). Then $(W, S) = (W_1 \times W_2, S_1 \sqcup S_2)$ is Coxeter system and $(\mathcal{C}, \delta) = (\mathcal{C}_1 \times \mathcal{C}_2, \delta_1 \times \delta_2)$ is a building of type (W, S). For example, if each $(\mathcal{C}_i, \delta_i)$ is the building corresponding to a tree T_i, then $(\mathcal{C}_1 \times \mathcal{C}_2, \delta_1 \times \delta_2)$ is a building corresponding to $T_1 \times T_2$; the associated Coxeter group is $D_\infty \times D_\infty$. Each apartment in $T_1 \times T_2$ is isomorphic to the euclidean plane with its square tiling.

It is proved in [1, Prop. 5.23] that the conditions **(WD1)**, **(WD2)**, **(WD3)** in Definition 4.87 are equivalent to the following two conditions on a chamber system $\mathcal{C}$, equipped with a function $\delta : \mathcal{C} \times \mathcal{C} \to W$.

- Each s-equivalence class has at least two elements.
- Given a reduced expression $\mathbf{s}$ for an element $w \in W$, there is a gallery of type $\mathbf{s}$ from C to D if and only if $\delta(C, D) = w$.

(This is the definition of a building that is given in [198].)

An *automorphism* of the building is a self-bijection of $\mathcal{C}$ which preserves s-equivalence classes for each $s \in S$. Equivalently, it is a self-bijection which preserves Weyl distance.

Definition 4.95 A subgroup G of Aut($\mathcal{C}$) is *chamber-transitive* if it is transitive on $\mathcal{C}$. It is *strongly transitive* if it is transitive on the set of pairs $(\mathcal{A}, C)$, where $\mathcal{A}$ is an apartment in $\mathcal{C}$ and $C \in \mathcal{A}$ (cf. [1, §6.1.1]). (In fact, it is not necessary to use all apartments in the definition of strongly transitive; $\mathcal{A}$ need only belong to a certain "system of apartments" satisfying the classical axioms for a building, cf. [1, §6.1].)

Suppose G is a chamber-transitive group of automorphism of the building $\mathcal{C}$. Choose a chamber C in $\mathcal{C}$. For each $T \in \mathcal{S}$, let G_T denote the stabilizer of $R_T(\mathcal{C})$, the spherical residue containing C. In particular, $G_\emptyset := B$ is the stabilizer of C, and for each $s \in S$, G_s is the stabilizer of the $\{s\}$-residue containing C. As in Example 4.86, $\mathcal{C} \cong \mathcal{C}(G, B, (G_s)_{s\in S})$, the chamber system associated to this family of subgroups. As promised earlier we get the following theorem.

Theorem 4.96 *With notation as above, suppose G is a chamber-transitive group of automorphisms of a building $\mathcal{C}$. Then $\{G_T\}_{T\in\mathcal{S}^{\mathrm{op}}}$ is a simple complex of groups over $\mathcal{S}^{\mathrm{op}}$. Its development $D(G, \mathcal{S}^{\mathrm{op}})$ can be identified with the poset of spherical residues $R(\mathcal{C})^{\mathrm{op}}$. Moreover, G is the direct limit of the $\{G_T\}_{T\in\mathcal{S}^{\mathrm{op}}}$.*

Proof The point is that since G is chamber-transitive, it acts on the poset $R(\mathcal{C})^{\mathrm{op}}$ with strict fundamental domain $\mathcal{S}^{\mathrm{op}}$. So, $\{G_T\}_{T\in\mathcal{S}}$ is a developable simple complex of groups. Since $|\mathcal{S}^{\mathrm{op}}|$ is simply connected, it follows that the geometric realization of $R(\mathcal{C})$ is simply connected and that the canonical homomorphism from the direct limit to G is an isomorphism (see Theorem A.7 (vi)′ in Appendix A.2).

Remark 4.97 In Theorem 4.96 we get the same direct limit G if we only use the smaller collection of those G_T where T ranges over the spherical subsets with $\mathrm{Card}(T) \leq 2$.

Remark 4.98 It turns out that if G is strongly transitive on a thick building, then it inherits the structure of a BN pair (also called a "Tits system", cf. [1, Thm. 6.56]). This concept will not be defined in this book.

Given $C \in \mathcal{C}$, the *combinatorial ball of radius n about C* is the set $B_C(n) := \{D \in \mathcal{C} \mid l(\delta(C, D)) < n\}$. Here $l : W \to \mathbb{N}$ denotes word length. There is a natural topology on the group $\mathrm{Aut}(\mathcal{C})$ of automorphisms of $\mathcal{C}$: an open neighborhood of $1 \in \mathrm{Aut}(\mathcal{C})$ is the set of automorphisms which fix each element of $B_C(n)$ for some $n \in \mathbb{N}$ and $C \in \mathcal{C}$. Since $B_C(1) = \{C\}$ we see that the stabilizer of C in $\mathrm{Aut}(\mathcal{C})$ is an open subgroup of $\mathrm{Aut}(\mathcal{C})$. When $\mathcal{C}$ is locally finite, $\mathrm{Aut}(\mathcal{C})$ is a locally compact, totally disconnected, topological group. As such, it has a Haar measure. A closed subgroup $G \subset \mathrm{Aut}(\mathcal{C})$ inherits a topology and a Haar measure. A subgroup $\Gamma \subset G$ is a *lattice* if it is discrete and if G/Γ has finite measure. It is a *uniform* lattice if G/Γ is compact. A discrete subgroup $\Gamma \subset \mathrm{Aut}(\mathcal{C})$ is a uniform lattice if and only if $\mathcal{C}/\Gamma$ is finite.

Example 4.99 (Projective Spaces) To an $(n+1)$-dimensional vector space V over a field $\mathbb{F}$ we associate a building $\mathcal{C}(V)$ as follows. A *chamber* C in $\mathcal{C}(V)$ is a maximal chain $V_1 < V_2 < \cdots < V_n$ of proper nonzero subspaces. As we shall explain in the next subsection, the chambers are the top-dimensional simplices in an $(n-1)$-dimensional simplicial complex Δ (called the *spherical realization of* $\mathcal{C}(V)$). A k-simplex of Δ is a chain of subspaces $V_1 < \cdots < V_{k+1}$ of length $k+1$. Inclusion of one chain into another corresponds to inclusion of one simplex as a face of the other. A codimension one face (i.e., a facet) of a chamber C is obtained by omitting one of the subspaces in the chain. Let us say that the facet is *type i* if it is obtained by omitting the i-dimensional subspace V_i. The Coxeter system (W, S) associated to $\mathcal{C}(V)$ has Coxeter group W the symmetric group on $n+1$ letters and $S = \{s_1, \ldots, s_n\}$, where s_i is the transposition $(i, i+1)$. In other words, the diagram for (W, S) is $\mathbf{A}_n$. Two chambers are s_i-equivalent if they share a common codimension-one face of type i. So, two distinct chambers belong to the same $\{s_i\}$-residue if and only if the intersection of the corresponding maximal chains is the chain obtained by omitting the ith element of each. More generally, a subset $T < S$ determines a face of codimension Card(T) in the simplex corresponding to C, namely, this face is the intersection of the codimension one faces indexed by the s_i that lie in T. The building $\mathcal{C}(V)$ is called the *projective space* of V. There is an obvious action of the group $PGL(V)$ on $\mathcal{C}(V)$. An apartment in $\mathcal{C}(V)$ corresponds to a choice of basis $\{e_1, e_2, \ldots, e_{n+1}\}$ (actually by the set of lines determined by the basis elements). Such a basis gives us a maximal chain C of subspaces $\langle e_1 \rangle < \cdots < \langle e_1, \ldots, e_n \rangle$, where $\langle e_1, \ldots, e_k \rangle$ means the subspace spanned by the first k basis vectors. The symmetric group W acts by permuting the basis vectors and the union of the chambers wC, $w \in W$, is a copy of the spherical Coxeter complex for (W, S) called an "apartment" of $\mathcal{C}$ (cf. Sect. 4.2.1). With its natural piecewise spherical metric, each apartment is isometric to the round sphere $\mathbb{S}^{n-1}$, triangulated as the barycentric subdivision of the boundary of an n-simplex (or rather its image under radial projection onto $\mathbb{S}^{n-1}$). In the classical realization of this building the number of chambers in each rank one residue is $q+1$, where $q = \text{Card}\,\mathbb{F}$ (where $q+1$ is the cardinality of the projective line over $\mathbb{F}$). Thus, $\mathcal{C}(V)$ is a finite simplicial complex if and only if $\mathbb{F}$ is a finite field.

When $\dim V = 3$ and $\mathbb{F}$ is the field $\mathbb{F}_2$ with 2-elements, the classical realization of $\mathcal{C}(V)$ is the Fano plane pictured in Fig. 4.2 below.

4.4.2 *Geometric Realizations of Buildings*

Given a Coxeter system (W, S), let $K = |\mathcal{S}^{\text{op}}|$ be its fundamental chamber. Even though $\mathcal{C}$ need not admit a chamber-transitive group action, the definition of the

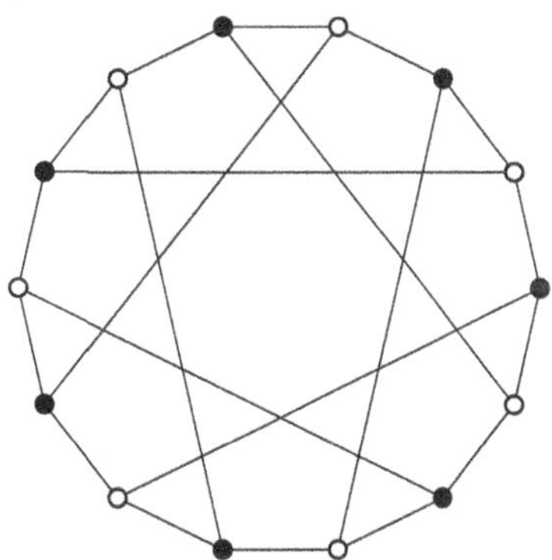

Fig. 4.2 Figure of incidence graph of Fano plane

basic construction can be modified to work for $\mathcal{C}$. The definition of $D(\mathcal{C}, K)$ is similar to that of the Davis–Moussong complex $\Sigma(W, S)$ in Sect. 4.2.3 or to the Deligne complex $\Lambda(W, S)$ in Sect. 4.3.3. To wit, $D(\mathcal{C}, K) = (\mathcal{C} \times K)/\sim$, where the equivalence relation $\sim$ is given by

$$(C, x) \sim (C', x') \iff x = x', \text{ and } C, C' \text{ belong to the same } S(x)\text{-residue.} \tag{4.21}$$

(As before, $S(x) = \{s \in S | x \in K_s\}$.) The polyhedron $D(\mathcal{C}, K)$ is the *standard realization of* $\mathcal{C}$. Since K is the geometric realization of $\mathcal{S}^{\mathrm{op}}$, $D(\mathcal{C}, K)$ can be identified with $|R(\mathcal{C})^{\mathrm{op}}|$, the geometric realization of the poset of spherical residues. Similarly, if X is a model chamber for (W, S) as in Remark 4.89, then one can define the *X-realization* of $\mathcal{C}$ by essentially the same formula: $D(\mathcal{C}, X) = (\mathcal{C} \times X)/\sim$.

As in Sect. 4.2.6, K is partitioned into Coxeter blocks B_T, with $T \in \mathcal{S}$, where B_T means the union of all simplices in the order complex of the interval $[T, \emptyset]$ of $\mathcal{S}^{\mathrm{op}}$ (i.e., $[T, \emptyset] = \{T' \in \mathcal{S} \mid T \geq T' \geq \emptyset\}$). Furthermore, each block B_T is identified with a convex polytope in some euclidean space. (Recall that B_T is isomorphic to the intersection of a Coxeter zonotope with a W_T-sector.) This gives a natural piecewise euclidean metric on K, as well as, the *natural piecewise euclidean metric on* $D(\mathcal{C}, K)$. Moreover, each apartment in $D(\mathcal{C}, K)$ is isometric to $D(W, K) = \Sigma(W, S)$, with its natural CAT(0) metric.

Remark 4.100 The calculation in [87] of the compactly supported cohomology of the standard realization, $D(\mathcal{C}, K)$, uses several different realizations of $\mathcal{C}$.

Theorem 4.101 (cf. [77]) *The natural piecewise euclidean metric on the standard realization of any building is* CAT(0).

The proof of this theorem basically follows from the fact that each apartment is a Davis–Moussong complex $\Sigma(W, S)$, which is CAT(0). The details of the proof will occupy the rest of this subsection.

First, we need to understand links in $\Sigma(W, S)$. Suppose (W_T, T) is a spherical Coxeter system. Then W_T acts on the unit sphere $\mathbb{S}^{m-1}$ in the geometric representation on $\mathbb{R}^m$, $m = \mathrm{Card}(T)$. A fundamental domain for the action on $\mathbb{S}^{m-1}$ is the fundamental spherical simplex $\sigma(T)$ defined in Sect. 4.2.1. Thus, $\mathbb{S}^{m-1} = D(W_T, \sigma(T))$. The set of "mirrors" of $\sigma(T)$ is the set its facets $\{\sigma(T)_t\}_{t \in T}$. Suppose $\mathcal{C}(T)$ is a spherical building of type (W_T, T). The *classical realization* of $\mathcal{C}(T)$ is the basic construction applied to $\mathcal{C}(T)$ with fundamental domain $\sigma(T)$:

$$D(\mathcal{C}(T), \sigma(T)) = (\mathcal{C}(T) \times \sigma(T))/\sim, \tag{4.22}$$

with its natural piecewise spherical metric induced from $\sigma(T)$ where the gluing relation $\sim$ is defined as in (4.21). We shall write $\sigma(\mathcal{C}(T))$ for $D(\mathcal{C}(T), \sigma(T))$. In this classical realization, each apartment of a spherical building is isometric to the round sphere $D(W_T, \sigma(T)) = \mathbb{S}^{m-1}$. More precisely, if $\mathcal{A}$ is an apartment in $\mathcal{C}(T)$, then the map induced by $\rho_C|_{\mathcal{A}}$ (from Definition 4.90) induces an isometry from the classical realization of $\mathcal{A}$ to $\mathbb{S}^{m-1}$. N.B. The classical realization of $\mathcal{C}(T)$ is *not* the same as the standard realization; the standard realization of $\mathcal{C}(T)$ uses for fundamental domain the Coxeter block B_T so that each apartment becomes isometric to the Coxeter zonotope $Z(W_T, T)$ while in the classical realization it is the image of a radial projection of $\partial Z(W_T, T)$ onto a sphere. So, for spherical buildings the standard realization is the cone on the classical realization.

Example 4.102 As in Example 4.93, suppose $\mathcal{C}$ is a generalized m-gon with its natural piecewise spherical structure. Then $\sigma(\mathcal{C})$ is a graph, each edge has length π/m and each apartment is isometric to a circle of length 2π. Moreover, $\Sigma(\mathcal{C})$ is CAT(1) (since its girth is $2m$) and its diameter is π (since its combinatorial diameter is m).

Example 4.103 As in Example 4.99, suppose $\mathcal{C}$ is a projective plane over the finite field $\mathbb{F}_q$ of order q. The classical realization $\sigma(\mathcal{C})$ is a graph; each apartment is the boundary of a hexagon; and since the link of each vertex is a projective line over $\mathbb{F}_q$, exactly $q+1$ edges meet at each vertex. (See Fig. 4.2 following Example 4.99 for the case $q = 2$.) If $\mathcal{C}$ is projective n-space, then each apartment is isomorphic to $\mathbb{S}^{n-1}$ triangulated as the barycentric subdivision of the boundary of an n-simplex.

The standard cellulation of $\Sigma(W, S)$ by Coxeter zonotopes has a subdivision into Coxeter blocks (see Definition 4.28). Let v_T be a vertex corresponding to the center of a Coxeter zonotope $Z(W_T, T)$. Then the link of v_T in $\Sigma = \Sigma(W, S)$ decomposes as a spherical join:

$$\mathrm{Lk}(v_T, \Sigma) = \mathbb{S}^{m-1} * \mathrm{Lk}(T, |\mathcal{S}|). \tag{4.23}$$

Here $\mathbb{S}^{m-1}$ is the link of v_T in the zonotope and $\mathrm{Lk}(T, |\mathcal{S}|) = |\mathcal{S}_{>T}| = \mathrm{Lk}(v_T, K_T)$ is the link of v_T in the stratum of a chamber corresponding to T. ($\mathrm{Lk}(T, \mathcal{S})$ also means the link of the simplex corresponding to T in $L(W, S)$.) Since $|\mathcal{S}_{>T}|$ is a link in the metric flag complex $L(W, S)$, it is also a metric flag complex (see [82, Lemma I.7.3]); so, by Moussong's Lemma 4.18, it is CAT(1). A similar analysis holds for the link of a vertex v_T in the geometric realization $|\mathcal{C}| = D(\mathcal{C}, K)$ except that in (4.23), the round sphere $\mathbb{S}^{m-1}$ must be replaced by the classical realization of a building $\mathrm{Res}_T(C)$. The formula in (4.23) then should be replaced by

$$\mathrm{Lk}(v_T, |\mathcal{C}|) = \sigma(\mathcal{C}(T)) * \mathrm{Lk}(T, |\mathcal{S}|), \tag{4.24}$$

where $\sigma(\mathcal{C}(T)) = D(\mathrm{Res}_T(C)^{\mathrm{op}}, \sigma(T))$. In other words, the link of a vertex in the standard realization of a building is the spherical join of the classical realization of a spherical building with a metric flag complex. So, to show that $\mathrm{Lk}(v_T, |\mathcal{C}|)$ is CAT(1) it suffices to prove the following lemma.

Lemma 4.104 (cf. [77]) *If $\mathcal{C}(T)$ is a spherical building of type (W_T, T), then its classical realization $\sigma(\mathcal{C}(T))$ is* CAT(1). *Its diameter is π.*

In fact, this lemma is a corollary of Theorem 4.101. Indeed, a neighborhood of the vertex v_T in $|\mathcal{C}(T)|$ is isometric to a neighborhood of the cone point in $\mathrm{Cone}(\sigma(\mathcal{C}(T))$. By Lemma 2.8, v_T has a CAT(0) neighborhood if and only if $\sigma(\mathcal{C}(T))$ is CAT(1).

Remark 4.105 On the other hand, if Lemma 4.104 holds for all spherical residues in $\mathcal{C}$, then, by the Link Condition (from Definition 2.10), $|\mathcal{C}|$ is NPC. Since it also is not hard to see that $|\mathcal{C}|$ is simply connected, it might seem that the way to prove Theorem 4.101 is to first prove Lemma 4.104 by an independent argument. (One needs to show that the spherical realizations of such links have no short geodesic loops. This should follow from the fact that each spherical apartment is a round sphere.) However, the proof Theorem 4.101 from [77] which is outlined below, is via a direct argument that is independent of the Link Condition.

In [60] Charney and Lytchak gave the following metric characterization of thick spherical buildings.

Proposition 4.106 (Charney-Lytchak, [60]) *A connected, piecewise spherical complex σ of dimension $n \geq 2$ is the classical realization of a thick spherical building if and only if*

- *σ is* CAT(1),
- *each $(n-1)$-cell is contained in at least three n-cells,*
- *the link of every k-cell with $k \leq n-2$, is connected, and*
- *the link of each $(n-2)$-cell has diameter π.*

Sketch of Proof of Theorem 4.101 The proof in [77] follows a standard argument given in [40, Ch. VI §3] in the case of euclidean buildings. An important consequence of Definition 4.87 is one of the classical axioms for a building: any two chambers are contained in a common apartment. For the standard realization, this implies that any two points x and y in $|\mathcal{C}|$ are contained in the realization of some apartment $\mathcal{A}$. For C a chamber in an apartment $\mathcal{A}$, we have the function $\rho_{C,\mathcal{A}} : \mathcal{C} \to \mathcal{A}$, the retraction onto $\mathcal{A}$ centered at C from Definition 4.90. Its geometric realization is denoted $\overline{\rho}_{C,\mathcal{A}} : |\mathcal{C}| \to |\mathcal{A}|$ (or simply by $\overline{\rho}$). If x, y are two points in $|\mathcal{A}|$, then $\overline{\rho}$ takes a geodesic segment connecting them in $|\mathcal{C}|$ to a geodesic segment in $|\mathcal{A}|$. Using the fact that $|\mathcal{A}|$ is CAT(0) one argues that the geodesic segment in $|\mathcal{C}|$ from x to y is equal to the geodesic segment in $|\mathcal{A}|$. Moreover, $\overline{\rho}$ is distance decreasing.

If x, y, z are vertices of a triangle in $\mathbb{E}^2$ and $p_t = tx + (1-t)y$ is a point on the segment from x to y, then it is an easy exercise to see that

$$d^2(z, p_t) = (1-t)d^2(z, x) + td^2(z, y) - t(1-t)d^2(x, y),$$

where $d^2(x, y)$ denotes the square of the euclidean distance. It follows that for a triangle in a geodesic space with vertices x, y, z, the CAT(0) inequality of Sect. 2.1.1 is equivalent to the above formula with the equality replaced by the inequality:

$$d^2(z, p_t) \leq (1-t)d^2(z, x) + td^2(z, y) - t(1-t)d^2(x, y) \tag{4.25}$$

Now choose an apartment $\mathcal{A}$ with $x, y \in |\mathcal{A}|$ and a chamber C so that $p_t \in |C|$. Let $\overline{\rho} : |\mathcal{C}| \to |\mathcal{A}|$ be the geometric realization of $\rho_{C,\mathcal{A}}$. Then

$$\begin{aligned} d^2(z, p_t) &= d^2(\overline{\rho}(z), p_t) \\ &\leq (1-t)d^2(\overline{\rho}(z), x) + td^2(\overline{\rho}(z), y) - t(1-t)d^2(x, y) \\ &\leq (1-t)d^2(z, x) + td^2(z, y) - t(1-t)d^2(x, y) . \end{aligned}$$

So, the CAT(0)-inequality holds for the triangle $[x, y, z]$.

As we showed in Theorem 4.7 of Sect. 4.1.3, the fact that $\Sigma(W, S)$ is CAT(0) implies that the $K(\pi, 1)$-Question for $W\mathcal{S}^{\mathrm{op}}$ has a positive answer. Similarly, if $\mathcal{C}$ is any building with a chamber-transitive automorphism group G, then Theorem 4.101 shows that the $K(\pi, 1)$-Question for the simple complex of groups associated to G has a positive answer. We state this as the following.

Theorem 4.107 (The $K(\pi, 1)$-Question for Chamber-Transitive Buildings) *Suppose $\mathcal{C}$ is a building with a chamber-transitive automorphism group G. Let $G\mathcal{S}^{\mathrm{op}}$ be the associated simple complex of groups. Then the $K(\pi, 1)$-Question for $G\mathcal{S}^{\mathrm{op}}$ has a positive answer. In particular, the aspherical realization of this simple complex of groups is $D(G, |\mathcal{S}^{\mathrm{op}}|) \times_G EG$, which is homotopy equivalent to BG.*

One also can use the argument in the proof of Theorem 4.101 to decide when the standard realization of a building has a piecewise hyperbolic, CAT(-1) metric. By Moussong's Theorem 4.23. $\Sigma(W, S)$ has a CAT(-1) structure if and only if W is word hyperbolic. Analogously, we get the following.

Theorem 4.108 (cf. [77, Remark 11.8]) *Suppose $\mathcal{C}$ is a building of type (W, S), with W word hyperbolic. Then its standard realization $|\mathcal{C}|$ ($= D(\mathcal{C}, K)$) can be given a piecewise hyperbolic metric that is* CAT(-1).

For example, if W is a reflection group on $\mathbb{H}^2$ generated by reflections across the faces of a hyperbolic polygon, then $\mathcal{C}$ is called a *Fuchsian building*. In the right-angled case these Fuchsian buildings were constructed and studied by Bourdon [29, 30], as well as Remy [194–196], Gaboriau–Paulin [122] and Haglund–Paulin [142].

4.4.3 Regular Right-Angled Buildings

Here we expand on some of the material in Sect. 3.1.2.

Suppose $\{E_s\}_{s\in S}$ is a family of discrete, usually finite, spaces indexed by S. In Sect. 3.1.1, given a simplicial complex L with vertex set S, we defined a cube complex Z_L as the polyhedral product:

$$Z_L = \prod^L \operatorname{Cone} \mathbf{E}$$

(cf. (3.15)), where $\operatorname{Cone}\mathbf{E}$ means the S-tuple $((\operatorname{Cone} E_s, E_s, *_s))_{s\in S}$. When L is a flag complex, Z_L is NPC by Theorem 3.14; so, its universal cover $\tilde{Z}_L$ is CAT(0). At the end of Sect. 3.1.1 we promised to show that $\tilde{Z}_L$ was the standard realization of a regular RAB. We do so now.

As was shown earlier, if each E_s is $\{\pm 1\}$, then $\tilde{Z}_L$ is the Davis–Moussong complex $\Sigma(W, S)$ where (W, S) is the RACS associated to the graph L^1.

We assume each E_s has at least two elements. By Example 4.92, E_s is a rank 1 building associated to the Coxeter system $(\mathbf{C}_2, \{s\})$, where $\mathbf{C}_2 = \langle s\rangle$ is the group of order 2 generated by s. As in Example 4.94, $\prod_{s\in S} E_s$ is a product building of type $((\mathbf{C}_2)^S, S)$. Its standard realization is the cube complex,

$$Z_{\Delta(S)} = \prod_{s\in S} \operatorname{Cone}(E_s) = \prod^{\Delta(S)} \operatorname{Cone}\mathbf{E},$$

which has fundamental chamber $\square^S$ $(:= \prod_{s\in S}[0,1])$. The polyhedral product Z_L is a cubical subcomplex of $Z_{\Delta(S)}$. If $\mathcal{C}' = \prod_{s\in S} E_s$ denotes the set of chambers in the product building, then Z_L is tessellated by copies of the fundamental chamber K $(= K_L)$, and, as in formula (3.15) of Sect. 3.1.2,

$$Z_L = D(\mathcal{C}', K) = (\mathcal{C}' \times K)/\sim$$

is the K-realization of the product building $\mathcal{C}'$.

For any chamber $D \in \mathcal{C}'$, put $K(D)$ equal to the image of $D \times K$ in $(\mathcal{C}' \times K)/\sim$; and call it a *chamber of* Z_L. In other words, Z_L is the realization of the product building using as model chamber the chamber K for the RACS (W, S) associated to L. Let $v_\emptyset$ denote the vertex of K corresponding to $\emptyset$. The point $c_D = D \times v_\emptyset$ is called the *center* of the chamber $K(D)$. Let $K^\circ := K - \partial K = \{x \in K \mid S(x) = \emptyset\}$ be the interior of K and let $K^\circ(D) = D \times K^\circ$

Next we describe the set of chambers $\mathcal{C}$ for the putative building structure on the universal cover $\tilde{Z}_L$. Note that Z_L is not simply connected (since the 2-skeleton of L is not equal to the 2-skeleton of $\Delta(S)$). Put $\pi = \pi_1(Z_L)$. (If each E_s is given the structure of a group G_s and $\Gamma = \prod_{L^1} G_s$ is the graph product, then, as explained in Sect. 3.1.2, π is the kernel of the natural epimorphism $\Gamma \to \sum G_s$.) Let $p : \tilde{Z}_L \to Z_L$ be the projection map of the universal cover. Since K° is the open cone on ∂K, it is simply connected; hence, $p^{-1}(K^\circ(D))$ is a disjoint union of copies of K°, each projecting homeomorphically onto $K^\circ(D)$. The closure of such a copy is called a *chamber* of $\tilde{Z}_L$. The set of chambers of $\tilde{Z}_L$ is denoted by $\mathcal{C}$. As before, $\tilde{Z}_L \cong (\mathcal{C} \times K)/\sim$. Identifying $\mathcal{C}'$ with the set of centers of chambers in Z_L and $\mathcal{C}$ with the set of centers of chambers in $\tilde{Z}_L$, we see that $\mathcal{C} = p^{-1}(\mathcal{C}')$. Chambers C, D in $\mathcal{C}$ are *s-adjacent* if they project to s-adjacent chambers in $\mathcal{C}'$. This gives $\mathcal{C}$ the structure of a chamber system over S.

Definition 4.109 A right-angled building $\mathcal{C}$ of type (W, S) is *regular* if for each $s \in S$, any two $\{s\}$-residues in $\mathcal{C}$ have the same cardinality, q_s.

More generally, a building $\mathcal{C}$ whose type is an arbitrary Coxeter system (W, S) is *regular* if for each spherical subset $T \in \mathcal{S}$, any two T-residues in $\mathcal{C}$ are isomorphic.

Theorem 4.110 (cf. [77, Thm. 5.1] and [83, Thm. 2.5]) *The chamber system $\mathcal{C}$ defined above is a regular right-angled building of type (W, S), where (W, S) is the* RACS *associated to L^1.*

Sketch of Proof To show that $\mathcal{C}$ is a building we must define a Weyl distance $\delta : \mathcal{C} \times \mathcal{C} \to W$ satisfying the axioms in Definition 4.87. The definition of δ is the obvious one. Suppose $(C, C') \in \mathcal{C} \times \mathcal{C}$. Choose a gallery of minimal length $C_0, C_1, \dots, C_k$ from $C = C_0$ to $C' = C_k$. Let $\mathbf{s} = (s_1, \dots, s_k)$ be its type and let $w(\mathbf{s}) = s_1 \cdots s_k$ be the corresponding element of W. The Weyl distance δ is defined by $\delta(C, C') = w(\mathbf{s})$. We must show that element $w(\mathbf{s}) \in W$ is independent of the choice of gallery (so that $w(\mathbf{s})$ is well-defined) and that δ satisfies Axioms **(WD1)**, **(WD2)** and **(WD3)** in Definition 4.87. The verification of these axioms in [83, Thm. 5.1] uses the fact that if $\mathcal{C}'$ is the product building and $p : \mathcal{C} \to \mathcal{C}'$ is induced by the covering projection, then for each spherical subset $T \in \mathcal{S}(W, S)$, p takes a spherical residue of type T in $\mathcal{C}$ isomorphically onto a corresponding T-residue in $\mathcal{C}$. By combining this with Tits' solution to the Word Problem for Coxeter groups [82, §3.4], the theorem follows.

So, the proof of Theorem 4.110 came down to the fact that the geometric realization of the chamber system $\mathcal{C}$ is simply connected and that $\mathcal{C}$ has T-residues which are spherical buildings for any $T \in \mathcal{S}(W, S)$ with $\text{Card}(T) \leq 3$. In fact, this argument gives a general principle which is stated as a result of Tits in the next subsection as Theorem 4.114. This result will be used again in the following subsections.

We continue our discussion of the regular RABs constructed in Theorem 4.110. Suppose each E_s is finite with $\text{Card}(E_s) = q_s + 1$. The RAB, $\mathcal{C}$, from Theorem 4.110 then has parameters $(q_s)_{s \in S}$. (In fact, the q_s can be arbitrary nonzero cardinal numbers ≥ 1.) According to the next result each regular RAB is determined by its S-tuple of parameters.

Proposition 4.111 (Haglund–Paulin [142, Prop. 1.2]) *Suppose $\mathcal{C}_1$ and $\mathcal{C}_2$ are two regular* RAB*s of the same type and with the same parameters $(q_s)_{s \in S}$. Then $\mathcal{C}_1$ is isomorphic to $\mathcal{C}_2$.*

Remark 4.112 The regular RAB of Theorem 4.110 can be seen to admit a chamber-transitive automorphism group (cf. [49]). In fact, let G_s denote the symmetric group on E_s and let B_s be the isotropy subgroup in G_s at the base point $*_s$. Put $B = \sum B_s$ and for each subset T of S, put $G_T = \sum_{s \in T} J_s(T)$, where

$$J_s(T) = \begin{cases} G_s \,, & \text{if } s \in T; \\ B_s \,, & \text{if } s \notin T. \end{cases}$$

(cf. Example 3.19. In other words, $G_T := \sum_{s \in T} J_s(T)$ (where $G_\emptyset = B$). Note that $J_s(T)/B = G_s/B_s$. If (W, S) is the RACS associated to L^1 and $\mathcal{S}^{op}$ is the dual poset of spherical subsets, then $\{G_T\}_{T \in \mathcal{S}^{op}}$ is a simple complex of groups over $\mathcal{S}^{op}$ (cf. Example 3.19). In Sect. 3.1.2a *generalized graph product* (or "relative graph product") Γ was defined as the direct limit of this simple complex of groups. The group Γ is chamber-transitive on $\mathcal{C}$. (In fact, the same is true if we replace G_s by any group which is transitive on E_s and B_s by the isotropy subgroup at $*_s$.) For more about automorphisms of RABs, see [49].

Trees of Buildings In contradistinction to Proposition 4.111, in certain cases when (W, S) is not right-angled, in [142] Haglund and Paulin have found examples where there are uncountably many isomorphism classes of regular buildings of type (W, S), each with the same S-tuple of parameters and with isomorphic spherical residues. The construction uses the notion of a tree of buildings, developed in [142]. Suppose that (W, S) splits as an amalgamated product:

$$W = W_+ *_{W_0} W_- \,, \qquad S = S_+ \cup S_- \,, \qquad S_0 = S_+ \cap S_- \,, \tag{4.26}$$

giving Coxeter systems (W_+, S_+), (W_-, S_-) and (W_0, S_0). Assume W_0 infinite. Let K_+, K_- and K_0 be the corresponding fundamental chambers. Let $\mathcal{C}$ be a chamber-transitive building of type (W, S). Then $\mathcal{C}$ can be given the structure of a tree of buildings as follows. Define a bipartite tree T with vertices the $S_\pm$ residues colored by $\{+, -\}$. The edges are the S_0-residues. The edge corresponding to $\mathrm{Res}_{S_0}(C)$ is incident to $\mathrm{Res}_{S_+}(C)$ and to $\mathrm{Res}_{S_-}(C)$. For a given $C \in \mathcal{C}$, let $\mathcal{C}_0$, $\mathcal{C}_+$ and $\mathcal{C}_-$ be the corresponding residues. Then $\mathcal{C}$ is formed by gluing together copies of $\mathcal{C}_\pm$ for each vertex of T along a copy of $\mathcal{C}_0$. If, up to isomorphism, there is more than one way to do the gluing, then for each edge one can try to reglue the vertex buildings along the edge buildings. To this end consider the set $\mathcal{F}$ of pairs (f_+, f_-) where $f_+ : \mathcal{C}_0 \to \mathcal{C}_+$ and $f_- : \mathcal{C}_0 \to \mathcal{C}_-$ are injections of buildings of type (W, S_0). Let $G_\pm$ and G_0 denote $\mathrm{Aut}(\mathcal{C}_\pm)$ and $\mathrm{Aut}(\mathcal{C}_0)$. The group $G_+ \times G_- \times G_0$ acts on $\mathcal{F}$ in an obvious fashion and if there is more than one $(G_+ \times G_- \times G_0)$-orbit on $\mathcal{F}$, then for each edge there is more than one way to accomplish the gluing. So, when there are at least two $(G_+ \times G_- \times G_0)$-orbits, there are uncountably many isomorphism classes of such trees of buildings over T (since there are uncountably many functions from $\mathrm{Edge}(T)$ to a set with at least two elements). Haglund–Paulin [142] prove that when (W, S) is a group generated by reflections across the faces of a certain 3-dimensional hyperbolic polytope K, then under certain conditions there is more than one $(G_+ \times G_- \times G_0)$-orbit on $\mathcal{F}$. In the Haglund–Paulin examples each vertex building is type $\mathbf{A}_3$ (in particular, the Coxeter system (W, S) for $\mathcal{C}$ is not right-angled).

Proposition 4.113 (Haglund–Paulin [142, Cor. 4.11]) *There are uncountably many isomorphism classes of buildings of type (W, S), where (W, S) is a certain 3-dimensional hyperbolic reflection system. Each of these examples is a tree of buildings over the Bass-Serre tree T for a splitting of (W, S) as in (4.26). So, in these examples all of the buildings have isomorphic chambers, S-tuple of parameters, and spherical residues, yet only one of them admits a chamber-transitive automorphism group.*

Similarly, in [122] Gaboriau and Paulin show that there are uncountably many Fuchsian buildings with L a $2k$-gon with its edges labeled 3.

4.4.4 A Local Approach to Buildings

Suppose $\mathcal{C}$ is a gallery-connected chamber system over S. Let (W, S) correspond to the Coxeter matrix. Suppose (W, S) is a Coxeter system with Coxeter matrix $M = (m(s,t))_{(s,t)\in S\times S}$. We say that $\mathcal{C}$ is *type* (W, S) if for each unordered pair $\{s, t\} \in \mathcal{S}$, each $\{s, t\}$-residue of $\mathcal{C}$ is a generalized $m(s, t)$-gon. (More usual terminology is that the chamber system has *type* M.) A chamber system of type (W, S) need not be a building since its standard realization need not be simply connected and hence, need not satisfy the conclusion of Theorem 4.101. For example, if $\mathcal{C}$ is a building of type (W, S) and Γ is a nontrivial torsion-free group of automorphisms of $\mathcal{C}$, then the chamber system $\mathcal{C}/\Gamma$ will not be a building. However, the following theorem of Tits says that this is essentially the only obstruction.

Theorem 4.114 (cf. Tits [220, 221] or [198, Cor. 4.10]) *A chamber system $\mathcal{C}$ of type (W, S) is a building if and only if*

(i) *each spherical residue is a building, and*
(ii) *the geometric realization of its poset of spherical residues, $|R(\mathcal{C})^{\mathrm{op}}|$, is simply connected.*

Remark 4.115 As stated, Theorem 4.114 is vacuous for a chamber system $\mathcal{C}$ whose type is a spherical Coxeter system. Indeed, since the entire chamber system is a spherical residue of type S, item (i) already requires it to be a building. So, suppose $\mathcal{C}$ is a chamber system of type (W, S) where (W, S) is a spherical Coxeter system of rank ≥ 3. Recall that the classical realization $\sigma(\mathcal{C})$ is homeomorphic to the order complex of the poset $R^{\mathrm{prop}}(\mathcal{C})$ of all spherical residues of form $\mathrm{Res}_T(C)$ where C is a chamber and T is a proper subset of S. One might speculate that, when the rank is greater than 3, a version of Theorem 4.114 should hold for the classical realization if one uses $R^{\mathrm{prop}}(\mathcal{C})$ rather $R(\mathcal{C})$. To wit, if (i) for each proper spherical subset T, each T-residue is a building and (ii) $|R^{\text{prop}}(\mathcal{C})|$ is simply connected, then $\mathcal{C}$ should be a building. (Note that when $\mathcal{C}$ is a spherical building of rank ≥ 3, $\sigma(\mathcal{C})$ is simply connected since each apartment is a sphere of dimension ≥ 2.) However, this is false for rank 3: there are examples of simply connected chamber systems of type $\mathbf{C}_3$ which are not buildings (type $\mathbf{C}_3$ is a refinement of saying that the Coxeter diagram is type $\mathbf{B}_3$). In other words, $|R^{\text{prop}}(\mathcal{C})|$ is simply connected but $\mathcal{C}$ is not a building. (See [198, Example 2, pp. 50–51].) The point is that $|R^{\text{prop}}(\mathcal{C})|$ is a 2-dimensional simplicial complex tessellated by $(2, 3, 4)$ spherical triangles so that the link of each vertex is a generalized polygon and even though $|R^{\text{prop}}(\mathcal{C})|$ is simply connected, it is not a building.

Definition 4.116 Suppose $\mathcal{C}$ and $\mathcal{C}'$ are chamber systems of type (W, S) and $\varphi : \mathcal{C} \to \mathcal{C}'$ is an adjacency-preserving map of chamber systems. Then φ is a *covering* if its restriction to each proper spherical residue is an isomorphism. (This implies that the standard realization of φ is a covering projection.)

Theorem 4.117 (cf. Ronan [198, Cor. 4.10]) *Suppose (W, S) is a spherical Coxeter system of rank ≥ 3 and that $\mathcal{C}$ is a chamber system of type (W, S). When* $\text{Card}(S) = 3$ *and $\mathcal{C}$ is the type of the root system* $\mathbf{C}_3$, *assume further that the classical realization of $\mathcal{C}$ is covered by a building. Then $\mathcal{C}$ is a building if and only if*

(i) *for each proper subset $T < S$, each T-residue* $\text{Res}_T(C)$ *is a spherical building, and*
(ii) $|R^{\text{prop}}(\mathcal{C})|$ *is simply connected.*

By definition, in a chamber system of type (W, S), any rank two spherical residue is a building. When $\text{Card}(S) = 3$, if we assume as in Theorem 4.117 that $\mathcal{C}$ is covered by a building then item (ii) implies that it is a building. hence, condition (ii) means that all spherical residues of rank ≤ 3 are buildings. Moreover, instead of requiring that these T-residues are buildings, one can get away with the condition that there are no type $\mathbf{C}_3$ residues and that the classical realization $\sigma(\text{Res}_T(C))$ is simply connected for each spherical residue of rank 3. Indeed, this follows from a version of Theorem 4.114 for the poset $R^{\text{prop}}(\mathcal{C})$ of proper spherical residues whenever (W, S) is a spherical Coxeter system. In [220] there is a replacement for Theorem 4.114 with the hypothesis that every rank three spherical residue of type $\mathbf{C}_3$ is covered by a building and the conclusion is that "the universal 2-cover of $\mathcal{C}$ is a building," where "universal 2-cover" means a chamber system $\tilde{\mathcal{C}}$ that $|R(\tilde{\mathcal{C}})|$ is simply connected as is the classical realization of each spherical residue of rank ≥ 3.

Buildings as Simple Complexes of Groups In Theorem 4.96 we demonstrated that every building with chamber-transitive automorphism group yields a simple complex of groups. In this paragraph we show the converse.

Suppose $G\mathcal{S}^{\text{op}} = (G_T)_{T \in \mathcal{S}^{\text{op}}}$ is a simple complex of groups over $\mathcal{S}^{\text{op}}$. Put $B = G_\emptyset$ and suppose that for each $T \in \mathcal{S}^{\text{op}}$, G_T is a chamber-transitive automorphism group of a building $\mathcal{C}_T$ (where $\mathcal{C}_T := G_T/B$ of type (W_T, T)).

Let $G\mathcal{S}^{\mathrm{op}}$ be the simple complex of groups $(G_T)_{T\in\mathcal{S}^{\mathrm{op}}}$ and let G be its direct limit. By Lemma 4.104 the classical realization, $\sigma(\mathcal{C}_T)$, of the spherical building $\mathcal{C}_T$ is CAT(1). This means that the simple complex of groups $G\mathcal{S}^{\mathrm{op}}$ is nonpositively curved in the sense of [35]; hence, it is developable [35, Thm. 4.17, p. 562]. Since the fundamental chamber $|\mathcal{S}^{\mathrm{op}}|$ $(= K(W, S))$ is contractible, it follows that the basic construction $D(G, K)$ is simply connected, hence, is $CAT(0)$ and so, is contractible. The set of chambers $\mathcal{C} := G/B$ obviously has the structure of a chamber system of type (W, S). Since each spherical residue is a spherical building and since $D(G, K)$ is simply connected, we can use Theorem 4.114 to deduce the following result of [122]. It is a converse to Theorem 4.96

Theorem 4.118 (Tits [221], Gaboriau-Paulin [122, Thm. 0.1]) *Suppose $\mathcal{C}$ is chamber system of type (W, S) defined by the simple complex of groups $G\mathcal{S}^{\mathrm{op}} = (G_T)_{T\in\mathcal{S}^{\mathrm{op}}}$ as above. Then $\mathcal{C}$ is a building. Moreover, the direct limit G is a chamber-transitive group of automorphisms.*

Roughly, this theorem means that we can construct a building by specifying it locally. In other words, a family of spherical buildings over $\mathcal{S}^{\mathrm{op}}$ can be glued together to get a building of type (W, S). However, one must be more careful: each of the local buildings must be equipped with a chamber-transitive group action and these local groups must fit together to give a complex of groups. This begs the question of how do we find such a complex of groups $(G_T)_{T\in\mathcal{S}^{\mathrm{op}}}$, in particular, what group B should correspond to $T = \emptyset$. A good way to produce such a complex of groups is to use the method of pullbacks described in the next subsection. That is, start with a building $\mathcal{C}'$ of type (W', S') and a simplicial map $f : L(W, S)) \to L(W', S')$ and then use f to get a non-standard realization of $\mathcal{C}'$ over $K(W, S)$. Then take the universal cover to get a building $\mathcal{C}$ of type (W, S). Equivalently, we can use f to pull back a complex of groups $\mathcal{G}'$ over $\mathcal{S}(W', S')^{\mathrm{op}}$ to get a complex of groups $f^*(\mathcal{G}')$ over $\mathcal{S}(W, S))^{\mathrm{op}}$. For example, if f is an inclusion, then $f^*(\mathcal{G}')$ has the same local groups as $\mathcal{G}'$; however, since some edges of L' are not in the image of f, the direct limit of $f^*(\mathcal{G}')$ will be larger than the direct limit for $\mathcal{G}'$.

4.4.5 Pullbacks

In Sects. 3.1 and 4.4.3 we explained a technique for constructing RACSs and RABs. This technique involves first taking a polyhedral product (of rank one buildings) and then passing to the universal cover. In this subsection we will explain a generalization of this technique which was developed in [83]. It works for arbitrary Coxeter systems, not just the right-angled ones. Given a building of type (W', S') and a function $f : S \to S'$ satisfying certain properties, first we define a new Coxeter system (W, S); secondly, by using the fundamental chamber for (W, S) rather than the one for (W', S'), we define a nonstandard, non-simply connected, realization of $\mathcal{C}'$; and thirdly, by taking the universal cover of this nonstandard realization of $\mathcal{C}'$ we get a new building $\mathcal{C}$ of type (W, S). In the right-angled case, the first step is analogous to passing from the product $(\mathbf{C}_2)^S$ of cyclic groups of order 2 to a RACG; the second step is analogous to forming the polyhedral product $Z_L = \prod^L \mathrm{Cone}\,\mathbf{E}$ as in Sect. 3.1.1; the third step corresponds to taking the universal cover of Z_L.

Suppose that (W', S') is a Coxeter system with nerve L', that L is another simplicial complex with vertex set S, and that $f : S \to S'$ defines a simplicial map $L \to L'$ whose restriction to each simplex is injective. Let $(m'(s', t'))_{(s',t') \in S' \times S'}$ denote the Coxeter matrix of (W', S'). Define a new Coxeter matrix $(m(s, t))_{(s,t) \in S \times S}$ by

$$m(s,t) := \begin{cases} 1 & \text{if } s = t, \\ m'(f(s), f(t)) & \text{if } \{s,t\} \in \mathrm{Edge}(L), \\ \infty & \text{otherwise,} \end{cases} \tag{4.27}$$

and let (W, S) be the resulting Coxeter system. In other words, if $m' : \mathrm{Edge}(L') \to \{2.3, \dots\}$ is the edge labeling defining (W', S'), then (W, S) is defined by the edge labeling m given by composing m' with $f|_{\mathrm{Edge}(L)}$, i.e., $m = m'f : \mathrm{Edge}(L) \to \{2, 3, \dots\}$. The map of generating sets $S \to S'$ extends to a homomorphism $\varphi_f : W \to W'$. Note that the homomorphism φ_f is surjective if and only if f is surjective.

First we consider the construction in the case of the Coxeter system (W, S). Let $K(L')$ denote the fundamental chamber for (W', S'), i.e., $K(L')$ is the geometric realization of the order complex of $\mathcal{S}(L')^{\mathrm{op}}$. Similarly, if $\mathcal{S}(L)$ denotes the poset of simplices in the simplicial complex L, then let $K(L) = |\mathcal{S}(L)^{\mathrm{op}}|$ be the dual chamber (i.e., $K(L)$ is the fundamental chamber for a W-action on $D(W, K(L))$). The piecewise spherical metric on L' induces one on L: the length of the edge $\{s, t\}$ is $\pi - \pi/m(s, t)$. The space $D(W, K(L))$ need not be the Davis–Moussong complex: a necessary and sufficient condition for this to be true is that $L = L(W, S)$, i.e., that L is a metric flag complex (cf. Definition 4.13). From now on, we assume this to be the case. The map f can be used to define a new stratification of $K(L)$ indexed by $\mathcal{S}(L')$:

$$K(L)_{s'} := \coprod_{s \in f^{-1}(s')} K(L)_s, \qquad \text{and for each } T' \in \mathcal{S}(L'),$$

$$K(L)_{T'} := \coprod_{T \in f^{-1}(T')} K(L)_T. \tag{4.28}$$

Let $D(W', K(L))$ be the result of applying the basic construction to this new stratification of $K(L)$. If $f|_{L^1}$ is an embedding onto the full subgraph of the 1-skeleton of L' spanned by $f(S)$, then W is isomorphic to the special subgroup $W'_{f(S)}$ of W' and $D(W, K(L))$ is a contractible subcomplex of $D(W', K(L))$. If this is not the case, then $W \neq W'$ and $D(W', K(L))$ is not simply connected. (Indeed, a necessary condition for $D(W', K(L))$ to be simply connected is that $K(L)_{s'}$ be connected for each $s' \in S'$ and that $K(L)_{\{s',t'\}} \neq \emptyset$ whenever $m'(s',t') \neq \infty$, see [82, Prop. 8.2.11]).) When $f(S) = S'$, the natural φ_f-equivariant map, $D(W, K(L)) \to D(W', K(L))$, is the projection map of the universal covering. Similarly, for A' the Artin group associated to (W', S'), the complex $D(A', K(L))$ is a quotient of the Deligne complex $D(A, K(L))$ $(= \Lambda(W, S))$.

More interesting is the case of buildings. If $\mathcal{C}'$ is a building of type (W', S') and $K(L)$ has stratification defined by (4.28), then $D(\mathcal{C}', K(L))$ is a nonstandard realization $\mathcal{C}'$. Let $\tilde{D}$ be the universal cover of $D(\mathcal{C}', K(L))$. The inverse image of a chamber of $D(\mathcal{C}', K(L))$ in $\tilde{D}$ is a disjoint union of copies of $K(L)$ (the inverse image is isomorphic to the product of $K(L)$ with the fundamental group of $D(\mathcal{C}', K(L))$). Such a copy of $K(L)$ in $\tilde{D}$ is called a *chamber* and the set of such chambers is denoted $\mathcal{C}$. Note that $\mathcal{C}$ has an obvious structure of a chamber system over S. The main result of [83], which was previously used in the proof of Theorem 4.110, is the following.

Theorem 4.119 (cf. [83, Thm. 2.5] As Well As Theorem 4.110) *The universal cover $\tilde{D}$ of $D(\mathcal{C}', K(L))$ is the standard realization of a building $\mathcal{C}$ of type (W, S).*

The building $\mathcal{C} = f^*(\mathcal{C}')$ is called the *pullback* of $\mathcal{C}'$ via f.

Remark 4.120 When $\mathcal{C}'$ admits a chamber-transitive automorphism group G', the pullback $f^*\mathcal{C}'$ could also be defined by using the method of simple complexes of groups explained at the end of the previous subsection. Let $B' < G'$ be the stabilizer of a given chamber $C' \in \mathcal{C}'$. For $R \leq S'$, let G'_R be the stabilizer of the R-residue containing C'. By Theorem 4.96, $\mathcal{G}' = (G'_R)_{R \in \mathcal{S}(W',S')^{\mathrm{op}}}$ is a simple complex of groups over $\mathcal{S}(W', S')^{\mathrm{op}}$. We can pull this back to get a simple complex of groups $G\mathcal{S}^{\mathrm{op}} = (G_T)_{T \in \mathcal{S}^{\mathrm{op}}}$, defined by $G_T = G'_{f(T)}$. Let G denote the direct limit of $G\mathcal{S}^{\mathrm{op}}$. There is an obvious simple morphism ψ' from this simple complex of groups to G'. The development of $(G_T)_{T \in \mathcal{S}^{\mathrm{op}}}$ with respect to ψ' is the complex $D(\mathcal{C}', K(L))$ discussed above and this leads to a homomorphism $G \to G'$ with kernel $\pi_1(D(\mathcal{C}', K(L))$. (See Appendices A.1 and A.2 for definitions of terms in the previous two sentences.)

For example, the simplicial map $f : L \to L'$ required in the definition of pullback could be one of the following:

(a) a coloring (defined below), where $L' = \Delta(S')$ is a simplex,
(b) a covering projection,
(c) the inclusion of a subcomplex, or
(d) a covering space of a subcomplex.

Definition 4.121 A *coloring* of a simplicial complex L is a simplicial map $f : L \to \Delta(S')$ whose restriction to each simplex is injective. The elements of S' are the *colors*. (The map f gives a coloring of the vertices of L by elements of S'.)

Coxeter Systems and Buildings of Type FC Suppose (W', S') is a spherical Coxeter system so that $L' = \Delta(S')$ is the simplex on S'. Also suppose L is a flag complex with vertex set S and that $f : L \to \Delta(S')$ is a coloring. The resulting Coxeter system (W, S) is type FC (see Definition 4.29). When $W' = (\mathbf{C}_2)^{S'}$ the complex $D(W', K(L))$ plays the role of the polyhedral product P_L $(= ([-1, 1], \{\pm 1\})^L)$ of (3.3). The natural map $D(W, K(L)) \to D(W', K(L))$ is the universal cover. Since $W \to W'$ is injective on spherical special subgroups, $\pi_1(D(W', K(L))) = \ker(W \to W')$ is a torsion-free subgroup of finite index in W. Since $D(W, K(L))$ $(= \Sigma(W, S))$ is CAT(0), the polyhedron $D(W', K(L))$ is NPC.

If $\mathcal{C}'$ is a spherical building of type (W', S') and $f : L \to \Delta(S')$ is a coloring, then $\mathcal{C} = f^*\mathcal{C}'$ is a regular building of type (W, S). While it is not true that every spherical Coxeter system (W', S') can occur as the type of a thick, locally finite spherical building, many of them can occur. For example, the irreducible Coxeter systems with diagrams $\mathbf{A}_n$, $\mathbf{B}_n$, $\mathbf{D}_n$, $\mathbf{G}_2$, $\mathbf{F}_4$, $\mathbf{E}_6$, $\mathbf{E}_7$, and $\mathbf{E}_8$ all can occur. For spherical buildings of rank two whose Coxeter group is a dihedral group of order $2m$ (i.e., for generalized m-gons), a theorem of Feit–Higman [117] asserts that thick, finite generalized m-gons exist only for $m \in \{2, 3, 4, 6, 8\}$ (cf. Example 4.93 and Theorem 4.130 below). The classical examples of spherical buildings will be discussed in more detail in Sect. 4.4.8. In the next subsection such thick spherical buildings will be used as inputs for the pullback construction to get many new examples of buildings.

4.4.6 *Using Pullbacks to Construct Examples*

Here are two questions with which we will be concerned in this subsection. To simplify the discussion, suppose all buildings are locally finite and thick.

Question 4.122 Which Coxeter systems (W, S) occur as the type of a thick, locally finite building?

In other words, given a finite simplicial graph L^1, for which labeling $m : \text{Edge}(L^1) \to \{2, 3, 4, 6, 8\}$ does there exist a thick, locally finite building $\mathcal{C}$ with Coxeter system defined by m? For example, one can ask this question when L^1 is a k-circuit.

Question 4.123 If $\mathcal{C}$ is a thick, regular building of type (W, S), then which S-tuples $(q_s)_{s \in S}$ can occur as parameters?

If $\mathcal{C}$ is a regular building of rank one, then, by Example 4.92, its parameter can be an arbitrary integer $q \geq 1$. If $\mathcal{C}$ has rank two, then it is a thick, generalized m-gon as in Example 4.93 and, by the Feit–Higman Theorem [117], $m \in \{2, 3, 4, 6, 8\}$. If $m = 2$, then $\mathcal{C}$ is a product of rank one buildings and there are no restrictions on the parameters. If $m = 3$, then $q_s = q_t$. When $m = 4$, 6 or 8 there are severe restrictions on the parameters (q_s, q_t) (see [198, Thm. 3.4, p. 39]). Suppose $\mathcal{C}$ is a locally finite, thick, irreducible spherical building of rank ≥ 3. By Tits' classification of finite spherical buildings of rank ≥ 3, which was discussed in the beginning of

Sect. 4.4, $\mathcal{C}$ is essentially the building of an algebraic group over a finite field $\mathbb{F}_q$. The Coxeter matrix of such an irreducible spherical Coxeter system of rank ≥ 3 must satisfy the crystallographic condition that no $m(s, t)$ can be equal to 5; in other words, all edges of the defining graph must be labeled 2, 3 or 4; so, $\mathbf{H}_3$ and $\mathbf{H}_4$ do not occur as subdiagrams of the Coxeter diagram. If the diagram is connected and the label 4 does not occur, then all generators $s \in S$ are conjugate and the q_s are all equal, their common value q is the order of $\mathbb{F}_q$ (a prime power). If the label 4 occurs, then it only occurs once, the diagram is $\mathbf{C}_n$ or $\mathbf{F}_4$ and there are two conjugacy classes of Coxeter generators in S; hence, there are two parameters q_s and q_t. (The root system $\mathbf{C}_n$ has the same Coxeter diagram as B_n.) For $\mathbf{C}_n$ the possibilities for (q_s, q_t) are (q, q), (q, q^2), (q^2, q) or (q^2, q^3). For $\mathbf{F}_4$ the possibilities for (q_s, q_t) are (q, q) or (q, q^2). (See [198, Appendix 6] or [53].) So, for a finite, irreducible spherical building, the parameters are determined by the finite field $\mathbb{F}_q$; in particular, both parameters are a power of the characteristic of $\mathbb{F}_q$. On the other hand, when $\mathcal{C}$ is reducible, then $(W, S) = (W_1 \times \cdots \times W_k, \bigsqcup S_i)$ with $k > 1$, $\mathcal{C} = \mathcal{C}_1 \times \cdots \times \mathcal{C}_k$ and the parameters for $\mathcal{C}_i$ and $\mathcal{C}_j$, with $i \neq j$ are independent of one another: for example, they can be associated to different primes. (When the parameters of $\mathcal{C}$ are associated to different primes we say that $\mathcal{C}$ is of "mixed type.") One of the main conclusions to be drawn from this subsection is that many buildings of mixed type exist, even when the building is not a product as we will see in the following example.

Example 4.124 (dim L = **1, So** $\mathcal{C}$ **Could Be a Fuchsian Building**) Suppose $\mathcal{C}'$ is a generalized m-gon, i.e., $(W', S') = (D_m, \{s', t'\})$ and L' is the 1-simplex $\Delta(S')$. If L is any bipartite simplicial graph, then it admits a coloring $f : L \to \Delta(S')$. For example, L could be a $2k$-gon or a finite tree. So, for any bipartite graph L, there is a labeling of Edge(L) where each edge is labeled by the same integer m. If $m \in \{2, 3, 4, 6, 8\}$, we can choose $\mathcal{C}'$ to be a thick generalized m-gon. Hence, if L is any bipartite graph, there is a building $\mathcal{C}$ such that each rank two spherical residue is isomorphic is the generalized m-gon $\mathcal{C}'$. In other words, the link of each vertex of $K(L)$ is isomorphic to the 1-dimensional piecewise spherical complex $\sigma(\mathcal{C}')$. (A vertex of $K(L)$ corresponds to the midpoint of an edge of L.) Suppose L is a $2k$-gon. When $2k = 4$, we require $m \geq 3$. Then $K(L)$ can be realized as a convex $2k$-gon in $\mathbb{H}^2$; so, $\mathcal{C}$ is a Fuchsian building. Such buildings need not be associated to any algebraic group or Kac–Moody group. For example, there are "non-Desarguesian" projective planes, i.e., generalized 3-gons that satisfy the axioms for a projective plane but are not isomorphic to a projective plane over any finite field or skew field and such generalized 3-gons can occur as links of vertices.

Example 4.125 (More Examples with dim L = 1) We can get many more examples with dim L = 1 by starting with (W', S') a Coxeter system of rank > 2. The Coxeter system (W', S') could be spherical (in which case $L' = \Delta(S')$) or possibly a euclidean Coxeter system (in which case $L' = \partial\Delta(S')$). To further simplify the discussion let us suppose L is a circuit of length n so that $K(L)$ is an n-gon. A simplicial map $f : L \to \Delta(S')$ is the same thing as a closed edge path, possibly with backtracking. If the diagram of (W', S') is $\widetilde{\mathbf{A}}_{n-1}$, then $\partial\Delta(S')$ contains an n-circuit with each edge labeled 3. Let $f : L \to \partial\Delta(S')$ be an embedding whose image is this n-circuit. If $\mathcal{C}'$ is a building of type (W', S'), then over each edge of $f(L)$ we have a rank two spherical residue with diagram $\mathbf{A}_2$, i.e., a projective plane. In particular, L can be an n-gon for n odd with each edge labeled 3. The resulting building $\mathcal{C}$ over $K(L)$ will be Fuchsian provided $n > 3$. In this fashion we can find examples of buildings with L an n-gon whose edges are labeled fairly randomly by elements of $\{2, 3, 4\}$. We note that when $\mathcal{C}'$ is an irreducible spherical building corresponding to an algebraic group over a finite field $\mathbb{F}_q$ or when it is an irreducible euclidean building corresponding to a discrete valuation ring with residue field $\mathbb{F}_q$, then $\mathcal{C}'$ and the resulting pullback $\mathcal{C} = f^*\mathcal{C}'$ will have thickness depending on the field $\mathbb{F}_q$.

If (W', S') is not irreducible, then different factors can be algebraic groups over finite fields of different characteristics. For example, the numbers $q_s + 1$ could be the cardinalities of projective lines over completely different fields. So, the S'-tuple, $(q_s)_{s\in S'}$, need not be a constant function of s. For example, suppose $(W', S') = (\sum W_i, \bigsqcup S_i)$ is a direct sum of dihedral groups W_i, $i \in \{1, \dots, k\}$, where $W_i = D_{m_i}$ is finite dihedral group of order $2m_i$ generated by $S_i = \{s_i, t_i\}$. Its nerve $L' = L(W', S')$ is a join of 1-simplices, $\sigma_1 * \cdots * \sigma_k$. Each edge of L' is labeled either by an m_i or by 2. Suppose $\mathcal{C}' = \mathcal{C}_1 \times \cdots \times \mathcal{C}_k$, where $\mathcal{C}_i'$ is a generalized m_i-gon. For $i \neq j$ the thickness parameters q_{s_i} and q_{s_j} are independent of each other (in fact if $m_i \neq 3$, then q_{s_i} and q_{t_i} also need not be equal). Hence, the S'-tuple $(q_s)_{s'\in S'}$ for $\mathcal{C}'$ need not be constant. Given a flag complex L and a simplicial map $f : L \to L'$, the labeling of Edge L' pulls back to a labeling of Edge L and we get a building $\mathcal{C}$. In this way, when L is a graph we can produce any labeling of Edge L by $\{2, 3, 4, 6, 8\}$ provided that whenever $\{e, f\}$ is a pair of adjacent edges if $m(e) \geq 3$, then $m(f) = 2$. For example, if L is a $2k$-circuit we can produce a labeling where every other edge is labeled by 2 and where the thicknesses at any two vertices, which are connected by an edge that is not labeled by 2, are different.

Example 4.126 (dim $L = 2$ **and** $K(L)$ **Is a Simple** 3**-Polytope)** Suppose L is a flag triangulation of S^2 and that $f : \operatorname{Vert} L = S \to \Delta(S')$ is a 3-coloring where $S' = \{s_0, s_1, s_2\}$ is the vertex set of the 2-simplex $\Delta(S')$. One way to obtain such an L together with a coloring f is to let L be the barycentric subdivision of the boundary complex of any 3-dimensional polytope and define $f : S \to \{s_0, s_1, s_2\}$ to be the coloring which assigns to the barycenter of a k-face, the color s_k. Another possibility is to take L to be the boundary complex of an octahedron. The edges of $\Delta(S')$ are $\{s_0, s_1\}$, $\{s_0, s_2\}$ and $\{s_1, s_2\}$. The coloring induces a function Edge $L \to \operatorname{Edge}(\Delta(S'))$. Suppose (W', S') is a spherical Coxeter system of rank three corresponding to some edge labeling $m' : \operatorname{Edge}(\Delta(S')) \to \{2, 3, 4\}$. (In other words, the possible off-diagonal entries of the Coxeter matrix for (W', S') are 2, 3 or 4.) Composing this with the edge labeling induced by the coloring, we get a labeling $m : \operatorname{Edge} L \to \operatorname{Edge}(\Delta(S'))$ and a corresponding Coxeter system (W, S) with fundamental chamber $K(L)$. The natural stratification on $K(L)$ is combinatorially equivalent to a 3-dimensional simple convex polytope. If $\mathcal{C}'$ is a thick spherical building of type (W', S'), then the resulting pullback $\mathcal{C}$ will be a building over

$K(L)$ such that each spherical residue of rank three is isomorphic to $\mathcal{C}'$. Since L is a barycentric subdivision, all empty 4-circuits are made of edges mapping to $\{s_0, s_2\}$. Hence, if $m'(s_0, s_2) \neq 2$, then L^1 contains no 4-circuit with each of its edges labeled by 2. It then follows from Andreev's Theorem (see Remark 4.22) that $K(L)$ can be realized as a convex polytope in $\mathbb{H}^3$ so that the dihedral angle along the edge dual to $e \in \operatorname{Edge} L$ is $\pi/m(e)$. When $m'(s_0, s_2) \neq 2$, W will be a cocompact reflection group on $\mathbb{H}^3$. So, if $\mathcal{C}'$ is a spherical building of type (W', S'), then the resulting building $\mathcal{C}$ $(= f^*\mathcal{C}')$ is of type (W, S) and has a piecewise hyperbolic, CAT(-1) metric on its standard realization so that each apartment is isometric to $\mathbb{H}^3$ (see Theorem 4.108). So, 3-colorable flag triangulations of S^2 yield many examples of hyperbolic buildings.

The Four Color Theorem enters the discussion here. Indeed, any triangulation of S^2 admits a 4-coloring $f : L \to \partial\Delta^3(S')$, where $S' = \{0, 1, 2, 3\}$ and by choosing (W', S') to be either a spherical or affine Coxeter group, one gets many examples of buildings $\mathcal{C}$ of type (W, S). Most of these also will be hyperbolic buildings (provided the triangulation contains no empty 4-circuit with each edge labeled by 2, cf. Example 4.16). Similar examples of hyperbolic buildings with fundamental chamber a hyperbolic polytope are explored by Haglund and Paulin in [141, 142].

Example 4.127 (dim L = 3 **and** $K(L)$ **Is a Simple** 4**-Polytope)** This is a continuation of Example 4.36 in Sect. 4.2.8. Suppose J is a flag triangulation of S^2 with vertex set T. Suspend J to get a triangulation L of S^3. Call the suspension vertices s_+ and s_- and let $S = T \cup \{s_+, s_-\}$. Then $K(L)$ is combinatorially equivalent to a convex polytope, namely, the prism on the simple polytope J^* dual to J. Let (W', S') be the spherical Coxeter system of rank 4 with Coxeter diagram $\mathbf{D}_4$. Label each edge of J by 2 and each edge of $L - J$ by 3. If J has no empty 4-cycles (for example J could be the boundary complex of an icosahedron), then L has no empty 4-cycles with each edge labeled 2. As in Example 4.36, if this is the case, the corresponding Coxeter system (W, S) is word hyperbolic. Suppose we can find a function $g : \operatorname{Vert}(J) \to \{s_0, s_1, s_2\}$ which is a 3-coloring of J. Extend this to a 4-coloring of L, call it $f : S \to S'$, by sending the suspension vertices s_+ and s_- to the central vertex s_3 of $\mathbf{D}_4$. Define the Coxeter group W to be the pullback of W' via f. Finally, let $\mathcal{C}'$ be a spherical building of type $\mathbf{D}_4$ and let $\mathcal{C} = f^*\mathcal{C}'$ be the pulled back building over $K(L)$. Although it may not be possible to 3-color J while simultaneously requiring it to have no empty 4-cycles, by using the method of simple complexes of groups discussed below, we can find a hyperbolic building isomorphic to $\mathcal{C}$ in which each of its spherical residues has type corresponding to a subdiagram of $\mathbf{D}_4$.

Example 4.128 (Branched Covers) We describe some further examples of type (a) in the list preceding Definition 4.121. Suppose L' is the nerve of a Coxeter system (W', S') and that $f : L \to L'$ is a covering space. The simplicial structure on L' lifts to a simplicial structure on L: a simplex of L is a component of the inverse image of a simplex of L'. We can pull back the labeling on Edge(L') to get a labeling on Edge(L) and hence, a Coxeter system (W, S) where $S = f^{-1}(S')$.

By construction, $L \leq L(W, S)$; an easy argument shows that $L(W, S) \leq L$ and hence, $L(W, S) = L$ (this uses the fact that each triangle in L projects to a triangle in L'). Since $K(L') = \text{Cone } L'$ and $K(L) = \text{Cone } L$, the natural projection $K(L) \to K(L')$ is a branched cover (branched at the cone point). The stratification of $K(L')$ indexed by $\mathcal{S}(W', S')$ lifts to one on $K(L)$ with the same index set. Furthermore, $K(L)$ also has a natural stratification indexed by $\mathcal{S}(W, S)$. As in the previous examples, $D(W, K(L))$ and $D(W', K(L'))$ are identified with the Davis–Moussong complexes $\Sigma(W, S)$ and $\Sigma(W', S')$; the natural map $D(W, K(L)) \to D(W', K(L))$ is the universal cover and $D(W, K(L)) \to D(W', K(L'))$ is a branched cover. Suppose $\mathcal{C}'$ is a building of type (W', S') with standard realization $D(W', K(L'))$. By Theorem 4.119, the universal cover of $D(\mathcal{C}', K(L))$ is the standard realization of a building $\mathcal{C}$ of type (W, S).

Lattices in Automorphism Groups We suppose $\mathcal{C}'$ is a building of type (W', S') and that $\mathcal{C}$ is the building of type (W, S) constructed in Theorem 4.119 via the pullback construction. If G' is an automorphism group of $\mathcal{C}'$, then $G' \curvearrowright D(\mathcal{C}', K(L)))$, where $K(L)$ is stratified as in (4.28). Let G be the group of all lifts of elements of G' to $D(\mathcal{C}, K(L))$. Then G is a group of automorphisms of $\mathcal{C}$ and we have a short exact sequence:

$$1 \to \pi \to G \to G' \to 1,$$

where $\pi = \pi_1(D(\mathcal{C}', K(L))$. (All this should be reminiscent of the construction in Sect. 3.1.2 of graph products using RABs.) If G' is chamber-transitive on $\mathcal{C}'$, then G is obviously chamber-transitive on $\mathcal{C}$. More generally, the sets of orbits $\mathcal{C}'/G'$ and $\mathcal{C}/G$ are naturally bijective with one another. If $\mathcal{C}'$ is locally finite and $G' \leq \text{Aut}(\mathcal{C}')$ is a lattice, then $G \leq \text{Aut}(\mathcal{C})$ is also a lattice.

The full automorphism group of a building constructed via a pullback is a locally compact group, usually not discrete and usually uncountable. By using ideas of Tits [217], it is proved in [141] that the full automorphism group is often a virtually simple group (as an abstract group).

Lemma 4.129 (cf. [83, Cor. 2.12]) *Suppose, as in the last two paragraphs of Sect. 4.4.5, that (W', S) is a spherical Coxeter system, that $f : L(W, S) \to \Delta(S)$ is inclusion of a subcomplex, that $\mathcal{C}'$ is a finite spherical building of type (W', S) and that $\mathcal{C} = f^*\mathcal{C}'$. Then $\pi = \pi_1(D(\mathcal{C}, K(L)))$ is a torsion-free uniform lattice in* Aut($\mathcal{C}$).

Proof Since $\mathcal{C}'$ is finite, the trivial subgroup is a lattice in Aut($\mathcal{C}'$); hence, its lift is a lattice in Aut($\mathcal{C}$). Since $\mathcal{C}$ is CAT(0), any finite subgroup of Aut($\mathcal{C}$) fixes a spherical residue and hence, projects isomorphically to a finite subgroup of Aut($\mathcal{C}'$). Hence, π is torsion-free.

4.4.7 Generalized Polygons

If $(W, \{s, t\})$ is a Coxeter system of rank two, then W is isomorphic to a dihedral group D_m (whose diagram is $\mathbf{I}_2(m)$) where m is the order of st. This subsection continues Example 4.93. A spherical building of rank two whose type is D_m is a called *generalized m-gon*. An equivalent definition is that a bipartite graph Ω is the geometric realization of a generalized m-gon if and only if its girth is $2m$ and its diameter is m. We assume Ω is regular so that its parameters $(q_s, q_t)_{s,t \in S}$ depend only on the types s and t of the vertices. If $m = 2$, then Ω is a complete bipartite graph. The valence of such a vertex is either $q_s + 1$ or $q_t + 1$. For example, if $m = 2$ and $(q_s, q_t) = (2, 2)$, then Ω is the utilities graph $K_{3,3}$. If the order m of st is odd, then $q_s = q_t$. The "free construction" of Tits shows that thick generalized m-gons exist for any $m \geq 2$. (See [198, Exercise 21, p. 37].) However, if we want Ω to be finite, then the following theorem imposes severe restrictions on m.

Theorem 4.130 (Feit–Higman [117]) *Suppose a finite generalized m-gon $\mathcal{C}$ is finite and thick. Then $m \in \{2, 3, 4, 6, 8\}$.*

If $m = 2$, then $\mathcal{C}$ is a RAB; so, the parameters q_s and q_t can be arbitrary positive integers (e.g., see Sect. 4.4.3). When m is odd, s and t are conjugate in the dihedral groups and we must have $q_s = q_t$. For $m > 2$ there are additional heavy restrictions on q_s and q_t. For example there is the following result of Feit–Higman (cf. [82, Thm. 18.3] or [198, Thm. 3.4]).

Theorem 4.131 (Feit–Higman [117]) *Suppose $\mathcal{C}$ is a finite, thick, generalized m-gon, with $m \in \{3, 4, 6, 8\}$. Then there are the following restrictions on the parameters q_s and q_t:*

$$\begin{cases} \text{for } m = 4, & \frac{q_s q_t (q_s q_t + 1)}{q_s + q_t} \in \mathbb{Z}, \\ \text{for } m = 6, & q_s q_t \text{ is a perfect square,} \\ \text{for } m = 8, & 2 q_s q_t \text{ is a perfect square.} \end{cases}$$

Moreover, for $m = 4$ or 8, $q_s \leq q_t^2$ and $q_t \leq q_s^2$, while for $m = 6$, $q_s \leq q_t^3$ and $q_t \leq q_s^3$.

The finite, thick generalized m-gons which correspond to spherical buildings over finite fields are precisely the "Mouffang polygons." (See[225].)

4.4.8 The Basic Algebraic Examples

In this subsection we consider the main examples of buildings associated to algebraic groups and Kac–Moody groups. Surveys of this material can be found in [115], [196], [142]. These examples can be used as inputs for the pullback construction in the previous subsections. In each case the building will have a chamber-transitive automorphism group.

Irreducible Spherical Buildings of Rank ≥ 3 In the beginning of Sect. 4.4, we discussed the result of Tits [218] on the classification of thick spherical buildings. Roughly, this says that if $\mathcal{C}$ is an irreducible spherical building of rank at least 3, then $\mathcal{C}$ is isomorphic to the building associated to an algebraic group over a field $\mathbb{F}$. Here we are mainly interested in the case where $\mathbb{F} = \mathbb{F}_q$ is a finite field of order q. (This implies that the classical realization of the building is a finite simplicial complex.) The Coxeter diagrams $\mathbf{A}_n$, $\mathbf{C}_n$, $\mathbf{D}_n$ correspond to classical groups $GL_n(q)$, $O_n(q)$, $Sp_{2n}(q)$, where the q means the field is $\mathbb{F}_q$. There also are twisted versions of such groups which go by symbols such as ${}^2\mathbf{A}_n(q)$ $(= U_{2n}(q)$ or $U_{2n}(q))$ and ${}^2\mathbf{D}_{n+1}(q)$ $(= O^-_{2n+2})$. The quotient of such a group by its center turns out to be a simple group. These "simple groups of Lie type" play a dominant role in the theory of finite simple groups (see Carter [53]). There are also versions of such algebraic groups corresponding to remaining irreducible spherical Coxeter diagrams of rank ≥ 3 other than $\mathbf{H}_3$ and $\mathbf{H}_4$. The construction of algebraic groups over $\mathbb{F}_q$ for the diagrams, $\mathbf{G}_2$, $\mathbf{F}_4$, $\mathbf{E}_6$, $\mathbf{E}_7$, $\mathbf{E}_8$ does not follow automatically from the theory of simple complex Lie algebras. Rather it depends on a result of Serre which states that the structure constants for these Lie algebras can be taken to be integers. It follows that the Lie algebras can be defined over $\mathbb{F}_q$. Descriptions of the spherical buildings over finite fields can be found in Ronan's book [198, Ch. 8] and lists in [198, appendix 6] or [122, Table1]. (See Fig. 4.2 for a picture of the projective plane corresponding to $GL_3(\mathbb{F}_q)$.)

As we shall see below, Serre's result leads to Kac–Moody Lie algebras, as well as to Kac–Moody groups over $\mathbb{F}_q$ (the analogs of the Chevalley groups) and eventually to Kac–Moody buildings.

Affine Buildings To say that $\mathcal{C}$ is an *affine building* means that its type is a euclidean Coxeter system. If (W, S) is an irreducible euclidean Coxeter system, then $\mathcal{S}(W, S)^{\mathrm{op}}$ can be identified with the face poset of the boundary complex of a simplex. Hence, the fundamental chamber K is a simplex. It follows that when the euclidean Coxeter system is irreducible, the standard realization of a building of that type is a simplicial complex. For example, any tree without terminal vertices is the realization of an affine building of type $(D_\infty, \{s, t\})$ (i.e., its type is $\tilde{\mathbf{A}}_1$). Affine buildings are often associated to a field with valuation, such as $\mathbb{Q}_p$ (the p-adic completion of the rationals) or $\mathbb{F}_q((t))$ (Laurent series over $\mathbb{F}_q$). If the residue field is finite, then the building will be locally finite. Next, we describe the classical example of an affine building.

Example 4.132 (The Building $\tilde{\mathbf{A}}_n(\mathbb{F}, v)$ (cf. [198, §9.2])) Suppose $\mathbb{F}$ is a field with a discrete valuation $v : \mathbb{F}^* \to \mathbb{Z}$, that $A = \{x \in \mathbb{F} \mid v(x) \geq 0\}$ is the *ring of integers*, that $\pi = \{x \in \mathbb{F} \mid v(x) \geq 1\}$ is the maximal ideal of A and that $k = A/\pi$ is the residue field. Let V be an $(n+1)$-dimensional $\mathbb{F}$-vector space. A *v-lattice* $\Lambda < V$ is a free A-submodule which generates V as a vector space. Two lattices are *equivalent* if one is a multiple of the other by a power of π. The equivalence class of Λ is denoted $[\Lambda]$. The standard realization of the building $\tilde{\mathbf{A}}_n(\mathbb{F}, v)$ will be an n-dimensional simplicial flag complex. Its vertices are the equivalence class of lattices. Equivalence classes $[\Lambda]$ and $[\Lambda']$ are connected by an edge if for some $\Lambda \in [\Lambda]$ and $\Lambda' \in [\Lambda']$ we have $\pi\Lambda < \Lambda' < \Lambda$. The standard realization of $\tilde{\mathbf{A}}_n(\mathbb{F}, v)$ is the order complex of this poset. (Hence, it is a flag complex.) A chamber C is a top-dimensional simplex, that is, $C = \{[\Lambda_0], \dots, [\Lambda_n]\}$, where we can choose representatives so that $\pi\Lambda_n < \Lambda_0 < \cdots < \Lambda_n$. This yields a building of type $\tilde{\mathbf{A}}_n$. The group $SL(V)$ $(\cong SL_{n+1}(\mathbb{F}))$ acts on it and the action is chamber-transitive. The case of SL_2, when the building is a tree, is one of the principal topics of Serre's book [207]. Suppose $C = \{[\Lambda_0], \dots, [\Lambda_n]\}$ is a chamber in $\tilde{\mathbf{A}}_n(\mathbb{F}, v)$ where $\pi\Lambda_n < \Lambda_0 < \cdots < \Lambda_n$. Then $\overline{V} = \Lambda_n/\pi\Lambda_n$ is an $(n+1)$-dimensional vector space over k and each $\Lambda_i/\pi\Lambda_n$ is a vector subspace. Moreover, $\Lambda_0/\pi\Lambda_n < \cdots < \Lambda_n/\pi\Lambda_n$ is a chamber in the spherical building for $SL(\overline{V})$, i.e., in the projective space discussed in Example 4.99. One can then choose a basis $e_1, \dots, e_n$ for V so that $\Lambda_n = \langle e_0, \dots, e_n\rangle A$, $\Lambda_{n-1} = \langle \pi e_0, e_1 \dots, e_n\rangle A$, $\Lambda_0 = \langle \pi e_0, \dots, \pi e_{n-1}, e_n\rangle A$. Any codimension-one face of $\{[\Lambda_0], \dots, [\Lambda_n]\}$ is obtained by omitting one of the $[\Lambda_i]$. The link of such a codimension one face is a projective line over k. So, if $k = \mathbb{F}_q$, then each parameter of $\tilde{\mathbf{A}}_n(\mathbb{F}, v)$ is q. In general, the link of an $(n-i)$-simplex is the classical realization of a spherical residue of rank i. It follows that the local isomorphism type of $\tilde{\mathbf{A}}_n(\mathbb{F}, v)$ is determined by the residue field k. For example, $\mathbb{Q}_p$ and $\mathbb{F}_p((t))$ have the same residue field, $\mathbb{F}_p$, and hence, are locally isomorphic. Using the fact that the fundamental chamber is a simplex, one can then show that the buildings $\tilde{\mathbf{A}}_n(\mathbb{Q}_p)$ and $\tilde{\mathbf{A}}_n(\mathbb{F}_p((t)))$ are isomorphic. For example, $\tilde{\mathbf{A}}_1(\mathbb{Q}_p)$ and $\tilde{\mathbf{A}}_1(\mathbb{F}_p((t)))$ are both regular trees of degree $p+1$. However, their natural chamber-transitive automorphism groups $SL_{n+1}(\mathbb{Q}_p)$ and $SL_{n+1}(\mathbb{F}_p((t)))$ are completely different.

Kac–Moody Buildings In order to construct a Kac–Moody group or a Kac–Moody building from a Coxeter matrix $(m(s,t))$ and a finite field $\mathbb{F}_q$, one first needs a "generalized Cartan matrix" (A_{st}). This means an $(S\times S)$-matrix with integer entries such that $A_{ss} = 2$ and $A_{st} \le 0$ for $s \ne t$. Furthermore, for $m(s,t) = \infty$, $A_{st} \le -2$, while for $m(s,t) \ne \infty$, $A_{st}A_{ts} = 4\cos^2 \frac{\pi}{m(s,t)}$ (cf. (4.15) in Sect. 4.2.9). The integrality condition then forces:

$$m(s,t) \in \{2, 3, 4, 6, \infty\}, \quad \text{for } s \ne t. \tag{4.29}$$

Kac [161] and Moody then use a generalized Cartan matrix together with Serre's result first to define a root system and then an infinite dimensional Lie algebra over $\mathbb{Z}$ (called a "Kac–Moody algebra"). Finally, in analogy with the construction of Chevalley groups, the difficult construction of Kac–Moody groups and their buildings was accomplished by Tits [223, 224] in the 1980's. The condition in (4.29) rules out certain Coxeter systems from occurring as the type of a Kac–Moody building, since in the labeled graph (L^1, m) no edge can be labeled by 5 or by an integer ≥ 7. Nevertheless, whenever the Coxeter matrix $(m(s,t))$ satisfies (4.29), one can find a corresponding generalized Cartan matrix (A_{st}). Thus, Kac–Moody buildings provide a method for constructing locally finite building whose type can be a fairly arbitrary Coxeter systems (W, S). Other references include [194–196].

Suppose $\mathcal{C}$ is a Kac–Moody building over a finite field and that G is the corresponding Kac–Moody group. Some affine buildings are Kac–Moody buildings. In particular, if $\mathcal{C}$ is an affine building defined by an algebraic group over the field $\mathbb{F}_q((t))$ with discrete valuation, then the algebraic group as well as the building are Kac–Moody. With its natural topology, the automorphism group G of a Kac–Moody building usually will not be discrete. In fact, if $\mathcal{C}$ is not affine, then *there is no uniform lattice in* G. On the other hand, the theory of twin buildings provides us with an embedding $G \hookrightarrow G \times G$ and an action $G \curvearrowright \mathcal{C} \times \mathcal{C}$ so that the image of G is an irreducible (non-uniform) lattice in $G \times G$. A high point in this area is the theorem of Caprace and Rémy in [52] that when the Kac–Moody building is irreducible and not affine, then there is a finite index subgroup of G which, when divided by its center, is simple (as an abstract group). So, there is a close analogy with the theory of finite Chevalley groups: irreducible spherical buildings lead to finite simple groups while irreducible, non-affine, Kac–Moody buildings yield infinite simple groups. The structure of the proof in [52] is similar to that of the main theorem in Burger–Mozes [43, 45] which states that simple groups occur as irreducible lattices in automorphism groups of the product of two trees (cf. Sect. 2.4.4).

Remark 4.133 It is observed in [52] that some of these Kac–Moody groups G have Kazhdan's Property (T). Hence, there exist infinite simple groups with Property (T).

Part IV
More on NPC Cube Complexes

Chapter 5
General Theory of Cube Complexes

Two important concepts are explained in this chapter. First, there is Sageev's definition [202] of a "pocset" or "abstract half-space system" and his result explaining how one can construct a CAT(0) cube complex from such a half-space system. (See also [181, 182, 197].) This construction was first used by Sageev [200] to prove that given a group G and a collection of "sufficiently deep codimension-one subgroups" one can construct a CAT(0) cube complex with G-action. The hyperplanes in this CAT(0) cube complex correspond to cosets of the codimension-one subgroups. The quotient of the CAT(0) cube complex by G is an NPC cube complex. This method undergirds many of the constructions of NPC cube complexes. The second important concept is the notion of Haglund–Wise [143] of what it means for an NPC cube complex to be "special." This definition rules out the occurrence of certain configurations of hyperplanes; for example, hyperplanes are not allowed to have self-intersections. Classifying spaces of RAAGs (or commutator subgroups of RACGs) have a universal property for special cube complexes: if B is an NPC special cube complex, then there is a locally isometric immersion of B into the classifying space of some RAAG, A_L, so, that $\pi_1(B)$ becomes a subgroup of A_L. (Here L^1 is the incidence graph of the set of hyperplanes in B.) The importance of this result is that RAAGs have strong separability properties which descend to subgroups. For example, any RAAG is residually finite and any word-quasiconvex subgroup of a RAAG is a virtual retract. The relevance of this to hyperbolic 3-manifold groups comes from work of Kahn–Marković (cf. Theorem 5.20), which shows that any such group contains many nice codimension-one subgroups (i.e., surface subgroups). After using Sageev's construction, it follows that any hyperbolic 3-manifold group is "virtually special," i.e., has a finite index subgroup which is the fundamental group of a compact special cube complex (cf. Theorem 5.21). These facts allowed Agol [3] to prove two of Thurston's conjectures about hyperbolic 3-manifolds: the Virtual Haken Conjecture and the Virtual Fibering Conjecture. These conjectures are discussed in Sect. 5.3.5.

M. W. Davis, *Infinite Group Actions on Polyhedra*, Ergebnisse der Mathematik
und ihrer Grenzgebiete. 3. Folge / A Series of Modern Surveys in Mathematics 77,
https://doi.org/10.1007/978-3-031-48443-8_5

Here is some more background. The fundamental groups of many closed hyperbolic n-manfolds can act freely and cocompactly on CAT(0) cube complexes. For $n = 3$, after passing to a subgroup of finite index, this was proved for all hyperbolic 3-manifold groups by Bergeron–Wise by using the Kahn–Marković result. (See Theorem 5.21.) For $n > 3$, it holds for certain arithmetic holds n-manifolds by a result of Bergeron-Haglund-Wise (cf. Theorem 5.24). In Sect. 5.3 we prove Haglund's result that word-quasiconvex subgroups of RACGs are virtual retracts.

We use the words "half-space" and "hyperplane" in several different ways in this chapter. First these are geometric notions that make sense in any CAT(0) cube complex. In this case we often confuse an open half-space with the set of vertices that lie in it. Secondly, there is the abstract notion of a "half-space system" or "wall set" (or "wall space"). The most abstract version is just a certain type of poset with an involution called "complementation." The poset with this additional structure is a "pocset." In these abstract versions a "hyperplane" is just a pair consisting of a half-space and its complement. In Sect. 5.1 we define the geometric notions of "hyperplanes" and "half-spaces" in CAT(0) cube complexes.

5.1 Abstract CAT(0) Cube Complexes: Half-Space Systems

In Sect. 5.1.1 we define geometric hyperplanes and half-spaces in a CAT(0) cube complex. In Sect. 5.1.2 we give abstract versions of these notions in terms of a "half-space systems" or "pocsets." Then we show how to recover a CAT(0) cube complex from a half-space system. In Sect. 5.1.3 we give conditions to insure that a group action on a CAT(0) cube complex will be proper and have compact quotient space. In Sect. 5.1.4 we discuss results of Bergeron and Wise for cubulating certain hyperbolic n-manifolds.

5.1.1 Hyperplanes in Cube Complexes

A *midcube* in $[-1, 1]^n$ is its intersection with a coordinate hyperplane $x_i = 0$ for some i with $1 \leq i \leq n$. The midcube $\{x_i = 0\}$ intersects all edges of $[-1, 1]^n$ that are parallel to the basis vector $\mathbf{e}_i \in \mathbb{R}^n$.

Throughout this section X denotes a cube complex; it is not assumed to be NPC. Two parallel edges of a square face in the cube complex X are said to be *square equivalent*. We extend this to an equivalence relation on all of Edge X.

Definition 5.1 A *hyperplane* in X dual to an equivalence class of edges is the union of all midcubes in X that intersect only the edges in the given equivalence class.

A more abstract definition simply would be to say that a *hyperplane* is an equivalence class of edges.

A hyperplane is *embedded* if it intersects each cube in at most one midcube. In other words, the hyperplane is embedded if it does not intersect any square in two adjacent edges. A hyperplane might have self-intersections. If so, this can remedied by "resolving" the hyperplane H to a cube complex $\hat{H}$, where $\hat{H}$ is formed from the disjoint union of all midcubes in H by gluing together two such whenever they share a common face. The natural map $\hat{H} \to H$ is then an immersion.

Each midcube in a cube has a normal bundle which is a trivial line bundle. These normal bundles glue together to give a normal bundle for each resolved hyperplane in the cube complex X. The hyperplane is *two-sided* if its normal bundle is trivial; otherwise, it is *one-sided*. Each hyperplane has an open tubular neighborhood in X which can be identified with the image an open interval bundle over $\hat{H}$.

In the next theorem we list some properties of (geometric) hyperplanes that are easy to verify in the case of a CAT(0) cube complex.

Theorem 5.2 (See Sageev [202] or Niblo–Reeves [182]) *Suppose Y is a* CAT(0) *cube complex. Then*

(i) *Each hyperplane is an embedded convex subset.*
(ii) *Each hyperplane has the structure of a* CAT(0) *cube complex.*
(iii) *Each hyperplane separates Y into two components.*

A (geometric) *half-space* in a CAT(0) cube complex Y bounded by a hyperplane H is one of the two components of $Y - H$.

Example 5.3 (Hyperplanes in $\mathbb{T}_L$ and P_L) . Consider the cube complexes of Sect. 3.1.1. Let L be a simplicial complex with vertex set I. Regard S^1 as the interval $[0, 1]$ with end points identified, i.e., S^1 is a 1-dimensional cube complex. We see that $(S^1)^I$ is a cube complex whose opposite face poset is bijective with the power set of I. A hyperplane of $(S^1)^I$ is a subtorus of the form $\{x_i = 1/2\}$. Suppose $\mathbb{T}_L$ is the polyhedral product of copies of S^1 defined by (3.5) (i.e., if $\mathbb{T}_L$ is the standard classifying space of the RAAG, A_L, then its hyperplane corresponding to $i \in I$ is defined by $\mathbb{T}_L \cap \{x_i = 1/2\}$. Note that this hyperplane does not separate $\mathbb{T}_L$. However, a lift of such a hyperplane does separate the universal cover $\widetilde{\mathbb{T}}_L$. (In fact, such hyperplanes separate in the 2^I-fold cover of $\mathbb{T}_L$ corresponding to the natural homomorphism $\pi_1(\mathbb{T}_L) \to (\mathbb{Z}/2)^I$. When L is a flag complex, this $(\mathbb{Z}/2)^I$-fold cover is the "pure Salvetti complex" of the RAAG, A_L.)

Let P_L be the subcomplex of $[-1, 1]^I$ defined in (3.3). Let $r_i \in (\mathbf{C}_2)^I$ be the reflection across the midcube $\{x_i = 0\}$ of $[-1, 1]^I$. Basically, the set of edges of $[-1, 1]^I$ that are orthogonal to $\{x_i = 0\}$ form a single parallel class, namely, the edge parallel to e_i; however, these edges might lie in several different square equivalence classes. In other words, although the intersection $\{x_i = 0\} \cap P_L$ is a union of midcubes of P_L, this intersection might not be connected. In this context we call $\{x_i = 0\} \cap P_L$ a "multiple hyperplane" of P_L. Each component of $\{x_i = 0\} \cap P_L$ is an actual hyperplane of P_L. Of course, such a hyperplane is embedded. For example, if L has no edges, then P_L is equal to the 1-skeleton of the n-cube $[-1, 1]^I$ and $\{x_i = 0\} \cap P_L$ consists of 2^{n-1} midpoints of edges. Since $\{0\}$ separates $[-1, 1]$, each hyperplane $\{x_i = 0\}$ separates the cube $[-1, 1]^I$ and hence, each multiple hyperplane separates P_L. Similarly, each hyperplane of the universal cover $\tilde{P}_L$ separates. So, properties (i) and (iii) of Theorem 5.2 hold for $\tilde{P}_L$. If L is a flag complex, then $\tilde{P}_L$ is CAT(0) and (ii) also holds.

Suppose Y is a CAT(0) cube complex with vertex set V. Each hyperplane of Y partitions V into two subsets h and h^c called *half-spaces*. Any vertex $v \in V$ is the intersection of all half-spaces h that contain it; moreover, for each vertex v and each partition $\{h, h^c\}$ of V into two half-spaces, v lies in exactly one of the two half-spaces. This suggests that the information in a CAT(0) cube complex can be encoded by a "system of half-spaces," as in the next subsection. This is explained by Niblo–Reeves in [183] based on earlier work of Roller [197] and Sageev [200, 202]. When Y is a tree, this theory was described earlier by Dunwoody [110]. There is also an equivalent formulation in terms of "walls" and "wall sets" due to Haglund–Paulin in [141]. These abstract notions are described in the next subsection.

5.1.2 Half-Space Systems and Sageev's Construction

Suppose Y is a CAT(0) cube complex with vertex set V. Each hyperplane of Y partitions V into two subsets h and h^c called *half-spaces*. Any vertex $v \in V$ is the intersection of all half-spaces h that contain it; moreover, for each vertex v and each partition $\{h, h^c\}$ of V into two half-spaces, v lies in exactly one of the two half-spaces. This suggests that the information in a CAT(0) cube complex can be encoded by a "system of half-spaces." This is explained by Niblo–Reeves in [182, 183] based on earlier work of Roller [197] and Sageev [200, 202]. When Y is a tree, this theory was described earlier by Dunwoody[110]. There is also an equivalent formulation in terms of "walls" and "wall sets" due to Haglund–Paulin in [141].

Here is the definition from [182]. The data for a *system of abstract half-spaces* consist of a triple $(\mathcal{H}, <, \iota)$ where $(\mathcal{H}, <)$ is a poset and ι is an order reversing involution of $\mathcal{H}$ denoted by $h \mapsto h^c$. The triple must also satisfy the following two conditions:

(1) For any two elements $h_1, h_2 \in \mathcal{H}$, there are only finitely many k such that $h_1 < k < h_2$.
(2) For any two distinct elements h, k of $\mathcal{H}$ at most one of the following holds:

$$h < k, \quad h < k^c, \quad h^c < k, \quad h^c < k^c.$$

A triple $(\mathcal{H}, <, \iota)$ satisfying (1) and (2) is a *half-space system*. (Following Roller [197], Sageev [202] calls such a triple $(\mathcal{H}, <, \iota)$ a "pocset," which stands for a "partialy ordered set with complementation"). Two half-spaces are *nested* if one of the four conditions in (2) holds and they are *transverse* if none of the conditions hold. (We are thinking of $\{h, h^c\}$ as the vertex sets of a complementary pair of geometric open half-spaces in a CAT(0) cube complex separated by a common bounding hyperplane $\overline{h}$. It makes sense to identify $\overline{h}$ with the pair $\{h, h^c\}$.) In other words, $\overline{h}$ can be identified with the orbit of a half-space h under the involution ι. Let $\overline{\mathcal{H}}$ denote the set of such orbits and let $\partial : \mathcal{H} \to \overline{\mathcal{H}}$ be the projection $h \mapsto \overline{h}$.

A more concrete version of a half-space system is given in [141] by using the equivalent notion of a "wall set." (See also [62, 151, 202, 234].) The definition is basically the same except that each half-space in $\mathcal{H}$ is required to be a subset of an underlying set Ω. A *wall* is then a partition of Ω into two subsets. A *wall set* is a pair $(\Omega, \mathcal{W})$ where $\mathcal{W}$ is a collection of walls. Two points of Ω are required to separated by finite, nonzero number of walls. One recovers the pocset $(\mathcal{H}, <, \iota)$ by defining the half-space determined by a wall to be one of the two elements h of the partition; the partial order is inclusion and the order-reversing involution is defined by $h^c = \Omega - h$.

A *vertex* v of half-space system $(\mathcal{H}, <, \iota)$ is a section of ∂, i.e., a function $v : \overline{\mathcal{H}} \to \mathcal{H}$ such that $\partial \circ v = \mathbb{I}$ and such that the following additional condition is satisfied: for distinct hyperplanes $\overline{h}$ and $\overline{k}$, $v(\overline{h}) \not< v(\overline{k})$. The vertex v *lies in a half-space* h, written as $v \in h$, if $v(\overline{h}) = h$. Sageev [202] uses the term of "ultrafilter" rather than "vertex." (Thus, an ultrafilter ω is a collection of half-spaces so that (a) if $\overline{h} = \{h, h^c\}$ is a hyperplane, then exactly one of h, h^c is in ω and (b) if $h \in \omega$ and $h < h'$, then $h' \in \omega$. So, given a wall set $(\Omega, \mathcal{W})$, an ultrafilter ω on $\mathcal{W}$ is a collection of half-spaces so that (a) and (b) hold (cf. [21, p. 58]). Each point $C \in \Omega$ determines a *principal ultrafilter* ω_C defined by the condition that $h \in \omega_C$ if and only if C lies in h. An ultrafilter ω is a *DCC ultrafilter* (or a *DCC vertex*) if it satisfies the *d*escending *c*hain *c*ondition. For example, consider the cubulation of the real line $\mathbb{R}$ by intervals of the form $[n, n+1]$. The hyperplanes are midpoints of edges of the form $n + \frac{1}{2}$. A half-space corresponding to such a midpoint $n + \frac{1}{2}$ is a component of the complement, $\mathbb{R} - \{n + \frac{1}{2}\}$, i.e., either $(-\infty, n + \frac{1}{2})$ (pointing to the left) or $(n + \frac{1}{2}, +\infty)$ (pointing to the right). The principal ultrafilter corresponding to a midpoint consists of all the half-spaces which contain it. There also are ultrafilters which are not DCC, say the ultrafilter consisting of all half-spaces pointing to the left or all half-spaces pointing to the right. These non-DCC ultrafilters should be thought of as boundary points (namely, $\pm\infty$) of the corresponding cubical structure on $\mathbb{R}$.

Example 5.4 (Half-Spaces in Coxeter Systems) Suppose (W, S) is a Coxeter system as defined in the beginning of Chap. 4. (This means, in particular, that S is a set of involutions which generate W.) Let

$$R = \{r \in W \mid r = wsw^{-1}, \text{ for some } w \in W \text{ and } s \in S\}. \tag{5.1}$$

An element of R is a *reflection*. One definition of a Coxeter system is that for each $r \in R$ the fix set of r separates $\mathrm{Cay}(W, S)$ (see [82, Chapter 3]). In other words, each reflection r defines a partition of W into two subsets h^r_+ and h^r_- where

$$h^r_+ = \{w \in W \mid l(w) < l(rw)\} \quad \text{and} \quad h^r_- = \{w \in W \mid l(w) > l(rw)\}. \tag{5.2}$$

So, we get a wall set $(\Omega, \mathcal{W})$ where $\Omega = W$ and $\mathcal{W} = R$ as in (5.1). The principal ultrafilters correspond to the vertices of $\text{Cay}(W, S)$, i.e., to W. The principal ultrafilter corresponding to the element $1 \in W$ is defined by $\omega_1 = h^r_+$. Not every vertex corresponds to a principal ultrafilter. A vertex which corresponds to a non-principal ultrafilter is a *phantom vertex*. For example, if $W = D_3$, the dihedral group of order 6, then there are 3 reflections and hence, 8 ($= 2^3$) vertices. Six vertices correspond to elements of W and the other two are phantom vertices.

Next we indicate how to recover a CAT(0) cube complex X from a half-space system. Heuristically, for each wall (i.e., hyperplane) $\overline{h}$, a vertex of a CAT(0) cube complex determines a choice of half-space bounded by $\overline{h}$, namely, the half-space that contains the vertex. For any two vertices v, w, the set $H(v, w)$ of *hyperplanes which separate them* is defined by $H(v, w) = \{\overline{h} \mid v(\overline{h}) \neq w(\overline{h})\}$. The vertices v, w are *adjacent* if $\text{Card}(H(v, w)) = 1$ in which case they are connected by an edge. This defines a graph Γ. Although this Γ might not be connected, it turns out that each of its connected components is the 1-skeleton of a CAT(0) cube complex. If v lies in h, then one can define a new section u of ∂ by putting $u(\overline{h}) = h^c$ and not changing its value on any other hyperplane. If v also lies in k and $k < h$, then $u(\overline{k}) = k < h = u(\overline{h})^c$; hence, u is not a vertex. On the other hand, if h is minimal among all half-spaces containing v, then the section u obtained by switching h with h^c is a vertex. If h_1, h_2 are a pair of minimal half-spaces both containing v, then we can switch them both simultaneously if and only if they are transverse. In this way we obtain a 4-circuit in Γ which we can then fill in with a square. More generally, if$h_1, h_2, \dots, h_n$ is a family of mutually transverse half-spaces each of which is minimal among the half-spaces containing v, then switching gives us an embedded copy of the 1-skeleton of an n-cube in Γ. Filling in all the n-cubes obtained in this way, we get a cube complex X.

Theorem 5.5 (Sageev [200, 201]) *Each component of the cube complex described above is* CAT(0).

So, vertices corresponding to DCC ultrafilters determine a path connected, CAT(0) cube complex X. The other ultrafilters, which are not DCC, lie in a different path components; they should be thought of as boundary points of X.

In terms of the wall set $(\Omega, \mathcal{W})$ there is an n-cube in X for each collection $\{\overline{h}_1, \dots, \overline{h}_n\}$ of n mutually transverse walls in $\mathcal{W}$.

The hyperplanes of a CAT(0) cube complex Y give partitions of its vertex set V into half-spaces. This defines a wall set $(V, \mathcal{W})$ (where $\mathcal{W}$ is the set of hyperplanes), as well as a half-space system $(\mathcal{H}, <, \iota)$. The resulting CAT(0) cube complex X $(= X(\mathcal{H}, <, \iota))$ is naturally identified with Y. There are no phantom vertices.

Example 5.6 (Codimension-One Submanifolds) We describe a classical situation where the cube complex is a tree. Suppose that M^{n+1} is an aspherical manifold and that $N^n < M^{n+1}$ is a π_1-injective submanifold of codimension one. Let $\tilde{M}$ be the universal cover of M and $\tilde{N}$ a component of the preimage of N in $\tilde{M}$. A translate of $\tilde{N}$ by an element of G $(= \pi_1(M))$ is called a *wall*. Each wall separates $\tilde{M}$, i.e., the complement of a wall consists of two open "half-spaces." Since the walls do

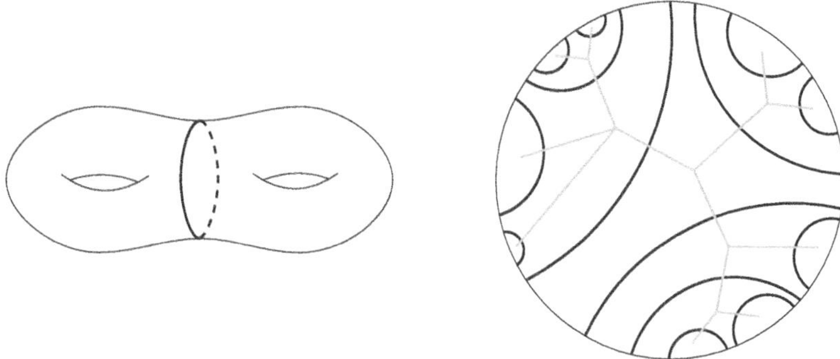

Fig. 5.1 Separating curve in a surface, its universal cover and the dual tree

not intersect, half-spaces cannot be transverse and we get a half-space system. The complement of the union of walls is a disjoint union of connected open regions, each of which is an intersection of half-spaces. The resulting cube complex X is 1-dimensional. The vertices of X correspond to the regions and each edge corresponds to the intersection of two regions, i.e., to a copy of $\tilde{N}$. The fact that $\tilde{M}$ is contractible implies that X is a tree. (See Fig. 5.1, which is copied from [21, Figure 6, p. 57].)

We can also get a half-space system by starting with an immersed codimension-one submanifold $f : N^n \looparrowright M^{n+1}$. Assume that f is a π_1-injective immersion and that the self-intersections are in general position. Also assume that the lift of f to the universal cover, $\tilde{N} \to \tilde{M}$, is an embedded submanifold (since $\tilde{M}$ is simply connected, $\tilde{N}$ necessarily separates $\tilde{M}$ into two "half-spaces." By considering all such lifts we get a family of codimension-one submanifolds intersecting transversely, a half-space system and then, a CAT(0) cube complex X. A group theoretic version of this will be is explained following Definition 5.7 below. An example of this is the important theorem of Kahn–Marković which says that every compact hyperbolic 3-manifold has (many) such π_1-injective immersed surfaces. (See Theorem 5.20 in Sect. 5.1.4 below.)

It follows from the work of Kahn–Marković [162] that the fundamental group π of any compact hyperbolic 3-manifold can act on a CAT(0) cube complex. Further arguments show that there are enough surface subgroups to find a π-action that is proper and cocompact. We will say more about this in Theorem 5.21 of Sect. 5.1.3, below.

Definition 5.7 A *codimension-one subgroup* of a finitely generated group G is a subgroup $H < G$ so that the quotient of the Cayley graph of G by H has at least two ends.

For example, if N is an immersed codimension-one submanifold of M as in Example 5.6, then $H = \pi_1(N)$ is a codimension-one subgroup of $G = \pi_1(M)$.

Example 5.8 (Codimension-One Subgroups) The construction in Example 5.6 of a CAT(0) cube complex associated to an immersed codimension-one submanifold can be extended to the construction of a CAT(0) cube complex for any codimension-one subgroup H of a group G (or more generally, a collection of codimension-one subgroups). In terms of the Cayley graph, to say that H is a codimension-one subgroup of G means that H separates $\mathrm{Cay}(G, S)$ into two "deep components," i.e., there is a number R such that the complement of an R-neighborhood of H in G has two components. Choose one of these components h and let h^c be its complement in G. (This is possible provided $\mathrm{Cay}(G, S)$ has extendible geodesic rays.) The translate of the partition $\{h, h^c\}$ by an element $g \in G$ is another partition of G corresponding to a complement of a neighborhood of gHg^{-1}. This gives G the structure of a wall set $(\Omega, \mathcal{W})$, where $\Omega = G$ and the set of walls $\mathcal{W}$ consists of all partitions $\{gh, gh^c\}$, where g ranges over a set of coset representatives for G/H. The resulting CAT(0) cube complex is denoted $X(G; H)$ (or more generally, $X(G; H_1, \dots, H_k)$ when H is replaced by a finite collection of codimension-one subgroups $\{H_1, \dots, H_k\}$.

Example 5.9 (Hyperplane Arrangements, Fans, Zonotopes, cf. Sects. 3.3.1, 4.2.2 or [235, Chapter 7]) Suppose $\mathcal{A} = \{H_1, \dots, H_n\}$ is an arrangement of finitely many linear hyperplanes in $\mathbb{R}^d$. These hyperplanes decompose $\mathbb{R}^d$ into a collection of polyhedral cones with cone points at the origin. This decomposition is the *fan* associated to $\mathcal{A}$, denoted $\mathrm{Fan}(\mathcal{A})$. A top-dimensional face in $\mathrm{Fan}(\mathcal{A})$ is a *chamber*. Each hyperplane $H \in \mathcal{A}$ separates $\mathbb{R}^d$ into two linear half-spaces h and h^c. Identify each hyperplane with the corresponding unordered pair of half-spaces $H = \{h, h^c\}$. In this way, $\mathcal{A}$ defines a half-space system $(\mathcal{H}(\mathcal{A}), <, \iota)$ in the above sense or equivalently, a wall set $(\Omega, \mathcal{W})$ with $\mathcal{W} = \mathcal{A}$ and Ω the set of chambers in $\mathrm{Fan}(\mathcal{A})$. A choice of chamber $C \in \Omega$ gives a principal ultrafilter ω_C on $\mathcal{A}$ defined as follows: if $H \in \mathcal{A}$ corresponds to a pair of half-spaces $H = \{h, h^c\}$, then $h \in \omega_C \iff C < h$. One says that h is a *positive half-space* with respect to C and that C is on the *positive side* of H. So, we can think of ω_C as a function $\mathcal{A} \to \{\pm 1\}$ defined by

$$\omega_C(H_i) = \begin{cases} +1, & \text{if } C \text{ is on the positive side of } H_i; \\ -1, & \text{if } C \text{ is on the other side of } H_i. \end{cases}$$

Any subset of $\mathcal{A}$ gives a collection of pairwise transverse hyperplanes; hence, any function $\mathcal{A} \to \{\pm 1\}$ (i.e., any element of $\{\pm 1\}^n$) defines a vertex in the corresponding CAT(0) cube complex X. It follows that X is the standard n-cube $[-1, 1]^n$. (My thanks to Katherine Goldman for explaining this result to me.) The poset of cones in $\mathrm{Fan}(\mathcal{A})$ is dual to the poset of faces of a convex polytope $Z(\mathcal{A})$ $(= Z)$ called the *zonotope* associated to $\mathcal{A}$. (See Sect. 3.3.1.) This polytope lies in the dual vector space $(\mathbb{R}^d)^*$. If $\mathcal{A}$ is essential, then $\dim Z = d$. (The arrangement is *essential* if the intersection of all the H_i is a single point, namely, the origin.) Here are two equivalent definitions of the zonotope Z (cf. [235, Definition 7.13]).

(a) A *zonotope* is the image of a cube $[-1, 1]^n$ under a linear map $\Phi : \mathbb{R}^n \to (\mathbb{R}^d)^*$. (This is Definition 3.71 in Sect. 3.3.1.)
(b) A *zonotope* is a Minkowski sum of intervals $[\varphi_1, -\varphi_1], \dots [\varphi_n, -\varphi_n]$ where $\varphi_i \in (\mathbb{R}^d)^* - \{0\}$. In other words,

$$Z = \{x \in (\mathbb{R}^d)^* \mid x = \sum_{i=1}^{n} x_i\varphi_i, \quad \text{where } -1 \leq x_i \leq 1\}. \tag{5.3}$$

To go from (b) to (a), let Φ be the $(d \times n)$-matrix with column vectors $\varphi_1, \dots, \varphi_n$. Then (5.3) is the same as putting $Z = \Phi([-1, 1]^n)$. To go from the arrangement $\mathcal{A}$ to the zonotope $Z(\mathcal{A})$, first choose a chamber $C \in \mathrm{Fan}(\mathcal{A})$. Each hyperplane $H_i \in \mathcal{A}$ is the kernel of a linear form $\varphi_i \in (\mathbb{R}^d)^*$. Let φ_i be such that it is positive on the positive side of H_i. The vertices of Z are dual to the chambers in $\mathrm{Fan}(\mathcal{A})$. Each chamber $D \in \mathrm{Fan}(\mathcal{A})$ determines a sign in $\{+, -\}$ for each φ_i and hence, a principal ultrafilter $\omega_D : \mathcal{A} \to \{\pm 1\}$. Such principal ultrafilters give the vertex set of Z. Whenever the number of chambers is less than 2^n, there are phantom vertices corresponding to the vertices of $[-1, 1]^n$ which project to interior points in Z. It follows that whenever Z is not isomorphic to $[-1, 1]^n$, it has phantom vertices. When $\mathcal{A}$ is a reflection arrangement associated to a finite Coxeter group we get a Coxeter zonotope which was discussed in more detail in Sect. 4.2.2. As a specific example of such a hyperplane arrangement, consider the case of p lines through the origin of $\mathbb{R}^2$ defined by linear forms $\varphi_1, \dots, \varphi_p$ in $(\mathbb{R}^2)^*$ in cyclic order. Then Z is a $2p$-gon with vertices, $\varphi_1, \dots, \varphi_p, -\varphi_1, \dots, -\varphi_p$. This shows that the 2-dimensional zonotopes are precisely the centrally symmetric polygons (cf. [235, Example 7.14]).

Example 5.10 (The Cube Complex $\tilde{P}_L$ and RACGs, cf. Example 5.3) Let L be a simplicial complex (not necessarily a flag complex) and let P_L be the associated cube complex defined in Sect. 3.1.1 and let $\tilde{P}_L$ be its universal cover. Then $W_L \curvearrowright \tilde{P}_L$, where W_L is the RACG associated to L^1. As in Example 5.3, the hyperplanes $\{x_i = 0\} \cap P_L$ lift to hyperplanes on $\tilde{P}_L$ and an associated half-space system $(\mathcal{H}, <, \iota)$. If L is a flag complex, then $\tilde{P}_L$ is CAT(0) and Sageev's realization of $(\mathcal{H}, <, \iota)$ is equal to $\tilde{P}_L$. On the other hand, if L is not flag, then Sageev's realization is $\tilde{P}_{\hat{L}}$ where $\hat{L}$ means the flag complex determined by L^1, i.e., $\hat{L}$ is the "flag completion" of L.

Example 5.11 (Reflection Groups) Half-space systems arise naturally in the theory of groups generated by reflections. To fix ideas let us suppose we are in the situation that was described in Sects. 3.2 and 4.2: W is a group generated by locally linear reflections on a manifold M. The fixed set of a reflection r is a codimension-one submanifold M^r called a *wall*. The complement $M - M^r$ has two components (called open *half-spaces*) that are interchanged by r. Let R denote the set of all reflections in W. Each component of $M - \bigcup_{r \in R} M^r$ is an "open chamber." A *chamber* is the closure of an open chamber. Choose a fundamental chamber K for

W on M. Let S be the set of reflections r whose walls intersect K in a codimension-one face. As in Sects. 3.2 and 4.2, R is the set of all conjugates of a elements of S by an elements of W, K is a strict fundamental domain for W on M and (W, S) is a Coxeter system. We get a wall set $(\Omega, \mathcal{W})$ where Ω is the set of chambers and $\mathcal{W} = \{M^r\}_{r \in R}$. This gives the same half-space system for (W, S) as in Example 5.4. Each chamber gives a principal ultrafilter on $\mathcal{W}$. The Davis–Moussong complex $\Sigma(W, S)$ of Sect. 4.2.3 is a nice geometric realization of (W, S). It is a CAT(0) cell complex where the cells correspond to the spherical special subgroups of W. Each such cell is a zonotope for the reflection arrangement corresponding to the given spherical subgroup as in Example 5.9. For each reflection $r \in R$, its fixed set Σ^r separates Σ into two half-spaces. As before, the closure of a component of $\Sigma - \bigcup_{r \in R} \Sigma^r$ is called a *chamber*. We get the same half-space system for (W, S) as before. Let $X(W, S)$ be the corresponding CAT(0) cube complex.

The facts in the next proposition are established in Niblo–Reeves [183].

Proposition 5.12 (Niblo–Reeves [183]) *Let $X(W, S)$ denote the* CAT(0) *cube complex associated to the Coxeter system (W, S)*

(a) *The cube complex $X(W, S)$ is finite dimensional (cf. [183, Corollary 2]).*
(b) *The complex $X(W, S)$ is locally finite (cf. [183, Theorem 3]).*
(c) *The W-action on $X(W, S)$ is proper (cf. [183, Lemma 5]).*

The dimension of the cell of $\Sigma(W, S)$ corresponding to the spherical subset $T \leq S$ is Card(T). Hence, the dimension of $\Sigma(W, S)$ is the maximum of Card(T), $T \in \mathcal{S}(W, S)$. For such a spherical subset T, let m_T denotes the number of hyperplanes in the reflection arrangement corresponding to W_T. By Example 5.9 the dimension of the corresponding cube in $X(W, S)$ is m_T. So, $\dim X(W, S) \leq n(W, S)$, where

$$n(W, S) = \max\{m_T \mid T \leq S \text{ is a spherical subset}\}. \tag{5.4}$$

Of course, when (W, S) is right-angled, $\dim X(W, S) = n(W, S) = \dim \Sigma(W, S)$.

In general, the W-action on $X(W, S)$ is not cocompact. For example, if W is the $(3, 3, 3)$ triangle group (a euclidean reflection group on $\mathbb{R}^2$), then one can show that $X(W, S)$ is the standard tiling of $\mathbb{R}^3$ by 3-cubes. Each 3-cube corresponds to a collection of 3 pairwise intersecting lines and two such collections lie in the same W-orbit if and only if they enclose equilateral triangles of the same area. Since there are arbitrarily large equilateral triangles in the $(3, 3, 3)$ tiling of $\mathbb{R}^2$, there are infinitely many W-orbits on 3-cubes. So, the W-action on $X(W, S)$ is not cocompact. More generally, suppose (W, S) is an irreducible euclidean Coxeter group of rank $d + 1$. It can be shown that $X(W, S)$ is the standard cubical tiling of $\mathbb{R}^{n(W,S)}$, where $n(W, S)$ is defined by (5.4).

Recall from Definition 4.41 that a Coxeter system (W, S) is 2-*spherical* if no entry of the Coxeter matrix $(m(s, t))_{(s,t) \in S \times S}$ is equal to ∞. When (W, S) is 2-*spherical* the following result of [46] shows that these irreducible euclidean reflection groups are the only obstruction to the W-action on $X(W, S)$ being cocompact.

Theorem 5.13 (Caprace [46]) *Suppose (W, S) is 2-spherical. Then the W-action on $X(W, S)$ is cocompact if and only if there is no special subgroup (W_T, T) which is an irreducible euclidean reflection group of rank ≥ 3.*

Caprace also shows that the condition in this theorem is equivalent to the condition that W has no subgroup isomorphic to a euclidean triangle group.

5.1.3 Geometric Actions on the Associated CAT(0) Cube Complexes

Definition 5.14 An isometric action of a group on a metric space is *geometric* if it is proper and cocompact.

Starting from a collection of codimension-one subgroups $\{H_1, \dots, H_k\}$ in a group G, Sageev [202] gives us a wall set $(\Omega, \mathcal{W})$ and a CAT(0) cube complex denoted by $X(\Omega, \mathcal{W})$ or by $X(G, H_1, \dots, H_k)$ so that $G \curvearrowright X(\Omega, \mathcal{W})$. Our goal in this section is to understand when the G-action on Sageev's CAT(0) cube complex is geometric.

First we deal with the question of when is $X(\Omega, \mathcal{W})$ finite dimensional. A subgroup $H \leq G$ satisfies the *bounded packing property* if for each positive real number D, there is a positive integer $N = N(D)$ bounding the cardinality of collections $\{g_1 H, \dots, g_n H\}$ of D-close cosets, cf. [234, Definition 7.2]. (Two cosets $g_1 H$ and $g_2 H$ are *D-close* if there are elements $x \in g_1 H$, $y \in g_2 H$, with $D(x, y) < D$.) One can define the bounded packing property for a finite collection $\{H_1, \dots H_k\}$ of subgroups of G in a similar fashion. Since a collection of transverse walls corresponds to a collection of D-close cosets, we see that the bounded packing property implies that each cube in $X(\Omega, \mathcal{W})$ has dimension $\leq N(D)$ for an appropriate choice of D. So, the next result is just a translation of the bounded packing property.

Lemma 5.15 (cf. Hruska–Wise [151, Corollary 3.31]) *Suppose a finitely generated group G has the bounded packing property with respect to finitely generated subgroups $H_1, \dots, H_k$. Then the associated* CAT(0) *cube complex $X(G, H_1, \dots, H_k)$ is finite dimensional.*

Remark 5.16 If the codimension-one subgroup $H < G$ corresponds to an embedded codimension-one submanifold $N \hookrightarrow M$ as in Example 5.6, then $X(G, H)$ is a tree; hence, one-dimensional. However, when M^3 is a certain graph manifold, Rubinstein and Wang have shown in [199] that there exists a π_1-injective immersed surface $N \looparrowright M^3$ such that any two translates of $\tilde{N}$ in $\tilde{M}^3$ have nonempty intersection. Hence, the associated cube complex is infinite dimensional.

We turn next to the question of properness. Suppose p, q are two points in a wall set Ω. (For example, if Y is a CAT(0) space and $\mathcal{W}$ a locally finite collection of convex hyperplanes which separate Y, then we could take Ω to be the complement in Y of $\bigcup_{H \in \mathcal{W}} H$.) Given any two points $p, q \in \Omega$, Let $\#(p, q)$ denote the number of hyperplanes in $\mathcal{W}$ which separate p from q. Note that if v, w are vertices of the CAT(0) cube complex $X(\Omega, \mathcal{W})$, then $\#(v, w)$ is the distance from v to w in the 1-skeleton of $X(\Omega, \mathcal{W})$ (i.e., in the ℓ^1-metric). Since this metric is quasi-isometric to the usual CAT(0) metric on $X(\Omega, \mathcal{W})$ we get the following lemma.

Lemma 5.17 (cf. [234, Lemma 7.13]) *The finitely generated group of isometries G acts (metrically) properly on $X(G, H_1, \dots H_k)$ if and only if for any vertex v, $\#(v, gv) \to \infty$ as $l(g) \to \infty$.*

Suppose Y is a CAT(0) space with a set of hyperplanes $\mathcal{W}$. One way to show that $\#(p, gp) \to \infty$ is to verify the following *linear separation property*. This states that there are constants $k_1, k_2 > 0$ so that $\#(p, q) \geq k_1 d_Y(p, q) - k_2$ for all $p, q \in Y$. The meaning of the linear separation property is that there is some sort of coarse fundamental domain for the G-action on Y which is bounded by a finite number of walls in $\mathcal{W}$. In the case of a word hyperbolic group, Bergeron and Wise use these notions to get the following criterion for a group to act properly on a CAT(0) cube complex.

Theorem 5.18 (Bergeron–Wise [18, Theorem 1.4]) *Let G be a word hyperbolic group. Suppose that for each pair of distinct points a, b in ∂G, there exists a quasiconvex codimension-one subgroup H so that ∂H separates a from b. Then there are a finitely many quasiconvex, codimension-one subgroups $H_1, \dots, H_k$ so that the G-action on the associated* CAT(0) *cube complex $X(G, H_1, \dots, H_k)$ is proper.*

For example, in the case of the reflection group action of W on the Davis–Moussong complex $\Sigma(W, S)$ (cf. Examples 5.11 and 5.4), there is a compact fundamental domain K which is bounded by the walls corresponding to the reflections of S. Suppose γ is a geodesic segment between points $p, q \in \Sigma$. Assume that p, q are "generic" in the sense that γ crosses no wall of codimension > 1. Each time γ crosses a wall it passes from a chamber wK to an adjacent chamber wsK for some $s \in S$; so, γ gives a sequence of adjacent chambers $K_1, \dots K_n$, i.e., an edge path of length n in the Cayley graph, $\mathrm{Cay}(W, S)$. So, when $q = gp$, $\#(p, gp)$ is the word length of g. Since the word distance in $\mathrm{Cay}(W, S)$ is quasi-isometric to the CAT(0) distance in Σ (cf. Sect. 4.2.3, as well as [82]) one sees that the linear separation property holds for $\Sigma(W, S)$.

On the other hand, consider the case of Example 5.6 of a connected π_1-injective embedded submanifold $N \hookrightarrow M$. To simplify the discussion suppose N^1 is an embedded curve in an orientable surface M^2. The walls are then the translates of $\tilde{N}$ in $\tilde{M}$. (See Fig. 5.1.) The corresponding cube complex Y is a tree and the quotient space $Y/\pi_1(M)$ is either an edge or a single loop depending on whether or not N separates M. These walls divide $\tilde{M}$ into regions and each such region is a lift of a component of $\hat{M}$ (which stands for M cut open along N so that $\hat{M}$ has either one or two components). In either case, the genus of a component M' of $\hat{M}$ is less than the genus of M, the induced homomorphism $\pi_1(M') \to \pi_1(M)$ is an injection onto an infinite subgroup of $\pi_1(M)$, and the corresponding region in $\tilde{M}$ is unbounded.

This region corresponds to a vertex of the tree and $\pi_1(M')$ is the isotropy subgroup at this vertex. Since this isotropy group is infinite, the $\pi_1(M)$-action on the tree Y is not proper.

Cocompactness The notion of a word hyperbolic group and a quasiconvex subgroup were defined in Sect. 2.1.3. A following result of [131], shows that in the presence of hyperbolicity, cocompactness is essentially automatic.

Lemma 5.19 (See [131] or [202, Theorem 2.1]) *Suppose $\{H_1, \dots, H_k\}$ is a collection of quasiconvex codimension-one subgroups in a word hyperbolic group G. Let $Y = X(G, H_1, \dots, H_k)$ be the resulting* CAT(0) *cube complex. Then Y/G is compact.*

5.1.4 Cubulating Hyperbolic Manifolds

How prevalent are examples of CAT(0) cube complexes with geometric group actions? The answer is not so clear. The tricky issue is to find enough codimension-one subgroups so that the group action on the corresponding cube complex is proper. As we saw in Sect. 3.1.1, one way to produce such examples is by using reflection groups. If (W, S) is a RACS with nerve L, then $\tilde{P}_L$ is a CAT(0) cube complex and the W-action is proper and cocompact. So, if $\Gamma < W$ is a torsion-free subgroup of finite index, then Γ acts freely and $\tilde{P}_L/\Gamma$ is a compact NPC cube complex. To be even more specific, if Γ is the kernel of the natural epimorphism $W \to (\mathbf{C}_2)^S$, then $\tilde{P}_L/\Gamma = P_L$, the polyhedral product defined by (3.6). One way to check that $\tilde{P}_L$ satisfies the linear separation property is to notice that any geodesic passing through the fundamental chamber K_L must intersect one of its walls and these are indexed by the finite set S. In the case of a (not necessarily right-angled) Coxeter group W, its action on the Davis–Moussong complex $\Sigma(W, S)$ (a space with convex walls) also has the linear separation property, since there is again a compact fundamental chamber. Hence, W acts properly on the associated CAT(0) cube complex $X(W, S)$. (However, as in Example 5.11, the W-action on $X(W, S)$ will not be cocompact unless Caprace's condition from Theorem 5.13 holds.)

In the remainder of this subsection we describe work on the question of when a closed hyperbolic n-manifold M^n $(= \mathbb{H}^n/\Gamma)$ is homotopy equivalent to a compact NPC cube complex. The method will be to find totally geodesic or nearly totally geodesic immersed codimension-one submanifolds which give codimension-one quasiconvex subgroups of $\Gamma = \pi_1(M^n)$. Sageev's construction then yields a CAT(0) cube complex Y. To insure properness of the $\pi_1(M^n)$-action on Y we need to be able to find sufficiently many codimension-one subgroups of $\pi_1(M^n)$. (To start, we need at least one.) In dimension 3 these subgroups are provided by the theorem of Kahn–Marković [162]. In dimension $n > 3$ there are results due to Bergeron-Haglund-Wise [17] that work for hyperbolic n-manifolds with arithmetic fundamental groups. Here the arithmeticity hypothesis is needed in order to get a large number of codimension-one subgroups.

The universal cover of M^n is identified with hyperbolic space $\mathbb{H}^n$ and $\Gamma = \pi_1(M^n)$ acts on it via isometries as well as on its sphere at infinity, $\partial\mathbb{H}^n \cong S^{n-1}$. The limit set of a hyperplane in $\mathbb{H}^n$ is a round $(n-2)$-sphere in S^{n-1}. A hyperplane $H < \mathbb{H}^n$ is a *Γ-hyperplane* if $\mathrm{Stab}_\Gamma(H)$ acts cocompactly on it. This implies that H projects to a totally geodesic immersed codimension-one submanifold $N^{n-1} \looparrowright M^n$, where $N^{n-1} = H/\mathrm{Stab}_\Gamma(H)$.

Dimension 3 Kahn–Marković [162] proved that the fundamental group of any closed hyperbolic 3-manifold M^3 contains a surface subgroup. In other words, M^3 contains a π_1-injective immersed closed surface of genus > 0. A π_1-injective immersed surface $N^2 \looparrowright M^3$ is *quasifuchsian* if the limit set of $\tilde{N}^2$ is a topologically embedded circle in S^2 and $\pi_1(N^2)$ is then called a *quasifuchsian subgroup* of Γ. It is known that every quasifuchsian subgroup of Γ is quasiconvex. (Indeed, a surface subgroup is either quasiconvex or normal, but the limit set of any infinite normal subgroup maps onto the whole sphere at infinity.) Here is the strong version of the Kahn–Marković result which shows that Γ has sufficiently many surface subgroups.

Theorem 5.20 (Kahn–Marković [162, Theorem 1.1]) *Let Γ be the fundamental group of a closed hyperbolic 3-manifold. For each round circle C in $S^2 = \partial\mathbb{H}^3$ there is a sequence of immersions $f_i : N_i \looparrowright M^3$ so that $(f_i)_*(\pi_1(N_i))$ is a quasifuchsian subgroup such that the limit sets $\partial \tilde{f}_i(\tilde{N}_i)$ converge pointwise to C.*

This gives enough codimension-one subgroups $H_1, \dots, H_k$ of the word hyperbolic group Γ so that Theorem 5.18 of Bergeron-Wise can be applied to get a geometric action on the resulting cube complex Y. (Since $\Gamma = \pi_1(M^3)$ is torsion-free, any proper Γ-action is necessarily free.) This gives the following basic result.

Theorem 5.21 ([18, Theorem 1.5]) *The fundamental group of any hyperbolic 3-manifold has a subgroup of finite index that acts freely and cocompactly on a* CAT(0) *cube complex. (In other words any hyperbolic 3-manifold has a finite cover that is homotopy equivalent to a compact NPC cube complex.)*

Arithmetic Hyperbolic Manifolds Borel [26] explains a method for using number theory to construct compact locally symmetric manifolds (equivalently, torsion-free uniform lattices in semisimple Lie groups). It follows from this construction that closed hyperbolic n-manifolds exist in each dimension n.

Example 5.22 (cf. [59, §6]) Let $F = \mathbb{Q}(\sqrt{d})$ be a totally real quadratic extension of the rationals and let A be the ring of integers in F. Choose an element $\varepsilon \in A$ so that ε is positive and its Galois conjugate $\overline{\varepsilon}$ is negative. Define a symmetric bilinear form φ on A^{n+1} by

$$\varphi(e_i, e_j) = \begin{cases} \delta_{ij}, & \text{if } 1 \le i, j \le n, \\ -\varepsilon, & \text{if } (i, j) = (n+1, n+1), \\ 0, & \text{otherwise,} \end{cases} \tag{5.5}$$

where $\{e_1, \ldots, e_{n+1}\}$ denotes the standard basis. Let φ_F and $\varphi_{\mathbb{R}}$ denote the induced bilinear forms on F^{n+1} and $\mathbb{R}^{n+1}$ and let $O(\varphi)$, $O(\varphi_F)$ and $O(\varphi_{\mathbb{R}})$ be the respective isometry groups. This gives an algebraic group $\mathbf{G}$, with $\mathbf{G}(\mathbb{R}) = O(\varphi_{\mathbb{R}}) = O(n, 1)$, $\mathbf{G}(\mathbb{Q}) = O(\varphi_{\mathbb{Q}})$, and with $O(\varphi)$ a uniform lattice in $bG(\mathbb{R})$. (That the lattice is uniform follows from the fact that if $\overline{\varphi}$ denotes the form obtained from φ by replacing ε by its Galois conjugate $\overline{\varepsilon}$, then $\overline{\varphi}_{\mathbb{R}}$ is positive definite). Since $\varphi_{\mathbb{R}}$ is type $(n, 1)$, $O(\varphi_{\mathbb{R}}) \cong O(n, 1)$, the isometry group of $\mathbb{H}^n$. (Use the quadratic form model of $\mathbb{H}^n$ as one sheet of the hyperboloid $\varphi_{\mathbb{R}}(x, x) = -\varepsilon$ in $\mathbb{R}^{n+1}$, where $x = (x_1, \ldots, x_{n+1}) \in \mathbb{R}^{n+1}$). There are many reflections in $O(\varphi)$, for example, the reflections across the hyperplanes, $(e_i)^{\perp}$, with $i \leq n$, or $(e_i - e_j)^{\perp}$, with $i, j \leq n$, as well as any conjugate of one of these reflections by an element $g \in O(\varphi)$. The subgroup W generated by these reflections is a Coxeter group acting properly on $\mathbb{H}^n$. Usually W will not be of finite index in $O(\varphi)$ and hence, will not have a bounded fundamental chamber on $\mathbb{H}^n$. However, in low dimensions, for certain fields F, W will have finite index in $O(\varphi)$. When this is the case, $O(\varphi)$ will act geometrically on the associated CAT(0) cube complex for W. Let Γ be a torsion-free subgroup of finite index in $O(\varphi)$. For example, Γ could be a congruence subgroup defined with respect to a suitable prime ideal in A. We are interested in hyperbolic manifolds of the form $M^n = \mathbb{H}^n/\Gamma$. (Call such M^n a *standard* arithmetic hyperbolic manifolds.) So, when W has finite index in $O(\varphi)$, M^n will be homotopy equivalent to a NPC cube complex. However, when W is not of finite index, the hyperplanes of reflection will not partition $\mathbb{H}^n$ into bounded fundamental domains and we cannot immediately invoke the linear separation property of the previous subsection to conclude that Γ acts properly on a CAT(0) cube complex. Nevertheless, as we shall see below, it does.

In [17, Prop. 3.4] Bergeron, Haglund and Wise prove the surprising result that every standard arithmetic hyperbolic closed n-manifold M^n is homotopy equivalent to a compact NPC cube complex. The context of [17] is slightly different from Example 5.22 in that hyperplanes of reflection do not appear explicitly. As before, F is a totally real number field, A is its ring of integers and φ_F is the symmetric bilinear form on F^{n+1}. Let $\mathbf{G}$ be the $\mathbb{Q}$-algebraic group obtained from $O(\varphi_F)$ by restriction of scalars from F to $\mathbb{Q}$. Then $\mathbf{G}(\mathbb{R}) = O(\varphi_{\mathbb{R}}) \cong O(n, 1)$, $\mathbf{G}(\mathbb{Q}) = O(\varphi_F)$, and $O(\varphi)$ is a uniform lattice in $G(\mathbb{R})$. Let Γ be a sufficiently deep congruence subgroup so that Γ is torsion-free. The resulting closed manifold $M^n = \mathbb{H}^n/\Gamma$ is a *standard* arithmetic hyperbolic n-manifold. For any hyperplane H in $\mathbb{H}^n$, let Γ_H denote its stabilizer in Γ. One says that H is a *Γ-hyperplane* if H/Γ_H is compact. Any totally positive, one-dimensional subpace of (F^{n+1}, φ_F) yields a Γ-hyperplane H. For such an H, the projection map $\mathbb{H}^n \to M^n$ descends to a totally geodesic immersion $H/\Gamma_H \looparrowright M^n$. Define a set of hyperplanes $\mathcal{H}(H)$ in $\mathbb{H}^n$ by

$$\mathcal{H}(H) := \{gH \mid g \in \Gamma\},$$

i.e., $\mathcal{H}(H)$ is a Γ-orbit of hyperplanes. For any finite subset $S < \mathbf{G}(\mathbb{Q})$, put

$$\mathcal{H}^S(H) = \{gH \mid g \in s\Gamma < \mathbf{G}(\mathbb{Q}) \text{ for some } s \in S\},$$

so that $\mathcal{H}^S(H)$ is a finite set of Γ-orbits of hyperplanes in $\mathbb{H}^n$. For $g \in \mathbf{G}(\mathbb{Q})$, gH/Γ_{gH} is a closed hyperbolic $(n-1)$-manifold called a *Hecke translate* of H/Γ_H. It maps to an immersed submanifold in M^n/Γ.

Proposition 5.23 (Bergeron-Haglund-Wise [17, Prop. 2.1]) *There is a finite subset S of $\mathbf{G}(\mathbb{Q})$ so that $\mathcal{H}^S(H)$ defines a linearly separated wall space structure on $\mathbb{H}^n$. Hence, Γ acts properly on the associated* CAT(0) *cube complex.*

More precisely, it is proved in [17] that there is a compact hyperbolic polytope P bounded by finitely many hyperplanes $g_1H, \dots, g_kH$ in $\mathcal{H}(H)$ so that P contains a fundamental domain for Γ (cf. the paragraph following Lemma 5.17).

Sketch of Proof Suppose $H_1', \dots, H_k'$ are hyperplanes in $\mathbb{H}^n$ bounding a simple polytope P' that contains a fundamental domain for Γ. The point is, that since $\mathbf{G}(\mathbb{Q})$ is dense in $\mathbf{G}(\mathbb{R})$, the $\mathbf{G}(\mathbb{Q})$-orbit of H is dense in the set of all hyperplanes in $\mathbb{H}^n$. Hence, we can find a finite set $S = \{g_1, \dots, g_k\} < \mathbf{G}(\mathbb{Q})$ so that $\{g_1H, \dots, g_kH\}$ is the set of hyperplanes spanned by the codimension-one faces of the desired polytope P.

Combining Proposition 5.23 with the result of Bergeron–Wise, Theorem 5.18, we get the following.

Theorem 5.24 (Bergeron-Haglund-Wise [17, Prop. 3.4] and also [145, §10]) *Suppose, as above, that a standard arithmetic hyperbolic group Γ is a torsion-free uniform lattice. Then Γ acts properly and cocompactly on a* CAT(0) *cube complex.*

This theorem only asserts that the hyperbolic manifold M^n $(= \mathbb{H}^n/\Gamma)$ is homotopy equivalent to an NPC cube complex and not that it is homeomorphic to one. In fact, this seems likely to be impossible in dimensions > 4. The reason is that, by Lemma 2.33, if the cubes of a cube complex Y are regular cubes in hyperbolic space of curvature $-\kappa$ for small positive κ, then the link of each vertex must be a flag complex satisfying the no $\square$-condition and by [82, Prop. I.6.6 of Appendix I] for $n > 4$, there is no such triangulation of S^{n-1}. So, it is unlikely that an NPC cube complex with word hyperbolic fundamental group could be a closed n-manifold for $n > 4$.

5.2 Special Cube Complexes

The theory of special cube complexes is due to Haglund and Wise [143–145]. A cube complex B is said to be *special* if the hyperplanes in B avoid certain pathological configurations (see Sect. 5.2.2 below). For example, each hyperplane must be embedded (i.e., self-intersections are not allowed). The main result of [143] is that the standard cube complexes for RAAGs have a universal property for NPC special cube complexes: if B is a NPC special cube complex, then there is a locally isometric immersion $f : B \looparrowright \mathbb{T}_L$, where $\mathbb{T}_L$ is the standard classifying space for a certain RAAG, A_L, defined by (3.5) in Sect. 3.1. It follows that f induces an injection $\pi_1(B) \hookrightarrow A_L$ and that on the level of universal covers it lifts to an isometric embedding $\tilde{B} \hookrightarrow \tilde{\mathbb{T}}_L$. This is important because RAAGs and RACGs have very strong separability properties. First of all, Coxeter groups have faithful linear representations. In fact, by using the Tits representation, one sees that any RACG is a subgroup of $SL(n, \mathbb{Z})$ for some integer n. Since any RAAG is commensurable to some RACG (cf. [93] or Sect. 5.3.1 below), the same holds for any RAAG. It follows that RAAGs are residually finite. Moreover, any word quasiconvex subgroup of a RACG or RAAG is separable. (The term "word quasiconvex subgroup" is defined in Definition 2.20; in the case of a subgroup of a RACG or RAAG it is always assumed to be with respect to the standard set of generators.) Hence, this property holds for $\pi_1(B)$, that is to say, for the fundamental group of an arbitrary NPC special cube complex. The definition of the immersion f is the obvious one. Let $\overline{\mathcal{H}}_B$ be the set of hyperplanes in B. Each $H \in \overline{\mathcal{H}}_B$ is dual to a square equivalence class of parallel edges. If H is two-sided, then we can choose an orientation for each edge in this equivalence class. The Artin group is defined by a graph L^1. The vertex set of L^1 is $\overline{\mathcal{H}}$ and vertices H, H' span an edge if and only if the hyperplanes H and H' intersect. The map $f : B^1 \to \mathbb{T}_L$ is defined by sending each vertex of B to the single vertex of $\mathbb{T}_L$ and each edge e in the equivalence class for H to the circle S^1_H corresponding to $H \in \operatorname{Vert} L$. The corresponding Artin generator is denoted a_H. For f to be well defined on B^1 we need each hyperplane to be 2-sided. The map f extends to the 2-skeleton of B: a square in B^2 with adjacent edges e, e' gives a pair of intersecting hyperplanes H and H' spanning a 2-torus, $S^1_H \times S^1_{H'}$, onto which the square $e \times e'$ maps. (The map fails to be an immersion precisely at the corner of each square that corresponds to a self-intersecting hyperplane.) Continuing with higher dimensional cubes, we get the map $B \to \mathbb{T}_L$.

Similarly, if B satisfies a slightly different version of specialness, one can find an isometric embedding $\tilde{B} \to \Sigma$ where Σ is the Davis–Moussong complex associated to some RACG, W_L, defined by a simplicial graph L^1 (cf. Sects. 4.2.3 and 3.1.1). On the level of quotient spaces the definition of the locally isometric immersion is not quite as direct as in the case of the RAAG, A_L. In order to define it we need a certain intermediate square complex $\mathrm{COX}(L^1)$, which we will define in the next subsection.

5.2.1 The Square Complexes COX(Γ) *and* ART(Γ)

Suppose Γ is a simplicial graph. Recall that the RACG, W_Γ, associated to Γ has a generator s_i for each $i \in \operatorname{Vert}(\Gamma)$ and relations $(s_i)^2$, for each $i \in \operatorname{Vert}(\Gamma)$ and a relation $[s_i, s_j]$ for each $\{i, j\} \in \operatorname{Edge} \Gamma$. Similarly, the RAAG, A_Γ, has a generator a_i for each $i \in \operatorname{Vert}(\Gamma)$ and a relation $[a_i, a_j]$ for each $\{i, j\} \in \operatorname{Edge} \Gamma$.The complex

ART(Γ) is the same as the 2-skeleton of the polyhedral product $\mathbb{T}_L$ from Sect. 3.1.1. (Here, as in the paragraph preceding Definition 2.32, L the flag completion of $\Gamma = L^1$.)

Next we give the definition of a square complex COX(Γ) from [143]. The universal cover UCOX(Γ) of COX(Γ) is closely related to the universal cover of the Davis–Moussong complex Σ for W_Γ defined in Sects. 3.1.1 and 4.2.3. The 1-skeleton of Σ is the Cayley graph of W_Γ and this is also equal to the 1-skeleton of the universal cover, UCOX(Γ). Each edge of Σ is labeled by some generator s_i, with $i \in \text{Vert}(\Gamma)$. Each 2-cell in Σ is a square with consecutive edges labeled $s_i s_j s_i s_j$ with $\{i, j\} \in \text{Edge}(\Gamma)$. The complex UCOX($\Gamma$) is constructed by replacing each such square by its double. (If $\Box_{ij}$ is such a square, then its double, $D\Box_{ij}$, is the *square pillow* formed by doubling $\Box_{ij}$ along its boundary.) There is a natural projection $f : \text{UCOX}(\Gamma) \to \text{COX}(\Gamma)$ which projects each pillow $D\Box_{ij}$ to $\Box_{ij}$. The group $W = W_\Gamma$ acts on its Cayley graph which is equal to the 1-skeleton, $\text{UCOX}^1(\Gamma)$. This extends to a free action on the 2-skeleton as follows: a conjugate of $s_i s_j$ acts on its doubled square $D\Box_{ij}$ by rotating the square $\Box_{ij}$ by π and then exchanging the two copies. Let W^+ be the subgroup of W consisting of the elements of even length. Put

$$\text{COX}(\Gamma) = \text{UCOX}(\Gamma)/W^+, \tag{5.6}$$

so that $\pi_1(\text{COX}(\Gamma)) = W^+$. The complex COX($\Gamma$) can be described independently of (5.6). Since there are two W^+-orbits of vertices on UCOX(Γ), the quotient square complex has two vertices v^+ and v^-. For each $i \in \text{Vert}(\Gamma)$ there is an edge e_i labeled s_i connecting v^- to v^+. Then, for each $\{i, j\} \in \text{Edge}\,\Gamma$, glue in a square $\Box$ along the edge path $e_i e_j e_i e_j$. This defines the 2-skeleton of COX(Γ). (N.B. The attaching map for the square is a degree two map from $\partial\Box$ onto the circle $e_i \cup e_j$. Hence, the image of $\Box_{ij}$ in COX(Γ) is a copy of $\mathbb{R}P^2$.)

5.2.2 *Definitions of Specialness*

We recall the definitions from [143, §3]. First, Haglund–Wise list the following five "pathologies" that can occur for a hyperplane H (or a pair of hyperplanes) in a cube complex B.

List of Pathologies

(1) H has *self-intersections*,
(2) H is *one-sided*,
(3) H is *directly self-osculating*,
(4) H is *indirectly self-osculating*,
(5) two hyperplanes are *inter-osculating* if they intersect and osculate.

Pictures of these pathologies can be found in [143, p. 1561, Figure 1], [234, §4.1] or in [21]. Self-intersection and the property of being one-sided were discussed in the beginning of Sect. 5.1.1. Hyperplanes H_1 and H_2 *osculate* if there are adjacent edges e_1, e_2 so that e_1, e_2 are not the edges of any square and so that e_i is dual to H_i. (Hence, the midcubes corresponding to e_1 and e_2 lie inside cubes which share the vertex $e_1 \cap e_2$). If H is two-sided, then an orientation of its normal bundle determines an orientation for each dual edge. After choosing an orientation for the normal bundle of H, we see there are two ways in which it can self-osculate at $v = e_1 \cap e_2$: it *directly self-osculates* at v if the orientation can be chosen so that v is the terminal vertex of both e_1 and e_2, and it *indirectly self-osculates* if v is the terminal vertex of one of the e_i and the initial vertex of the other. Finally H_1 and H_2 *inter-osculate* if they both intersect and osculate.

Definition 5.25 A simple cube complex B is *A-special* if the pathologies of type (1), (2), (3), (5) do not occur. It is *C-special* if pathologies (1), (5) do not occur, if its 1-skeleton B^1 a bipartite graph and if no hyperplane self-osculates. The cube complex B is *special* if pathologies of type (1), (3), (5) do not occur.

For example, the cube complex $\mathbb{T}_L$ from 3.1.1 (the Salvettti complex for A_L) is special.

Let B be a square complex and let Γ_B be the graph with a vertex for each hyperplane of B and with an edge between vertices whenever the corresponding hyperplanes intersect. We call Γ_B the *incidence graph* of B.

Definition 5.26 (*A-Typing* and *C-Typing*, cf. Haglund–Wise [143, Def. 3.14]) The hyperplanes in a square complex B are two-sided if and only if there is a combinatorial map $B^1 \to \mathrm{ART}(\Gamma_B)$ sending each of the oriented edges in the equivalence class corresponding to an oriented hyperplane H to the corresponding loop in $\mathrm{ART}(\Gamma_B)$. Since $\dim B \leq 2$, this map extends to a combinatorial map on 2-skeletons $\tau_A : B \to \mathrm{ART}(\Gamma_B)$ provided each hyperplane is embedded. The map τ_A is an *A-typing* of B. Similarly, B^1 is a bipartite graph if and only if there is a map $B^1 \to \mathrm{COX}(\Gamma_B)$ sending equivalent edges corresponding to some hyperplane to the corresponding edge of Γ_B. This extends to a combinatorial map $\tau_C : B \to \mathrm{COX}(\Gamma_B)$ called a *C-typing* of B.

In [143] typing maps are used to show that the fundamental group of a special cube complex B embeds into a RAAG (when B is A-special) or into a RACG (when it is C-special). The main result of [143] is the following.

Theorem 5.27 (Haglund–Wise [143, Theorem 4.2]) *Suppose B is a cube complex. As above, let $\Gamma = \Gamma_B$ be the incidence graph determined by the hyperplanes of B.*

(i) *If B is A-special, then there is a locally isometric immersion $B^2 \looparrowright \mathrm{ART}(\Gamma)$. On the level of universal covers this is covered by an isometric embedding and on the level of fundamental groups it gives an embedding $\pi_1(B) \hookrightarrow A_\Gamma$.*
(ii) *If B is C-special, then on the 2-skeleta there is a locally isometric immersion $\tau_C : B^2 \looparrowright \mathrm{COX}(\Gamma)$. This gives an embedding $\pi_1(B) \hookrightarrow W_\Gamma$ and for the*

universal cover of B an isometric embedding $\tilde{B} \to \Sigma_\Gamma$, where Σ_Γ is the Davis–Moussong complex for W_Γ. (In other words, if L is the flag completion of Γ, then then $\Sigma_\Gamma = \tilde{P}_L$ where $\tilde{P}_L$ is as in Sect. 3.1.1.)

Let $\tau_A : B^2 \to \mathrm{ART}^2(\Gamma)$ and $\tau_C : B^2 \to \mathrm{COX}(\Gamma)$ be the typing maps and let $\tau : B^2 \to X^2$ stand for either one of them. Here is a sketch of the proof of the above theorem in [143]. One first shows that τ_A is a "locally isometric immersion" on the level of 2-skeleta. For each vertex $v \in B^0$, τ restricts to a combinatorial map $\tau_v : \mathrm{Lk}(v, B^2) \to \mathrm{Lk}(v, X^2)$ of simple graphs. To say that τ is an *immersion* means that (a) each τ_v is an embedding (i.e., a homeomorphism onto its image). To say that τ is an *isometric immersion* means that (b) each τ_v is an isomorphism onto a full subgraph. Condition (a) follows from the fact that no two hyperplanes osculate. To verify (b) we need to show that if e_1 and e_2 are adjacent edges at v (corresponding to vertices of $\mathrm{Lk}(v, B^2)$) and if $\tau(e_1)$, $\tau(e_2)$ are adjacent edges of a square at $\tau(v)$, then e_1 and e_2 are adjacent edges of a square in X^2. So, (b) follows from the fact that the hyperplanes in B^2 dual to e_1 and e_2 do not inter-osculate. It follows that τ gives an isometric embedding $\tilde{\tau} : \tilde{B}^2 \to \tilde{Y}^2$, as well as an injection on fundamental groups. In the case of statement (ii) we compose $\tilde{\tau}$ with the natural projection $\mathrm{UCOX}(\Gamma) \to \Sigma^2_\Gamma$ to get a W_Γ-equivariant embedding $\tilde{B}^2 \to \Sigma^2_\Gamma$. Since B^2 is special, it follows from [144, Lemma 3.13] that B^2 is completable to an NPC cube complex (see Definition 2.32). So, we can assume that B is NPC and that its universal cover $\tilde{B}$ is CAT(0). Finally, the equivariant embeddings of 2-complexes then extend to give equivariant embeddings from $\tilde{B}$ to Σ_Γ or from $\tilde{B}$ to the universal cover of $\mathrm{ART}(\Gamma)$ (cf. [144, Lemma 3.13]).

Example 5.28 (The Polyhedral Product P_L Again, cf. Example 5.3) Let L^1 be a simplicial graph with vertex set $I = \{1, \dots, n\}$ and let L be its flag completion. Associated to L^1 we have the RACG, W $(= W_{L^1})$ with nerve L. We return to the NPC cube complex P_L from Sect. 3.1.1 and the study of its hyperplanes in Example 5.3. The cube complex P_L is a subcomplex of the n-cube, $[-1, 1]^n$. The universal cover of P_L is equal to the Davis–Moussong complex Σ_L of Sect. 4.2.3 and W is the group of all lifts of the $(\mathbf{C}_2)^n$-action on P_L. It follows that $\pi_1(P_L) = [W, W]$, the commutator subgroup of W. The 1-skeleton of $[-1, 1]^n$ is a bipartite graph (the coloring is defined using the parity of the distance from a vertex to a base vertex) and since $[-1, 1]^n$ and P_L have the same 1-skeleton, $(P_L)^1$ is also bipartite. Similarly, since $[-1, 1]^n$ is C-special, so is P_L (cf. Definition 5.26). We use Γ to denote the incidence graph of P_L. It is tempting to think that Γ is equal to L^1. This would be true if each multiple hyperplane of P_L were connected (see Example 5.3 for the definition of "multiple hyperplane"); however, since hyperplanes are required to be connected, the vertex set of Γ is not equal to I. For each $i \in I$, let $c(i)$ be the set of components of $\{x_i = 0\} \cap P_L$ and let $\hat{I} = \bigsqcup_i c(i)$. The edges of L^1 lift to edges in Γ. More precisely, for each $\{i, j\} \in \mathrm{Edge}\, L^1$ and each unordered pair $\{c, c'\}$, with $c \in c(i)$ and $c' \in c(j)$, we have that $\{c, c'\} \in \mathrm{Edge}\, \Gamma$. In other words, Γ is the polyhedral join of $\{c(i)\}_{i \in I}$ with respect to L^1 (see Definition 3.11). By Theorem 5.27 (ii), the C-typing map from the 2-skeleton of P_L $(= P_{L^1})$ to

COX(Γ) induces an injection $\pi_1(P_L) \to W_\Gamma$ and on the level of universal covers, an isometric embedding $\tilde{P}_{L^1} = \Sigma_{L^1} \hookrightarrow \Sigma_\Gamma$. Let $p : \hat{I} \to I$ be the natural map which takes $c(i)$ to i. Then p extends to a map of graphs $\Gamma \to L^1$ and then to an epimorphism $p : W_\Gamma \to W$. Once we choose a base vertex in P_{L^1} (or dually a fundamental chamber), we get a section ι of $p : \hat{I} \to I$, defined by $i \mapsto \hat{i}$ where $\hat{i}$ is the hyperplane in $c(i)$ dual to an edge containing the base vertex. This leads to a section $\iota : L^1 \to \Gamma$ and then to a section (also denoted ι) of $p : W_\Gamma \to W_L$. (Exercise: show that W_Γ is the semidirect product of W_L and a free group.) Moreover, there is a p-equivariant projection $\Sigma_\Gamma \to \Sigma_{L^1}$ which restricts to the identity map on Σ_{L^1}. So, after some work we see that the Haglund–Wise construction ends up exhibiting $\pi_1(P_L)$ as the commutator subgroup of W_L, as expected.

5.2.3 Virtual Specialness

A group G is *special* if it is the fundamental group of a special cube complex; it is *compact special* if the special cube complex is compact. It is *virtually special* if there is a subgroup of finite index $H \leq G$ which is the fundamental group of a special cube complex.

Proposition 5.29 ([144, Theorem 8.1]) *Let W be a finitely generated Coxeter group and let X be its Niblo–Reeves* CAT(0) *cube complex. Then there is a finite index torsion-free subgroup $H < W$ so that X/H is a special cube complex.*

So, Coxeter groups are virtually special.

Proposition 5.30 ([143, Prop. 3.10]) *Suppose B is a compact special cube complex. Then B has a finite cover which is both C-special and A-special.*

This is proved by showing that if B is a special cube complex, then it has a two-fold cover such that

(1) every closed path in B^1 has even combinatorial length (and hence, that B^1 is bipartite),
(2) if $H_1, \dots, H_k$ are hyperplanes in B, then there is a 2^k-fold cover so that any closed edge path intersects each H_i an even number of times (and hence, each H_i is two-sided), and
(3) no hyperplane indirectly self-osculates.

A corollary to Proposition 5.30 and Theorem 5.27 is that RAAGs and RACGs both have a universal property for fundamental groups of special cube complexes, as stated in following theorem.

Theorem 5.31 ([143]) *Suppose B is a compact special cube complex. Then there is a finite cover $B' \to B$ and a locally isometric immersion $B' \looparrowright X$ where X is either* *the standard classifying space for a* RAAG *or the quotient of the Davis–Moussong complex for a* RACG *by a suitable torsion-free subgroup.*

Remark 5.32 Behrstock-Hagen-Sisto [14] introduced the notion of a "hierarchically hyperbolic space" as a common framework for studying mapping class groups and cubical groups. They proved the universal cover of any compact special cube complex is hierarchically hyperbolic. Hence, any RACG is a hierarchically hyperbolic group (cf. [72, Section 3.4])

In [3], Agol proved the following important theorem. Earlier this had been a conjecture of Wise. The proof is an ingenious and complicated argument for constructing an appropriate finite-sheeted cover of an NPC cube complex with word hyperbolic fundamental group. A sketch of Agol's proof can be found in Bestvina's survey article [21, §8].

Theorem 5.33 (Agol [3], also see [202, Thm. 4.3]) *If a word hyperbolic group G acts properly and cocompactly on a* CAT(0) *cube complex, then G is virtually special.*

5.3 Separability

In [140, p. 168] Haglund calls a subgroup $H < G$ *word-quasiconvex* if there is a constant C so that any geodesic path in the Cayley graph of G joining two points of H has all its vertices of distance $\leq C$ from H. Thus, $H < G$ is word-quasiconvex if the geodesic in $\mathrm{Cay}(G)$ between any two points of H lies in some Hausdorff neighborhood of H in $\mathrm{Cay}(G)$. The group H is a quasiconvex subgroup if it is word-quasiconvex with respect to some set of generators of G. i.e., if $H \hookrightarrow G$ is a quasi-isometric embedding. (For the meaning of "quasi-isometric embedding", see Definition 2.19. It follows from a basic lemma about quasi-geodesics in a Gromov hyperbolic space, that, in a word hyperbolic group G that if $H < G$ is a quasi-isometrically embedded subgroup, then it is word-quasiconvex. The main result in this section is Haglund's Theorem 5.45 which states that a subgroup of a RACG is word-quasiconvex with respect to the standard set of generators, then it is a virtual retract (meaning that it is a retract of a subgroup of finite index in the RACG). In the next subsection we begin with a result of Davis–Januszkiewicz [93] that says that every RACG is commensurable with some RAAG.

5.3.1 RAAGs *Are Commensurable with* RACGs

Suppose Γ is a finite simplicial graph with vertex set I. As usual, let A_Γ and W_Γ be the RAAG and RACG associated to Γ. The generating sets of A_Γ and W_Γ are denoted by $\{a_i\}_{i\in I}$ and $\{s_i\}_{i\in I}$, respectively. Also, define C_Γ to be $[W_\Gamma, W_\Gamma]$, the commutator subgroup of W_Γ. Define new graphs $O\Gamma$ and $\Delta\Gamma$ with respective vertex sets $I \times \{-1, 1\}$ and $I \times \{0, 1\}$. In $O\Gamma$ we have two copies of Γ, one at level -1 and the other at level 1; vertices $(i, -1)$ and $(j, 1)$ are connected by an edge if and only if $\{i, j\} \in \text{Edge}(\Gamma)$. In $\Delta\Gamma$ the vertices at level 1 are also connected by edges to give a copy of Γ; all vertices at level 0 are connected to give the complete graph on $I \times 0$; finally, edges at different levels, $(i, 0)$ and $(j, 1)$, are connected whenever $i \neq j$.

Consider the RACGs, $W_{O\Gamma}$ and $W_{\Delta\Gamma}$. The subset of generators for these groups corresponding to the vertices in $I \times 1$ is again denoted by $S = \{s_i\}_{i \in I}$. The set of generators of $W_{O\Gamma}$ corresponding to vertices at level -1 are denoted by $T = \{t_i\}_{i \in I}$ while the set of generators for $W_{\Delta\Gamma}$ corresponding to vertices at level 0 is denoted by $R = \{r_i\}_{i \in I}$. Thus, $S \sqcup T$ is the set of generators for $W_{O\Gamma}$ and $S \sqcup R$ is the set of generators for $W_{\Delta\Gamma}$. Regard $(\mathbf{C}_2)^I$ as the subgroup $\langle R \rangle$ of $W_{\Delta\Gamma}$ and define four homomorphisms:

$$\begin{array}{llll} \varphi : W_{\Delta\Gamma} \to (\mathbf{C}_2)^I & \theta : W_{\Delta\Gamma} \to (\mathbf{C}_2)^I & \alpha : W_{O\Gamma} \to W_{\Delta\Gamma} & \beta : A_\Gamma \to W_{\Delta\Gamma} \\ \quad s_i \mapsto 1 & \quad s_i \mapsto r_i & \quad s_i \mapsto s_i & \quad a_i \mapsto s_i r_i \\ \quad r_i \mapsto r_i & \quad r_i \mapsto r_i & \quad t_i \mapsto r_i s_i r_i & \end{array}$$

Theorem 5.34 (Davis–Januszkiewicz [93, p. 230] and cf. [173, Prop. 5]) *The maps $\alpha : W_{O\Gamma} \to \ker(\varphi)$ and $\beta : A_\Gamma \to \ker(\theta)$ are isomorphisms. After identifying $W_{O\Gamma}$ and A_Γ with their images, we see that $W_{O\Gamma}$ and A_Γ are both normal subgroups of $W_{\Delta\Gamma}$ of index 2^n, where $n = \text{Card}(I)$. Moreover, $C_{\Delta\Gamma} = W_{O\Gamma} \cap A_\Gamma = C_{O\Gamma}$ is a normal subgroup in both $W_{O\Gamma}$ and A_Γ and it has index 2^n in both. So, the groups $C_{\Delta\Gamma}$, $W_{O\Gamma}$, A_Γ and $W_{\Delta\Gamma}$ are all commensurable with each other.*

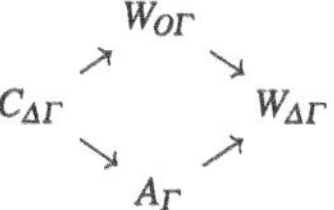

Theorem 5.34 means that $W_{\Delta\Gamma}$ can be written as a semidirect product in two different ways: as $W_{\Delta\Gamma} = W_{O\Gamma} \rtimes (\mathbf{C}_2)^I$ and as $W_{\Delta\Gamma} = A_\Gamma \rtimes (\mathbf{C}_2)^I$.

Remark 5.35 One consequence of this theorem is that there is no substantial difference between the two different definitions of special in Sect. 5.2.2. In either case we can prove Theorem 5.27. If B is an A-special cube complex, then some finite cover B' isometrically immerses in a compact quotient of the Davis–Moussong complex of some RACG and if B is C-special then it has a finite cover that isometrically immerses in the standard classifying space for some RAAG.

Remark 5.36 Let L, OL, ΔL be the flag complexes determined by Γ, $O\Gamma$, $\Delta\Gamma$, respectively. The complex OL is the polyhedral join over L of copies of S^0. In [105] it is called the *octahedralization* of L. Sometimes we replace Γ by L and instead write our groups as W_L, A_L, W_{OL} and $W_{\Delta L}$. Thus, $W_{O\Gamma}$ $(= W_{OL})$ is the graph product $\prod_\Gamma D_\infty$ over Γ of copies of the infinite dihedral group. Since $\mathbb{Z}$ is a subgroup of index 2 in D_∞, there is a natural inclusion $A_\Gamma \hookrightarrow W_{O\Gamma}$. However, in general, this inclusion will take A_Γ to a subgroup of infinite index in $W_{O\Gamma}$ so it does not induce the commensurability in Theorem 5.34.

Proof of Theorem 5.34 The proof of Theorem 5.34 is geometric. In the case of an $(n-1)$-simplex Δ and its octahedralization $O\Delta$ is the boundary complex of an n-octahedron and $K_{O\Delta}$ can be identified with the n-cube $[-1, 1]$. It follows that fundamental chamber K_{OL} for the Davis–Moussong complex Σ_{OL} of W_{OL} can be identified with a cubical subcomplex of $[-1, 1]^I$. If we identify opposite faces $K_{OL} \cap \{x_i = -1\}$ and $K_{OL} \cap \{x_i = 1\}$ we obtain $\mathbb{T}_L$, the standard model for the classifying space BA_L. Thus, K_{OL} also is a (non-strict) fundamental domain for the A_L-action on the universal cover of $\mathbb{T}_L$. (The standard CAT(0) cube complexes for W_{OL} and for A_L have the same underlying space but the cubical structures are slightly different. In the complex for W_{OL} a cube is a face of $[-1, 1]^I$, while in the complex for A_L a cube is a face of $[0, 2]^I$.) Subdivide $[-1, 1]^I$ into 2^n n-dimensional cubes. The intersection $K_{OL} \cap [0, 1]^I$ is not quite the standard fundamental chamber for $W_{\Delta L}$ but it is strata-preserving homotopy equivalent to it. So, it does no harm to assume that $K_{\Delta L} = K_{OL} \cap [0, 1]^I$ and that the Davis–Moussong complex for $W_{\Delta L}$ is a cubical subdivision of the Davis–Moussong complex for W_{OL}. Since $\langle R\rangle K_{\Delta L} = K_{OL}$ is the fundamental chamber for the W_{OL}-action, it follows that α is injective, hence, an isomorphism. Similarly, since $K_{\Delta L}$ is a fundamental domain for the action of $(\mathbf{C}_2)^I$ $(= \langle R\rangle)$ on $\mathbb{T}_L$, β is an isomorphism.

Corollary 5.37 *The* RAAG, A_Γ, *is linear; in fact, it is a subgroup of* $GL(N, \mathbb{Z})$ *for some integer* N.

Proof The RACG, $W_{\Delta\Gamma}$, is a subgroup of $GL(2n, \mathbb{Z})$, where $n = \mathrm{Card}(I)$, and A_Γ is a subgroup of index 2^n in $W_{\Delta\Gamma}$.

More Properties of RAAGs Besides being linear, the RAAG, A_Γ, has many other nice properties, for example:

- Any RAAG is biorderable and hence, locally indicable, cf. [63]. (A group G is *locally indicable* if every finitely generated subgroup admits an epimorphism onto $\mathbb{Z}$.)
- A_Γ is residually torsion-free nilpotent.
- The strong Atiyah Conjecture holds for A_Γ, cf. [173]. (This is a conjecture about the integrality of L^2-Betti numbers of spaces with fundamental group A_Γ. It implies the Zero Divisor Conjecture for $\mathbf{C}[A_\Gamma]$.)
- Any word-quasiconvex subgroup of A_Γ (with respect to the standard generating set) is a virtual retract. (See Corollary 5.46 in the next subsection.)

5.3.2 Convex Subsets of Coxeter Groups and Cube Complexes

Suppose (W, S) is a RACS and that $\Sigma(W, S) = D(W, K)$ is its Davis–Moussong complex thought of as being defined by the basic construction (3.9) of Sect. 3.1.1 rather than as a zonotopal cell complex (see Sect. 4.2.3). So, $D(W, K)$ is a union of chambers $D(W, K) = \bigcup_{w \in W} wK$. For each reflection $r \in R$, let Σ^r denote the fixed set of r on Σ $(= \Sigma(W, S))$, i.e., Σ^r is the wall associated to r. A *half-space bounded by* Σ^r is the closure of one of the two components of $\Sigma - \Sigma^r$. So, a half-space is a union of chambers in $D(W, K)$. If (W, S) is a RACS, then Σ is a CAT(0) cube complex and Σ^r is a hyperplane in the sense of Sect. 5.1.1.

Definition 5.38 Given a subset $C < D(W, K)$ let $\mathcal{H}(C)$ denote the set of half-spaces which contain it. The subset $C < D(W, K)$ is *convex* if $C = \bigcap_{\mathcal{H} \in \mathcal{H}(C)} \mathcal{H}$. The convex subset C is a *convex union of chambers* if $\mathcal{H}(C)$ does not contain any pair of opposite half-spaces. A bounding hyperplane Σ^r of a half-space containing C is a called *supporting hyperplane* of a convex union of chambers if it intersects some chamber of the closure $\overline{C}$ in a codimension-one face of that chamber.

Each element of $w \in W$ can be identified with the center point of the corresponding chamber wK.

Suppose W is a RACG. The center points of chambers are then the vertices of the corresponding cube complex. A subset $Y \leq W$ is *convex* if it is the set of centers of a convex union of chambers. An equivalent definition is to say that a subset Y of vertices of the cube complex is *convex* if it is a convex subset of the 1-skeleton of the cube complex. This means that if $y_0, y_1 \in Y$, then any intermediate vertex on any geodesic edge path from y_0 to y_1 also belongs to Y. Given a subset of $U < D(W, K)$, its *convex hull*, $\mathrm{Conv}(U)$, is the intersection of all closed half-spaces that contain U. If $U < W$ is a subset of vertices, then we shall also say that the convex hull of U is the set of vertices which are contained in $\mathrm{Conv}(U)$.

Suppose H is a finitely generated subgroup of W. Then $H \curvearrowright \mathrm{Conv}(H)$. The subgroup H is *convex cocompact* if the H-action on $\mathrm{Conv}(H)$ is cocompact. This is equivalent to the definition of word-quasiconvex subgroup given in the beginning of this section.

Remark 5.39 The term "convex cocompact" had a different meaning in Sect. 4.2.9; there it referred to a projective representation of a Coxeter group so that the action was cocompact on a convex subset of projective space.

5.3.3 Convex Cocompact Subgroups of RACGs

Suppose that (W, S) is a RACS and C is some convex union of chambers in the Davis–Moussong complex $D(W, K)$ $(= \Sigma(W, S))$. Let $\mathcal{H}$ be the subset of $\mathcal{H}(C)$ consisting of those half-spaces whose bounding hyperplanes intersect C in a codimension-one subset. Such a half-space is called a *supporting half-space* of the corresponding supporting hyperplane. In other words, $\mathcal{H}$ is a minimal subset of $\mathcal{H}(C)$ such that $C = \bigcap_{h \in \mathcal{H}} h$. Let $S(C)$ denote the set of reflections across the supporting hyperplanes for the half-spaces of $\mathcal{H}(C)$. Let $W(C) = \langle S(C) \rangle$ be the subgroup of W generated by $S(C)$. Taking intersections with walls corresponding to $S(C)$ we get a mirror structure on C. The mechanism driving the results of this subsection is the observation that $W(C)$ is a right-angled reflection group on $D(W, K)$ with strict fundamental domain C. More precisely, there is the following basic lemma.

Lemma 5.40 *The pair* $(W(C), S(C))$ *is a* RACS. *The inclusion* $C \hookrightarrow D(W, K)$ *induces a* $W(C)$*-equivariant homeomorphism* $D(W(C), C) \to D(W, K)$.

Haglund proves a cubical version of this lemma in [140, Prop. 3.21] for the cube complexes $\Sigma(W, S)$ and $\Sigma(W(C), S(C))$.

Remark 5.41 A version of Lemma 5.40 holds for general Coxeter groups; however, one needs to add a hypothesis. The hypothesis is that if distinct elements $s, t \in S(C)$ satisfy $m(s, t) \neq \infty$ and h_s and h_t are the corresponding half-spaces, then $h_s \cap h_t$ is a fundamental domain for the dihedral group $\langle \{s, t\} \rangle$. In other words, the corresponding walls make a "dihedral angle" of $2\pi/m(s, t)$ rather than $2\pi/km(s, t)$ for some k, with $1 < k < m$. (This is called the *angle condition.*) This condition is automatic when W is right angled for then the only possibility for $m(s, t)$ is 2. If the angle condition does not hold, then the translate of a supporting wall of C by an element of $W(C)$ may intersect the interior of C so that C will not be a fundamental chamber. Then the lemma then states that C is a strict fundamental domain for $W(C)$ on $D(W, K)$ if and only if C satisfies the angle condition.

Before stating the main result of this subsection, Theorem 5.45 below, we will warm up by proving the residual finiteness of any RACG. (Recall Definition 2.51 of Sect. 2.4.4: a group G is *residually finite* if for any $g \in G - \{1\}$, there is a finite index subgroup $H < G$ so that $g \in G - H$. In other words, the elements 1 and g have distinct images in G/H, or more geometrically, the vertices 1 and g of the Cayley graph have distinct images in the quotient $\mathrm{Cay}(G, S)/H$. (N.B. After replacing H by the intersection of all subgroups conjugate to it, we many assume that H is a *normal* subgroup of finite index in G.)

Proposition 5.42 (cf. [205, Lemma 3.4] and [140]) *Any* RACG *is residually finite.*

Proof Let W be a RACG and $w \in W - \{1\}$. Then we can find a convex union of chambers C so that the set $\mathrm{Chamb}(C)$ of chambers that are contained in C is finite and so that the chambers corresponding to 1 and w are both in $\mathrm{Chamb}(C)$. Since $\mathrm{Chamb}(C)$ is finite, the subgroup $W(C)$ has finite index in W. Since C is the orbit space of $W(C)$ on $D(W, K)$, it follows that 1 and w have distinct images in $W/W(C)$.

Remark 5.43 Since W has a faithful linear representation, Proposition 5.42 also follows from the standard result that finitely generated linear groups are residually finite.

Definition 5.44 A subgroup H of a group G is a *virtual retract* of G if there is a subgroup of finite index $G' \le G$ that contains H and a retraction $r : G' \to H$. (The corresponding notion for an immersed subspace $f : A \looparrowright X$ of X is that there is a finite-sheeted covering space X' of X, a lift of f to an embedding $f' : A \to X'$ and a retraction $r : X' \to f'(A)$.)

The following theorem of Haglund is the main result in this subsection.

Theorem 5.45 (cf. Haglund [140, Theorem A]) *Suppose W is a* RACG *and H is a convex cocompact subgroup (i.e., H is a word-quasiconvex subgroup of W with respect to the standard set of generators). Then H is a virtual retract of W.*

Sketch of Proof Put $C = \mathrm{Conv}(H)$. The group H acts on C. In particular, it permutes the supporting hyperplanes of C. Hence, it acts on $S(C)$. Since intersections of supporting hyperplanes are preserved, it acts on the nerve $L(W(C), S(C))$. Hence, H acts on $(W(C), S(C))$ by diagram automorphisms. We claim that the subgroup $HW(C) < W$ is a semidirect product. To see this, first note that taking the orbit space gives a retraction $D(W, K) = D(W(C), C) \to C$. Identifying $w \in W$ with the center point of the chamber wK, this retraction induces a map $r : W \to \mathrm{Chamb}(C) = W/W(C)$. Identify H with a subset of $\mathrm{Chamb}(C)$ and put $W' = r^{-1}(H)$. Then W' is a subgroup (in fact, $W' = HW(C)$) and we have a short exact sequence:

$$1 \to W(C) \longrightarrow W' \xrightarrow{r} H \to 1.$$

So, W' is the semidirect product, $W' = H \ltimes W(C)$. Since $\mathrm{Chamb}(C)/H$ is finite, W' has finite index in W, i.e., H is a virtual retract.

As we explained in Sect. 5.3.1, given a flag complex L, the RAAG, A_L, and the RACGs, W_{OL} and $W_{\Delta L}$, all act on the same CAT(0) space X. All three groups are commensurable; however, they give three different structures on X as a CAT(0) cube complex. (When L is a point, X is isomorphic to $\mathbb{R}$ (the universal cover of the circle) and $A_L = \mathbb{Z}$ acting with fundamental domain $[0, 2]$; the groups W_{OL} and $W_{\Delta L}$ are both infinite dihedral with fundamental chambers $[-1, 1]$ and $[0, 1]$, respectively.) In general, the group A_L is a subgroup of index 2^n in $W_{\Delta L}$. Hence, Theorem 5.45 has the following corollary.

Corollary 5.46 *Suppose A is a* RAAG *and H is a word-quasiconvex subgroup of A (with respect to the standard generating set), cf. [152]. Then H is a virtual retract of A.*

Proof Suppose $A = A_L$. For a base point take the cone point $0 \in K_{OL} \cap [0, 1]^I = K_{\Delta L}$. Consider $H(0)$, the H-orbit of 0. Let $C = \mathrm{Conv}(H)$ be the convex hull of $H(0)$ in $D(W_{\Delta L}, K_{\Delta L})$. As before, $H \curvearrowright C$ and we get a subgroup W' of finite index in $W_{\Delta L}$ and a semidirect product decomposition, $W' = H \ltimes W(C)$, where $W(C)$ is a reflection subgroup of $W_{\Delta L}$. Put $A' = W' \cap A_L$. It is easy to see that A' is index 2^n in W' and that we have a semidirect product decomposition $A' = H \ltimes A(C)$, where $A(C)$ is the kernel of the retraction of A' onto H.

Corollary 5.47 *Suppose B is a compact special cube complex B, then $\pi_1(B)$ is residually finite. In particular, virtually special groups are residually finite.*

More or less the same method proves the following.

Theorem 5.48 (cf. Haglund–Wise [143] or [21, Theorem 4.2]) *Suppose B and Z are compact special NPC cube complexes and $f : B \looparrowright Z$ is a locally isometric immersion. Then there is a finite cover Z' of Z and a lift of f to Z' so that Z' retracts onto the image of the lift of f. So, $\pi_1(B)$ is a virtual retract of $\pi_1(Z)$.*

Sketch of Proof The special case $Z = \mathbb{T}_L$, where $\mathbb{T}_L$ denotes the standard classifying space of a RAAG associated to an A-typing of B, was handled in Corollary 5.46 above. In the general case first use Theorem 5.27 (i) to immerse Z in the classifying space $X = \mathbb{T}_L$ of a RAAG (possibly after replacing Z by a finite cover) to get a finite cover $X' \to X$ and a lift to X' of the composition $B \to Z \to X$ so that X' retracts onto the image of B. Pull back X' to a cover Z' of Z. Then f lifts to Z' and Z' retracts onto the image of the lift of f.

5.3.4 Separability of Word-Quasiconvex Subgroups

Definition 5.49 A subgroup H of a group G is *separable in G* if for each $g \in G - H$ there is a subgroup of finite index $G' < G$ such that $H < G'$ and $g \notin G'$.

For example, G is residually finite if and only if the trivial subgroup is separable in G.

Lemma 5.50 ([140, Lemma 3.26]) *Assume H and G' are subgroups of a group G with G' of finite index and $H < G'$. If H is separable in G', then it is separable in G.*

Proof We must separate $g \in G - H$ from H by a finite-index subgroup $G'' < G$. If $g \notin G'$, take $G'' = G'$. If $g \in G'$, then let $G'' < G'$ be a finite-index normal subgroup which separates g from H.

Lemma 5.51 ([140, Lemma 3.27]) *Assume G is residually finite and that $r : G \to H$ is a retraction onto a subgroup H. Then H is separable in G.*

Proof Let $N = \ker(r)$ so that G is the semidirect product $H \ltimes N$. Let $g \in G - H$. Then we can write $g = hn$ with $r(n) = 1$ so that $h = r(g)$. Since $g \notin H, n \neq 1$. Since G is residually finite, there is a finite index normal subgroup $G' < G$ such that $n \notin G'$. Put $N' = N \cap G'$. Then the subgroup generated by H and N' is the semidirect product $H \ltimes N'$ with $g \notin G'$. Moreover, $[G : G'] = [N : N'] < \infty$.

Combining these two lemmas we get the following.

Proposition 5.52 ([140, Proposition 3.28]) *If H is a virtual retract of a residually finite group G, then H is separable in G.*

The next theorem results from combining Propositions 5.42 and 5.52 with Theorem 5.45.

Theorem 5.53 (Haglund [140, Theorem A]) *Any word-quasiconvex subgroup of a* RACG *is separable.*

Remark 5.54 Haglund proves the generalization of Theorem 5.45 for any graph product of finite groups: any word-quasiconvex subgroup of such a graph product is separable (cf. [140, Theorem B]).

Similarly, by using Corollary 5.46 we see that any word-quasiconvex subgroup of a RAAG is separable.

Corollary 5.55 *Suppose B is a compact A-special NPC cube complex. Then $\pi_1(B)$ is a separable subgroup of some* RAAG. *If B is C-special, then an index* 2 *subgroup of $\pi_1(B)$ is a separable subgroup of some* RACG.

Proof After replacing B by a finite cover, we can find a locally isometric immersion $B \looparrowright \mathbb{T}_L$ where X is the classifying space of a RAAG A_L. Passing to universal covers we get an isometric embedding $\tilde{B} \hookrightarrow \tilde{\mathbb{T}}_L$ and hence, an embedding of $\pi_1(B)$ as a subgroup of finite index in A_L (or equivalently, as a subgroup of the RACG, $W_{\Delta L}$). Thus, $\pi_1(B)$ is a convex cocompact subgroup of $W_{\Delta L}$. So, $\pi_1(B)$ is a word-quasiconvex subgroup and hence, by Theorem 5.53, a separable subgroup.

Since any quasi-isometrically embedded subgroup of a word hyperbolic group is quasiconvex, one can use Lemma 5.50 and the method for proving Theorem 5.48 to get the following.

Theorem 5.56 (Haglund–Wise [143, Theorem 7.3] or [21, Prop. 4.3]) *Suppose a word hyperbolic group G is the fundamental group of a compact cube complex. Then any quasiconvex subgroup of G is separable.*

Conversely, if B is a compact NPC cube complex such that $\pi_1(B)$ is word hyperbolic and separable on quasiconvex subgroups, then B is virtually special (cf. [143]).

The hypothesis that G is word hyperbolic is definitely necessary in Theorem 5.56. Indeed, the Burger-Mozes examples in Sect. 2.4.4 are NPC square complexes whose fundamental groups are not residually finite.

Remark 5.57 One of the main uses of subgroup separability is for converting π_1-injective immersions into embeddings. Indeed, suppose G is the fundamental group of an n-manifold M^n and that H is the image of the fundamental group of a codimension-one submanifold N^{n-1} under a π_1-injective immersion $f : N \looparrowright M$ in general position and with only double points. Let $P < M$ be a component of the set of double points and let α be a loop in $\mathrm{Im}(f)$ which starts at a basepoint $x_0 \in \mathrm{Im}(f)$, travels along one sheet of the immersion to a point in P and then returns to x_0 along the other sheet. Suppose H is separable. Then there is a subgroup G' of finite index in G so that $H < G'$ and $\alpha \notin G'$. Let M' be the finite-sheeted cover of M corresponding to G'. Then f lifts to an immersion $f' : N \to M'$ with no double points lying above P. Continuing to eliminate double points in this fashion, we eventually reach an embedding $N \hookrightarrow M$. (See [21, Figure 2] for a picture of this argument when $n = 1$.)

5.3.5 Applications to 3-Manifolds

In the late1970 s Thurston proposed a new vision of 3-manifolds: his Geometrization Conjecture (see [214]). This was analogous to the classical Uniformization Theorem for Riemann surfaces, which asserts that any such surface is conformally equivalent to one with a metric of constant curvature. The main part of the Geometrization Conjecture concerned hyperbolic 3-manifolds. It asserted that a closed 3-manifold M^3 has a hyperbolic structure if and only if it is aspherical and atoroidal (i.e., its fundamental group does not contain a copy of $\mathbb{Z} \times \mathbb{Z}$). (There is a similar version of this conjecture when M^3 is the interior of a compact manifold with tori as its boundary components.) Thurston proved this conjecture for Haken 3-manifolds. He also proved it in the case where M^3 is a fiber bundle over S^1 with "pseudo-Anosov" holonomy. Of course, the conjecture would also follow if virtual versions of either of these results could be proved, that is, after replacing M by a finite sheeted cover M'. This led Thurston to ask in [215, Questions 16 and 18] if every aspherical, atoroidal 3-manifold is virtually Haken and similarly, did it virtually fiber over S^1? Positive answer to these questions became known as the Virtual Haken Conjecture and the Virtual Fibering Conjecture. Although Thurston presented these conjectures as methods for proving geometrization, neither conjecture was known to hold even when the 3-manifold was hyperbolic. In 2002 and 2003 Perelman wrote three preprints in which he proved the full Geometrization Conjecture (including the Poincaré Conjecture!) by using the Ricci flow. This obviated the original motivation for proving the Virtual Haken Conjecture and the Virtual Fibering Conjecture and reduced these questions to the case of hyperbolic 3-manifolds. The culmination of the techniques developed in this chapter is Agol's proof in [3] of these two conjectures. These basically follow from his proof of Wise's Conjecture, Theorem 5.33. This is all explained in Bestvina's survey article [21].

The argument goes roughly as follows. First one uses Sageev's construction from Sect. 5.1.2 together with the Kahn–Marković Theorem 5.20 to show that the fundamental group G of a closed hyperbolic 3-manifold M^3 is the fundamental group of a finite NPC cube complex. By Theorem 5.33 there is a subgroup of finite index $G' < G$ so that the corresponding cover is special. So, quasiconvex subgroups of G are separable. Hence, as explained in Remark 5.57, each of surface subgroup of $\pi_1(M^3)$, which exists by the Kahn–Marković Theorem, lifts to an incompressible embedded surface in some finite sheeted cover M' of M. Hence, M is virtually Haken.

As for the Virtual Fibering Theorem, Agol defines the property of a group being RFRS and he showed that if the fundamental group of a hyperbolic 3-manifold is RFRS, then the 3-manifold virtually fibers over S^1; moreover, RAAGs have the

RFRS property. Using Theorem 5.33 again, he concluded M has a finite-sheeted cover that fibers over S^1.

Remark 5.58 The definition of a Haken 3-manifold involves the existence of an incompressible surface Σ^2 in M^3 (in fact, only one incompressible surface is needed). This means that we have a π_1-injective embedding $\Sigma^2 \hookrightarrow M^3$. In particular, $\pi_1(\Sigma^2)$ is a subgroup of $\pi_1(M^3)$. Conversely, the inclusion $\pi_1(\Sigma^2)$ into $\pi_1(M^3)$ is induced by a map $\Sigma^2 \to M^3$ and any such map is homotopic to an immersion $\Sigma^2 \looparrowright M^3$. Proving M^3 to be Haken involves two steps: (1) showing $\pi_1(M^3)$ has at least one surface subgroup and (2) changing the resulting immersed surface into an embedded surface. So, a crucial problem in 3-manifold theory is being able to solve (2).

Chapter 6
Hyperbolization

"Hyperbolization" refers to various functorial constructions of Gromov [136] for converting a cell complex J into an NPC polyhedron $\mathcal{H}(J)$. Several such procedures are discussed in this chapter. The result is usually an NPC cube complex. In Sect. 6.1.1 we discuss some properties which a hyperbolization procedure might satisfy. In Sect. 6.3 the product with interval procedure is explained, while in Sect. 6.4 Gromov's construction is explained. A procedure for constructing locally CAT(-1) spaces, called "strict hyperbolization," is explained in Sect. 6.5. Relative hyperbolization procedures are explained in Sects. 6.1.2 and 6.3.1. In Sect. 6.2 we give applications of hyperbolization and relative hyperbolization to constructions of NPC manifolds.

6.1 Properties of Hyperbolization

Hyperbolization constructions are defined by induction on dimension, roughly, as follows. Assuming the hyperbolization of the $(n-1)$-skeleton J^{n-1} has been defined, the hyperbolization of an n-cell σ of J is defined to be some NPC n-manifold boundary, whose boundary is the hyperbolization of $\partial\sigma$. Finally, the hyperbolization of J^n is defined to be the result of using functoriality to glue on hyperbolized n-cells to the hyperbolization of J^{n-1} in the same combinatorial pattern as the cells of J^n are glued onto J^{n-1}. Basic references for this material are [59, 91, 136].

In the original hyperbolization procedures of [136] and [91], the result was a locally CAT(0), piecewise euclidean polyhedron. The term "hyperbolization" is actually a misnomer for such a procedure. In Gromov's original vision, the output of hyperbolization should have been piecewise hyperbolic and locally CAT(-1) space. Such a procedure will be called a *strict hyperbolization*. Even without requiring the construction to be locally CAT(-1), the output of a true hyperbolization procedure

M. W. Davis, *Infinite Group Actions on Polyhedra*, Ergebnisse der Mathematik
und ihrer Grenzgebiete. 3. Folge / A Series of Modern Surveys in Mathematics 77,
https://doi.org/10.1007/978-3-031-48443-8_6

still should be an NPC space with word hyperbolic fundamental group. However, as is explained in [59, Corollaries 4.3 and 4.4], for $n \geq 4$, Gromov's original hyperbolization procedures in [136], as well as the procedures of [91], do not have such fundametal groups. In other words, the initial versions of hyperbolization were not strict.

Gromov said that his purpose in defining a hyperbolization procedure was simply to show that there was a simple method for constructing a large number of nonpositively curved (or negatively curved) spaces, e.g., given a polyhedron J, one can find an NPC polyhedron $\mathcal{H}(J)$ whose homology maps onto that of J. One motivation for finding such a construction was the Kan–Thurston Theorem of [163] which asserts that for any J one can find an aspherical polyhedron $\mathcal{A}(J)$ and a map $c : \mathcal{A}(J) \to J$ inducing an isomorphism on homology. In a Kan–Thurston construction one can take $\mathcal{A}(J)$ to be a finite complex whenever J is a finite complex.

6.1.1 Axioms for Hyperbolization

Let $\mathcal{C}_n$ be a category of cell complexes of dimension $\leq n$. A *morphism* in $\mathcal{C}_n$ from J_1 to J_2 is a cellular immersion $i : J_1 \to J_2$ such that i takes each cell of J_1 isomorphically onto a cell of J_2. For example, i could be an embedding of a subcomplex or a covering map. Let $\mathcal{PE}_n$ be the category of NPC piecewise euclidean polyhedra. A morphism in $\mathcal{PE}_n$ is a locally isometric immersion. A *hyperbolization* $\mathcal{H}$ is, first of all, a functor from a subcategory of $\mathcal{C}_n$ to $\mathcal{PE}_n$. In the initial cases that we discuss, $\mathcal{H}(J)$ is an NPC cube complex; however, its fundamental group will usually not be word hyperbolic group; often it will contain copies of $\mathbb{Z} \times \mathbb{Z}$. A *strict* hyperbolization procedure should take values in the category $\mathcal{PH}_n$ of n-dimensional, piecewise hyperbolic, locally CAT(-1) spaces so that the fundamental group of a result of a strict hyperbolization procedure will be a word hyperbolic group. Eventually, a method of [59] for constructing strict hyperbolizations will be explained in Sect. 6.5.

In what follows, except in Sect. 6.3.1 and in Sect. 6.5, $\mathcal{H}(J)$ will have the structure of an NPC cube complex. A consequence of functoriality is that if J_1 is a subcomplex of J_2, then $\mathcal{H}(J_1)$ is a totally geodesic subspace of $\mathcal{H}(J_2)$; hence, $\pi_1(\mathcal{H}(J_1)) \to \pi_1(\mathcal{H}(J_2))$ is injective. Another consequence is that if $G \curvearrowright J$, then there is an induced G-action on $\mathcal{H}(J)$.

There are several properties (or "axioms") which we might want to require the functor $\mathcal{H}$ to satisfy.

(H0) (*Preservation of local structure*). For each n-cell σ^n, its hyperbolization $\mathcal{H}(\sigma^n)$ is an NPC n-manifold with totally geodesic boundary $\mathcal{H}(\partial\sigma^n)$. Moreover, if $\mathrm{Lk}(\sigma, J)$ denotes the link of σ in J (so that there is a neighborhood of σ in J of the form $\sigma \times \mathrm{Cone}(\mathrm{Lk}(\sigma, J))$), then there is a neighborhood of $\mathcal{H}(\sigma)$ in $\mathcal{H}(J)$ of the form $\mathcal{H}(\sigma) \times \mathrm{Cone}(L')$ where L' is PL homeomorphic to $\mathrm{Lk}(\sigma, J)$.

(H1) There is a map $c : \mathcal{H}(J) \to J$ so that for each cell σ of J, c takes $\mathcal{H}(\sigma)$ to σ.
(H2) The hyperbolization of a point is a point.
(H3) For each cell σ of J, the manifold with boundary, $\mathcal{H}(\sigma)$, is orientable.

The map, $c : \mathcal{H}(J) \to J$ in **(H1)** is determined up to homotopy by functorality and by Axiom **(H0)**. First, for any 0-cell σ^0, $\mathcal{H}(\sigma^0) \to \sigma^0$ is the constant map. We can assume by induction on n that, for any n-cell σ, there is a map $c : \mathcal{H}(\partial\sigma) \to \partial\sigma$ unique up to homotopy. Since σ is contractible, this map extends to $c : \mathcal{H}(\sigma) \to \sigma$. Axiom **(H2)** implies that $c : \mathcal{H}(\sigma) \to \sigma$ has degree 1 modulo 2. If, in addition, **(H3)** holds, then $c : \mathcal{H}(\sigma) \to \sigma$ is a *degree one* map. By **(H0)**, **(H2)**, **(H3)** that the map $c : (\mathcal{H}(\sigma^n), \mathcal{H}(\partial\sigma^n)) \to (\sigma^n, \partial\sigma^n)$ is a degree one map between connected, orientable manifolds with boundary. It follows easily from this that $c : \mathcal{H}(J) \to J$ induces an epimorphism on homology. (Even when $\mathcal{H}(\sigma^n)$ is not orientable c is still an epimorphisn on homology with coefficients in $\mathbb{Z}/2$.) In particular, if J is an orientable manifold, then so is $\mathcal{H}(J)$ and $c : \mathcal{H}(J) \to J$ is a degree one map. In view of the Kan–Thurston Theorem [163] one might speculate about the possibility of requiring $c : \mathcal{H}(J) \to J$ to induce a homology isomorphism in each degree. However, Axiom **(H0)**, which requires that hyperbolization takes n-manifolds to n-manifolds, precludes this possibility. Indeed, since there is no aspherical surface with the same homology as S^2, it is impossible to define $\mathcal{H}(S^2)$ to be simultaneously NPC and a homology 2-sphere.

Here are three more possibilities for axioms.

(H4) If J is a smooth cellulation of a smooth manifold, then $\mathcal{H}(J)$ also is a smooth manifold.
(H5) For any cell σ, the tangent bundle of $\mathcal{H}(\sigma)$ is trivial.
(H6) Suppose J is a smooth or PL manifold, then $c : \mathcal{H}(J) \to J$ pulls back the stable tangent bundle of J to the stable tangent bundle of $\mathcal{H}(J)$. (This often follows from **(H5)**.)

Möbius Band Hyperbolization In Sect. 3.3.2 we described Gromov's procedure $X \mapsto \mathcal{H}(X)$ for converting a cubical complex X (or more generally a zonotopal complex) into an NPC cube complex, $\mathcal{H}(X)$, called the *Möbius band hyperbolization*. It satisfies Axioms **(H0)** **(H1)** **(H2)** and **(H4)**. Axiom **(H3)** fails completely: the Möbius band hyperbolization of d-cube is not orientable for $d > 1$.

Before describing in Sect. 6.4 a nonstrict hyperbolization procedure of Gromov [136] which satisfies all the axioms, we give some consequences of the existence of such a procedure.

Relative Hyperbolization Gromov [136, §3.4.C] proposed the idea of finding a relative version of a hyperbolization procedure $\mathcal{H}$. As explained in [91] and [96], the existence of a suitable relative version is a consequence of the previous axioms. Given a pair of cell complexes, (J, B), a *relative hyperbolization* procedure should functorially associate a space $\mathcal{H}(J, B)$ such that B is a subspace of $\mathcal{H}(J, B)$. Although it is not required that B be nonpositively curved, if it is, then $\mathcal{H}(J, B)$

should also be NPC. We list some axioms for relative hyperbolization below. As for the relative version of Axiom (**H0**), one might want to require that a regular neighborhood B in $\mathcal{H}(J, B)$ be homeomorphic to a regular neighborhood of B in J. However, let us stick to axioms that are easier to satisfy.

(**RH0**) If J is a manifold with boundary and B is a union of its boundary components, then $\mathcal{H}(J, B)$ is a manifold with boundary and B a union of boundary components of $\mathcal{H}(J, B)$.

(**RH1**) (*Asphericity*). If each component of B is aspherical, then $\mathcal{H}(J, B)$ is aspherical.

As we explain in the next subsection, a consequence of the existence of a hyperbolization procedure satisfying (**H0**), (**H1**) is the existence of a relative version of it. For the next axiom see [15].

(**RH2**) (*Relatively hyperbolic groups*), cf. [15]. If $\mathcal{H}(J, B)$ is induced from a strict hyperbolization procedure $\mathcal{H}$, then $\pi_1(\mathcal{H}(J, B))$ is a relatively hyperbolic group (in the sense of Bowditch [31] or Farb [114]) where the peripheral subgroups are the conjugates of the fundamental groups of the components of B.

Finally, if B has the structure of an NPC piecewise euclidean polyhedron and B is a subcomplex of a cell complex J (no metric on J is assumed), then we show in Sect. 6.3 how to construct an NPC polyhedron $\mathcal{H}(J, B)$ satisfying the following.

(**RH3**) Suppose B is an NPC polyhedron and that it is a subcomplex of J.Then the piecewise euclidean polyhedron $\mathcal{H}(J, B)$ is NPC and B is a totally geodesic subspace of it.

6.1.2 A Relative Version of Hyperbolization

In this subsection we show how the existence of a suitable hyperbolization procedure implies the existence of a relative hyperbolization version satisfying Axioms (**RH0**), (**RH1**), (**RH2**). The discussion follows [96].

Suppose $\mathcal{H}$ is a hyperbolization procedure satisfying (**H0**), (**H1**). For the purpose of simplifying notation, suppose it also satisfies (**H2**). We suppose that if $\mathcal{H}(\sigma)$ is defined for a cell σ, then it is also defined for Cone(σ). (We call such a cell σ *allowable*.) Note that this assumption precludes the possibility of using the Möbius band hyperbolization procedure or indeed, any procedure that is only defined for the subcategory of cube complexes. The point is that for any allowable cell σ we will need to be able to define $\mathcal{H}(\text{Cone}(\sigma))$. There is no problem if σ is a simplex; however, the cone on a d-cube is not a $(d+1)$-cube. Also suppose that for each cell complex J, $\mathcal{H}(J)$ is an NPC cube complex.

Let B be a subcomplex of a cell complex J and let $\{B_\alpha\}$ be its set of path components. Let CB denote the disjoint union of the Cone(B_α) and let $J \cup CB$ denote complex formed by attaching each Cone(B_α) to J along B_α. Hyperbolize to get $\mathcal{H}(J \cup CB)$. By Axiom **(H2)**, there is a unique point $b_\alpha \in \mathcal{H}(J \cup CB)$ corresponding to the cone point of Cone(B_α). By **(H0)**, each b_α has a small neighborhood in $\mathcal{H}(J \cup CB)$ homeomorphic to the open cone on B_α. Remove these neighborhoods to get a space $\mathcal{H}(J, B)$ called the *relative hyperbolization of J with respect to B*. Removing these neighborhoods destroys the nonpositive curvature of $\mathcal{H}(J \cup CB)$. So, $\mathcal{H}(J, B)$ will not be endowed with any nonpositively curved metric. (On the other hand, a different relative hyperbolization procedure described in Sect. 6.3.1 will yield an NPC complex.) Note that B_α, the base of Cone(B_α), is a subspace of $\mathcal{H}(J, B)$ homeomorphic to the link of b_α in $\mathcal{H}(J \cup CB)$.

Let $\overline{\mathcal{H}}(J \cup CB)$ denote the universal cover of $\mathcal{H}(J \cup CB)$ and let $\overline{\mathcal{H}}(J, B)$ denote the inverse image of $\mathcal{H}(J, B)$ in $\overline{\mathcal{H}}(J \cup CB)$. The key fact is that the inclusion $B_\alpha \hookrightarrow \mathcal{H}(J, B)$ induces an injection on fundamental groups (cf. Corollary 6.2 below). Let $\overline{b}_\alpha$ denote a lift of the cone point b_α to $\overline{\mathcal{H}}(J \cup CB)$.

Lemma 6.1 (cf. [96, Lemma 2.2]) *Let L_α $(= B_\alpha)$ be the link of a lifted cone point $\overline{b}_\alpha$ in $\overline{\mathcal{H}}(J \cup CB)$. Then $\overline{\mathcal{H}}(J, B)$ retracts onto L_α. Hence, $\pi_1(L_\alpha) \to \pi_1(\overline{\mathcal{H}}(J \cup CB))$ is injective.*

Proof Since $\overline{\mathcal{H}}(J \cup CB)$ is CAT(0), geodesic contraction provides a deformation retraction of $\overline{\mathcal{H}}(J \cup CB) - \overline{b}_\alpha$ onto L_α. This deformation retraction restricts to a retraction $\overline{\mathcal{H}}(J, B) \to L_\alpha$.

Since $\pi_1(L_\alpha) \to \pi_1(\overline{\mathcal{H}}(J \cup CB))$ is injective and since $\overline{\mathcal{H}}(J, B)$ is a covering space of $\mathcal{H}(J, B)$, we get the following corollary to Lemma 6.1.

Corollary 6.2 *For each component B_α of B, the homomorphism $\pi_1(B_\alpha) \to \pi_1(\mathcal{H}(J, B))$ is injective.*

Theorem 6.3 (cf. [96, Theorem 2.5]) *The relative hyperbolization $\mathcal{H}(J, B)$ is aspherical if and only if each component of B is aspherical.*

The proof of Theorem 6.3 uses the notion of "the universal branched cover" defined below.

Definition 6.4 Suppose X is a geodesic metric space and $\{X_i\}_{i\in I}$ is a locally finite collection of locally convex subspaces indexed by some set I. (The X_i are called the "strata" of X.) The *universal branched cover of X along the X_i* is the space $\tilde{X}$ formed by taking the metric completion of the universal cover of $X - \bigcup X_i$.

For example, if the geodesic metric space X is a surface and the subset B is a singleton, then $X - B$ is a punctured surface and its universal cover corresponds to the usual picture of the hyperbolic plane with the inverse image of B as cusps on the circle at infinity. The universal branched cover $\tilde{X}$ is obtained by filling in the cusps. However, the metric on the interior of $\tilde{X}$ is not that of the hyperbolic plane since geodesics need not extend to infinity: the distance from a point in the universal cover to a cusp point is finite.

We continue our discussion of $\mathcal{H}(J, B)$ and its properties. Let $\tilde{\mathcal{H}}(J \cup CB)$ denote the universal branched cover of $\mathcal{H}(J \cup CB)$ branched along the cone points b_α. (It is also the universal branched cover of $\overline{\mathcal{H}}(J \cup CB)$ branched along the cone points.)

The cubical structure on $\mathcal{H}(J \cup CB)$ induces one on $\tilde{\mathcal{H}}(J \cup CB)$. The link L_α of b_α in $\mathcal{H}(J \cup CB)$ is a subdivision of the link of the corresponding point in J. If $\tilde{b}_\alpha$ is a point in $\tilde{\mathcal{H}}(J \cup CB)$ lying above b_α, then its link $\tilde{L}_\alpha$ is a covering space of L_α. Since $\mathcal{H}(J \cup CB)$ is NPC, L_α is a simplicial flag complex. It follows that $\tilde{L}_\alpha$ also is a flag complex and hence, that $\tilde{\mathcal{H}}(J \cup CB)$ is a CAT(0) cube complex. Let $\tilde{\mathcal{H}}(J, B)$ denote the inverse image of $\mathcal{H}(J, B)$ in $\tilde{\mathcal{H}}(J \cup CB)$. The space $\tilde{\mathcal{H}}(J \cup CB)$ can be reconstructed from $\tilde{\mathcal{H}}(J, B)$ by gluing back a copy of Cone$(\tilde{L}_\alpha)$ for each cone point in the inverse image of b_α.

Proof of Theorem 6.3 The "only if" part follows immediately from Lemma 6.1. If B_α is aspherical, then its universal cover $\tilde{L}_\alpha$ is contractible. Since $\tilde{\mathcal{H}}(J \cup CB)$ is CAT(0), it is contractible. Since this contractible space is formed by attaching cones on contractible spaces to $\tilde{\mathcal{H}}(J, B)$, we see $\tilde{\mathcal{H}}(J, B)$ also is contractible.

6.2 Applications to Aspherical Manifolds

An important use of hyperbolization and relative hyperbolization is to convert examples of manifolds with certain properties into aspherical manifolds with similar properties. Before giving the details of any specific hyperbolization procedure, we first indicate how the axioms in the previous section can be used to find such examples of aspherical manifolds. References include [80, 82, 91, 96].

6.2.1 Bordisms of Aspherical Manifolds

Suppose $\mathcal{H}$ satisfies axioms **(H0)**–**(H6)**. Axiom **(H6)** implies that when J is a manifold, the map $c : \mathcal{H}(J) \to J$ pulls back the Pontrjagin classes and Stiefel-Whitney classes of J to those of $\mathcal{H}(J)$. In particular, if c has degree 1 mod 2, then the Stiefel-Whitney numbers of $\mathcal{H}(J)$ are the same as those of J and if c has degree one, then so are its Pontrjagin numbers. This means that when $\mathcal{H}(J)$ is a smooth orientable manifold, its bordism class is the same as that of J. A direct proof of this fact is provided in part (i) of the next theorem. Both parts of this theorem were suggested by Gromov [136, pp. 117–118].

Theorem 6.5 (cf. [91, Example 1g.1] and [96, Theorem 1.1])

(i) *Suppose M is a triangulable, closed n-manifold. Then M is bordant to an aspherical manifold M' (which can be taken to be an NPC cube complex).*
(ii) *Suppose B is a disjoint union of closed aspherical n-manifolds and that B is the boundary of a triangulable $(n+1)$-manifold J. Then B bounds an aspherical manifold.*

Proof

(i) Start with $M \times [0, 1]$ and attach Cone(M) onto $M \times 1$. Apply relative hyperbolization to get $\mathcal{H}(M \times [0, 1], M \times 1)$. It follows from **(H0)** that $\mathcal{H}(M \times [0, 1], M \times 1)$ is a manifold with boundary. By **(H2)**, it has two boundary components. One is the NPC cube complex $\mathcal{H}(M \times 0)$ and the other is $M \times 1$.
(ii) The manifold B is the boundary of the relative hyperbolization, $\mathcal{H}(J, B)$.

Remark 6.6 Theorem 6.5 applies to any bordism theory: unoriented, oriented, or even complex (co)bordism, as well as bordism of topological manifolds. In fact, Axiom (**H6**) implies that $c : \mathcal{H}(M) \to M$ is a normal map in the sense of surgery theory. Moreover, $\mathcal{H}(M \times [0, 1], M \times 1) \to (M \times [0, 1], M \times 1)$ is a normal bordism between c and the identity map on M.

6.2.2 Nontriangulable Aspherical Manifolds

In this subsection and the next we expand the examples of Sects. 3.2.1 and 3.2.3 in Sect. 3.2. We will use notions that were discussed Sect. 3.2.1 such as: PL manifold, polyhedral homology manifold, and generalized homology sphere.

Given a topological manifold M^n, in [166] Kirby–Siebenmann defined a cohomology class $\Delta(M^n) \in H^4(M^n; \mathbb{Z}/2)$, which is an obstruction for M^n to have a PL structure. For $n \geq 5$, it is the only obstruction.

Let $Q^4(E_8)$ be the smooth 4-manifold with boundary formed by plumbing together 8 copies of the tangent disk-bundle to S^2 according to the E_8 diagram. Its boundary Σ^3 is Poincaré's homology 3-sphere. The intersection form of $Q^4(E_8)$ is the symmetric bilinear form associated to E_8, i.e., it is an even form with signature 8. Triangulate $Q^4(E_8)$ and put $X^4(E_8) = Q^4(E_8) \cup \text{Cone}\, \Sigma^3$. It is a polyhedral homology manifold with one non-manifold point. In [121, p. 367] Freedman proved that any homology 3-sphere bounds a contractible topological manifold. It follows that $X^4(E_8)$ can be resolved to a homotopy equivalent topological 4-manifold $M^4(E_8)$ (the E_8-*manifold*) by the simple procedure of removing Cone Σ^3 and replacing it with a contractible manifold W^4 cobounding Σ^3. By Rohlin's Theorem $M^4(E_8)$ is not homotopy equivalent to a smooth or PL manifold. (In general, when M^4 is obtained by resolving the PL singularities of a polyhedral homology 4-manifold by this procedure, its Kirby–Siebenmann obstruction $\Delta(M^4)$ can be identified with the sum of the Rohlin invariants of the homology 3-spheres which occur as links of singular points.) By the 3-dimensional Poincaré Conjecture (that is, by Perelman's Theorem), every triangulation of a 4-manifold is a PL manifold; in particular, $M^4(E_8)$ cannot be triangulated. (This was proved earlier by Akbulut–McCarthy using Casson's invariant.) We apply hyperbolization to $X^4(E_8)$ to get the following result of [91].

Theorem 6.7 (Davis–Januszkiewicz [91, Theorem 5a.1])

(i) There are closed aspherical 4-manifolds that cannot be triangulated.
(ii) In each dimension ≥ 4 there are closed aspherical manifolds that admit no PL structure.

Proof

(i) Hyperbolize $X^4(E_8)$ to get a polyhedral homology manifold $\mathcal{H}(X^4(E_8))$ with only one non-manifold point (this uses (**H0**)). Resolve it to get a topological manifold N^4. Since N^4 is homotopy equivalent to $\mathcal{H}(X^4(E_8))$, it is aspherical. Its Kirby–Siebenmann invariant, $\Delta(N^4)$, is not zero; so, N^4 cannot be triangulated.
(ii) For any $k \geq 0$ the Kirby–Siebenmann obstruction of $N^4 \times T^k$ is nonzero.

Although there exist topological manifolds n every dimension ≥ 4 with no PL structure, until fairly recently it seemed possible that every manifold of dimension ≥ 5 might be triangulable. Galewski–Stern [126] and Matumoto [177] analyzed the obstructions to triangulability. Their analysis showed that the obstruction depends on properties of homology 3-spheres. Let θ_3^H be the abelian group of homology cobordism classes of integral homology 3-spheres. Rohlin's μ-invariant defines a homomorphism, $\mu : \theta_3^H \to \mathbb{Z}_2$. The obstruction to triangulating a manifold M^n is a cohomology class in $H^5(M^n; \ker \mu)$, namely, it is the image of the Kirby–Siebenmann obstruction $\Delta(M^n)$ under the Bockstein homomorphism associated to the coefficient sequence,

$$0 \longrightarrow \ker \mu \longrightarrow \theta_3^H \xrightarrow{\mu} \mathbb{Z}_2 \longrightarrow 0.$$

If every homology 3-sphere with nonzero μ-invariant had order 2 in θ_3^H, this Bockstein homomorphism would be the zero map. So, Galewski–Stern showed that in dimensions ≥ 5 triangulability came down to a question about homology 3-spheres: Is there a homology 3-sphere Σ^3 with $\mu(\Sigma^3) \neq 0$ which represents an element of order 2 in θ_3^H. Such a Σ^3 exists if and only if in dimensions ≥ 5 every manifold is triangulable. In [174] Manolescu proved that no such Σ^3 exists and hence, that the manifolds constructed by Galewski–Stern [127] are not triangulable. In [90] hyperbolization and relative hyperbolization are used to prove the following.

Theorem 6.8 (Davis–Fowler–Lafont [90]) *In each dimension $n \geq 6$ there are closed aspherical n-manifolds that cannot be triangulated.*

Proof In dimensions $n \geq 5$ the Galewski–Stern manifold has the form $P_1 \cup U \cup P_2$ where P_1 is a polyhedral homology manifold with boundary and P_2 is a PL manifold with boundary. Also, $\Delta(P_1) \neq 0$ and $\Delta(\partial P_1) = 0$. For $n \geq 6$ one can use Edwards' Theorem 3.33 (cf. [126, Theorem 1.5]) to show that P_1 can be chosen to be a topological manifold with boundary. The manifold U is then the mapping cylinder of a homeomorphism $\partial P_1 \to \partial P_2$. If $n = 5$, the polyhedral homology 4-manifold ∂P_1 might not be a topological 4-manifold and a different procedure must be used to choose U. For this reason our hyperbolization techniques do not work in dimension 5. In any case, when $n \geq 6$, the aspherical version of the Galewski–Stern manifold in [90] is again the union of 3 pieces, $P_1' \cup U \cup P_2'$, where $P_1' = \mathcal{H}(P_1)$. Since $\Delta(\partial P_1') = 0$, $\partial P_1'$ is a PL manifold; so, it also bounds a PL manifold P_2. The manifold P_2' is then defined to be the relative hyperbolization $\mathcal{H}(P_2, \partial P_1')$. As before, U is the mapping cylinder of a homeomorphism from $\partial P_1'$ to $\partial P_2'$, the manifold $P_1' \cup U \cup P_2'$ is aspherical and we see that the Bockstein of the Kirby–Siebenmann obstruction is nonzero. Details of this argument can be found in [90].

Remark 6.9 Since a hyperbolization functor is a procedure that applies to cell complexes, usually simplicial complexes, one cannot directly apply hyperbolization to a nontriangulable manifold. In the previous argument, we got around this problem as follows: the polyhedral homology manifold $\partial P_1'$ is a topological manifold with a non PL triangulation; moreover, $\partial P_1'$ admits a different structure as a PL manifold (giving two inequivalent triangulations of $\partial P_1'$). One then combines P_1', the relative hyperbolization P_2' and the mapping cyliner U of a homeomorphism $\partial P_1' \to \partial P_1'$ between these inequivalent triangulations.

6.2.3 More on NPC Manifolds not Covered by Euclidean Space

This subsection expands on the examples of Sect. 3.2.3 of aspherical polyhedral homology manifolds that are obtained by hyperbolization and that happen to be topological manifolds.

When is the universal cover of an NPC polyhedral homology manifold simply connected at infinity? Consider, for example, the polyhedral homology manifold $\mathcal{H}(X^4(E_8))$ constructed in the proof of Theorem 6.7. It has one non-manifold point whose link is Poincaré's homology 3-sphere Σ^3. The universal cover $\tilde{\mathcal{H}}$ of $\mathcal{H}(X^4(E_8))$ is not simply connected at infinity because its fundamental group at infinity is isomorphic to an inverse limit of free products, $\pi_1(\Sigma^3) * \cdots * \pi_1(\Sigma^3)$. To understand this, consider a ball of radius r in $\tilde{\mathcal{H}}$. It follows that ∂B_r is homeomorphic to the connected sum $\Sigma^3 \# \cdots \# \Sigma^3$ where there is one summand for each PL singular point in B_r. Hence, $\pi_1^\infty(\tilde{\mathcal{H}})$ is the inverse limit of free products as claimed, cf. the final paragraph of Sect. 3.2.1. (See [82, Example 9.2.7] for details of this argument.) Although the polyhedral homology manifold $\mathcal{H}(X^4(E_8))$ is not a topological 4-manifold, it is homotopy equivalent to the 4-manifold N^4 in the proof of Theorem 6.7. The universal cover $\tilde{N}^4$ of N^4 is proper homotopy equivalent to $\tilde{\mathcal{H}}$ and hence, also is not simply connected at infinity.

Next suppose we have a polyhedral homology manifold P^n, $n \geq 5$, whose PL singular set consists of a single edge e, meaning that the link of e is a homology $(n-2)$-sphere that is not simply connected. Further suppose that the link of each endpoint of e is simply connected. By Edwards' Theorem 3.33, P^n is a topological manifold. In particular, a regular neighborhood R of e is a contractible manifold. So, ∂R is a homology $(n-1)$-sphere. Its fundamental group may or may not be trivial. Examples where it is nontrivial can be found in [7] or [91, Remark 5.b.2] or in the proof of Theorem 3.4.1, where ∂R is the double of an acyclic manifold A (cf. Sect. 3.2.3). In fact, ∂R can be an arbitrary integral homology sphere.

Let $\mathcal{H}$ be any hyperbolization procedure satisfying (**H0**). Choose P^n as above with a singular edge e so that the boundary of a regular neighborhood of e is a non-simply connected homology sphere. Hyperbolize P^n to get $\mathcal{H}(P^n)$. Since P^n is a topological manifold, so is $\mathcal{H}(P^n)$. Let $\tilde{\mathcal{H}}(P^n)$ be its universal cover. The ball of radius r in $\tilde{\mathcal{H}}(P^n)$ contains many copies of e and hence, many copies of R. Arguing as in the case of the universal cover of $\mathcal{H}(X^4(E_8))$ in the final paragraph of Sect. 3.2.1, we see that ∂B_r is homeomorphic to a connected sum of copies of ∂R. So, $\tilde{\mathcal{H}}(P^n)$ is not simply connected at infinity and hence, not homeomorphic to $\mathbb{R}^n$. This gives the following theorem, which was stated earlier as Theorem 3.41 in Sect. 3.2.3.

Theorem 6.10 ([91, Theorem 5b.1]) *In each dimension $n \geq 5$, there exists a closed topological n-manifold M^n with the structure of an NPC cube complex so that its universal cover $\tilde{M}^n$ is not homeomorphic to $\mathbb{R}^n$. In particular, the manifold $\tilde{M}^n$ is a* CAT(0) *cube complex not homeomorphic euclidean space. By using a strict hyperbolization procedure for $\mathcal{H}$, as in Theorem 6.33 from Sect. 6.5, we get examples where $\tilde{M}^n$ is* CAT(−1).

We can easily arrange for P^n to be smoothable (for example, P^n can be chosen to be homeomorphic to a sphere). Then $M^n = \mathcal{H}(P^n)$ is also smoothable. (This is just the question of lifting the stable topological tangent bundle to a vector bundle; if the procedure $\mathcal{H}$ satisfies (**H6**), then a lift of the stable tangent bundle for P^n pulls back to a lift of the stable tangent bundle for $\mathcal{H}(P^n)$.)

6.3 The Product with Interval Hyperbolization Procedure

The hyperbolization procedure that is easiest to define is the "product with interval procedure." This construction, denoted by $J \mapsto \mathcal{I}(J)$, was first described in [91, §4b] and in [59, §2]. The definition of the construction is by induction on the skeleta of the cell complex J. Assuming $\mathcal{I}(J^{n-1})$ has been defined for all $(n-1)$-dimensional complexes J^{n-1}, given an n-cell σ, one defines $\mathcal{I}(\sigma)$ to be the product $\mathcal{I}(\partial\sigma) \times [-1, 1]$. One then glues this to $\mathcal{I}(J^{n-1})$ along the subcomplex $\mathcal{I}(\partial\sigma) \times \{\pm 1\}$. In the first version of this construction in [91] we started the induction with the 1-skeleton and, as with the Möbius band hyperbolization, defined $\mathcal{I}(J^1) = J^1$. A few years later Lowell Jones observed that there is a cleaner version of this construction if one started with the 0-skeleton, setting $\mathcal{I}(J^0) = J^0$, and then defined it inductively for the n-skeleton, $n \geq 1$ as above. This cleaner version is discussed in [80, §17] and in [82, §12.8]. After giving the definition we will see that $\mathcal{I}(J)$ is essentially the same as the polyhedral product construction P_L of (3.3) of Sect. 3.1, where L is the first derived neighborhood of a vertex in J. Moreover, this procedure can be relativized to give a procedure $\mathcal{I}(J, B)$, where J is a thickening of some nonpositively curved complex B. This is a relative hyperbolization procedure satisfying Axiom (**RH3**). As we shall see in Sect. 6.3.1 it gives a version of the Reflection Group Trick with nonpositive curvature.

The *product with interval procedure* is a sequence of functors $\mathcal{I}_0, \dots, \mathcal{I}_n, \dots,$ where $\mathcal{I}_n(J)$ an n-dimensional cube complex defined for cell complexes J of dimension $\leq n$. If $\dim J = 0$, put $\mathcal{I}_0(J) = J$. Assume by induction that $\mathcal{I}_{n-1}$ has been defined and let J be a complex of dimension $\leq n$. Put $\mathcal{I}_n(J^{n-1}) = \mathcal{I}_{n-1}(J^{n-1}) \times \{\pm 1\}$. For each $\sigma \in J^{(n)}$, put $\mathcal{I}_n(\sigma) = \mathcal{I}_{n-1}(\partial\sigma) \times [-1, 1]$. Finally, let $\mathcal{I}_n(J)$ be the result of gluing the $\mathcal{I}_n(\sigma)$ to $\mathcal{I}_n(J_{n-1})$ along the common subspace $\mathcal{I}_{n-1}(\partial\sigma^n) \times \{\pm 1\}$. It follows by induction that $\mathcal{I}_n(J)$ is a n-dimensional cubical complex. (Indeed, since $\mathcal{I}_n(\sigma) = \mathcal{I}_{n-1}(\partial\sigma) \times [-1, 1]$, we see that if $\mathcal{I}_{n-1}(\partial\sigma)$ is an $(n-1)$-dimensional cube complex, then $\mathcal{I}_n(\sigma)$ is an n-dimensional cube complex.) It follows from this definition that distinct vertices v_1, v_2 of J correspond to distinct components of $\mathcal{I}_n(J)$. Moreover, the topology of such a component of $\mathcal{I}_n(J)$ depends only on a regular neighborhood (i.e., the first derived neighborhood) of the vertex in J.

For example, if v is a vertex of a simple graph J^1, then the component of $\mathcal{I}_1(J^1)$ which contains $v \times \{\pm 1\}$ is formed by gluing on an edge $v \times [-1, 1]$ to $v \times \{\pm 1\}$ for each edge of J^1 which contains v. In other words, this component is the suspension of $\mathrm{Lk}(v, J^1)$. As another example, if v is the vertex of a 2-cell σ, then $\mathrm{Lk}(v, \partial\sigma) = S^0$; so, the component of $\mathcal{I}_1(\partial\sigma)$ which contains $v \times \{\pm 1\}$ is homeomorphic to S^1(where S^1 is cellulated as a digon). Hence, $\mathcal{I}_2(\sigma) \cong S^1 \times [-1, 1]$. For any vertex $v \in J^{(0)}$, we have that $\mathcal{I}_n(v) = v \times \{\pm 1\}^n \subset \mathcal{I}_n(J)$. So, for $n > 0$, $\mathcal{I}_n$ never satisfies (**H2**). If v is a vertex of an n-cell σ of J, then there is a unique component of $\mathcal{I}(J)$ containing $v \times \{\pm 1\}^n$. Henceforth, assume that $\dim(J) = n$ and that v is not an isolated vertex. Denote by $\mathcal{I}(J, v)$ the component of $\mathcal{I}_n(J)$ which contains $v \times \{\pm 1\}^n$.

Next we analyze and describe the cubical cell structure on $\mathcal{I}(J, v)$ in order to show that it satisfies the Link Condition of Definition 2.10 and hence, that it is NPC. The complex $\mathcal{I}(J, v)$ depends only on the poset $\mathcal{P}$ of cells which contain v. So, we may further assume that $J = \mathrm{Cone}(\mathrm{Lk}_v)$, the cone on another simplicial complex, Lk_v with v the cone point. Each i-cell in J that contains v is a join $v * \sigma$ for some $(i-1)$-cell $\sigma \in \mathrm{Lk}_v^{(i-1)}$. Let L be the order complex of $\mathcal{P}$ (= the barycentric subdivision of Lk_v). A simplex in L is a *flag* in $\mathcal{P}$, i.e., a totally ordered finite subset $f = \{\sigma_0, \dots, \sigma_k\}$ of $\mathcal{P}$. For each $f \in L$ and each integer i with $1 \leq i \leq n$, there is at most one $\sigma \in f$ with $\dim\sigma = i - 1$. Denote this cell by f_i. If there is no such cell, put $f_i = *_i$, where $*_i$ is just a symbol for a point that is disjoint from everything else. Let $A(f) = \{i \mid 1 \leq i \leq n, \ \ f_i \neq *_i\}$.

The cells in $\mathcal{I}(J, v)$ are determined by the poset $\mathcal{S}(L)$ of simplices f of L as follows. Let $t = (t_1, \dots, t_n) \in [-1, 1]^n$. Let $f \in L$ be a flag in Lk_v and let $k = \mathrm{Card}(A(f))$ (so that f corresponds to a $(k-1)$-simplex of L). Let $\square(f)$ be the union of k-dimensional faces of $[-1, 1]^n$ defined by the condition,

$$t_i \in \{\pm 1\}, \text{ if } i \notin A(f).$$

The conclusion is that the link of a copy of v in $\mathcal{I}(J, v)$ is identified with L, the barycentric subdivision of Lk_v. Since any barycentric subdivision is a flag complex, we conclude from Gromov's Lemma 2.28 that $\mathcal{I}(J, v)$ is an NPC cube complex. It is trivial to verify that $J \mapsto \mathcal{I}(J, v)$ satisfies all the axioms for a hyperbolization procedure except (**H2**) (since $\mathcal{I}_n(v)$ consists of 2^n points instead of a single point).

Proposition 6.11 *For a given vertex v of a cell complex J, the correspondence $J \mapsto \mathcal{I}(J, v)$ is a hyperbolization procedure satisfying all the axioms except (**H2**).*

The failure of Axiom (**H2**) for the product with interval procedure means that it cannot be used for the applications to manifolds given in Sects. 6.2.1 and 6.2.2.

Next we want to show that the product with interval procedure essentially coincides with our previous construction of right-angled reflection groups in Sects. 3.1 and 3.2. Let $\mathbf{C}_2 = \{\pm 1\}$. There is a natural action of $(\mathbf{C}_2)^n$ on $[-1, 1]^n$. Let $\square(f)$ be a $(\mathbf{C}_2)^n$-orbit of a k-dimensional face. In other words, $\square(f)$ is the disjoint union of 2^{n-k} copies of a k-cube. These orbits of cubes can be glued together to form $\mathcal{I}(J, v)$. In other words,

$$\mathcal{I}(J, v) = \Big(\bigsqcup_{f \in \mathcal{S}(L)} f \times \square(f) \Big) \Big/ \sim, \tag{6.1}$$

where the gluing relation $\sim$ is defined as follows. If $f \leq f'$, then $f \times \square(f)$ is naturally a subcomplex of $f' \times \square(f')$. Then, given flags f' and f'', the orbits of cubes $f' \times \square(f')$ and $f'' \times \square(f'')$ are glued together along the common subcomplex $(f' \cap f'') \times \square(f' \cap f'')$. The $(\mathbf{C}_2)^n$-action on $[-1, 1]^n$ induces $(\mathbf{C}_2)^n \curvearrowright \mathcal{I}(J, v)$ as a reflection group. There is a strict fundamental domain K defined by the equations: $t_i \geq 0$, for $1 \leq i \leq n$. So, as in (6.1), $K = \big(\bigsqcup f \times \square(f)\big)/\sim$. All this is reminiscent of the construction of the commutator cover of a RACG defined as a polyhedral product in Sect. 3.1.1. In fact, let $S = \mathrm{Vert}(L)$ be the vertex set of L. Since L is a barycentric subdivision it is a flag complex, and there is a RACS (W, S) with nerve equal to L. Each element $s \in S$ corresponds to a cell $\sigma(s)$ in Lk_v so there is a function $\varphi : S \to \{1, \dots, n\}$ defined by $\varphi(s) = \dim \sigma(s) + 1$ and a corresponding homomorphism $\phi : (\mathbf{C}_2)^S \to (\mathbf{C}_2)^n$ induced from φ and defined by $\phi(s) = r_{\varphi(s)}$, where $r_1, \dots, r_n$ are the standard reflections on $[-1, 1]^n$. (Here $(\mathbf{C}_2)^S$ means the direct product (or sum) of S copies of $\mathbf{C}_2$.) The map φ is a coloring as in Definition 4.121.

As in Sect. 3.1.1, $P_L = ([-1, 1], \{\pm 1\})^L$ denotes the polyhedral product. We have $(\mathbf{C}_2)^S \curvearrowright P_L$ and the lifts of this action give an action of a RACG, W, on the universal cover of P_L. This universal cover is the Davis–Moussong complex $\Sigma(W, S)$. The fundamental group of P_L is the commutator subgroup $[W, W]$, i.e., $\pi_1(P_L)$ is the kernel of the natural epimorphism $W \to (\mathbf{C}_2)^S$. Let $\mathcal{I} = \mathcal{I}(J, v)$ be the product with interval hyperbolization. The group $(\mathbf{C}_2)^S$ acts on P_L and $\mathcal{I}$ can be identified with the quotient space P_L/H, for $H = \ker[\phi : (\mathbf{C}_2)^S \to (\mathbf{C}_2)^n]$. So, $P_L \to \mathcal{I}$ is the projection map of a normal covering space with group of deck transformations H. In the language of Sect. 3.2.4 or [92], $\mathcal{I}$ is a "small cover" associated to the RACG, W. Since $\Sigma(W, S)$ is the universal cover of P_L, it is also the universal cover of $\mathcal{I}(\text{Cone}(\text{Lk}_v), v)$.

Theorem 6.12 *Suppose* $J = \text{Cone}(\text{Lk}_v)$ *for some* $(n-1)$*-dimensional cell complex* Lk_v*. Let* L *denote the barycentric subdivision of* Lk_v *and let* (W, S) *be the* RACS *with nerve* L*. Then*

(i) $\mathcal{I}(J, v)$ *is an NPC cube complex of dimension* n*. (ii) The universal cover of* $\mathcal{I}(J, v)$ *can be identified with* $\Sigma(W, S)$ *and its fundamental group with the kernel of* $\phi : W \to (\mathbf{C}_2)^n$*. (So,* $\pi_1(\mathcal{I}(J, v))$ *is a torsion-free subgroup of* W *of index* 2^n*.)*

In other words, $\mathcal{I}(J, v)$ is essentially the same construction as P_L when the flag complex L is given as the barycentric subdivision of another cell complex Lk_v. More precisely, $\mathcal{I}(J, v) = P_L/H$.

6.3.1 The Reflection Group Trick with Nonpositive Curvature

There is relative version of the product with interval construction that is compatible with nonpositive curvature. This allows us to define a relative hyperbolization satisfying Axiom (**RH3**). This version is explained here and elsewhere (in [82, §12.8] and [80, §17]).

Suppose that B is a connected, NPC cell complex and that B is a subcomplex of another cell complex J. No metric on J is assumed. For example, B could be an NPC cube complex and J could be any other cube complex containing B. Alternatively, after subdividing, we could assume that B is a simplicial complex with a piecewise euclidean NPC metric and that J is another simplicial complex containing B. We need to assume the following:

$$\text{For any cell } \sigma \text{ of } J, \sigma \cap B \text{ is either empty or a single cell of } B. \tag{6.2}$$

Let $\mathcal{P}(J, B)$ denote the poset of cells of J which have nonempty intersection with B and which are not contained in B. Define $\dim(J, B)$ to be the largest dimension of any $\sigma \in \mathcal{P}(J, B)$. For $\dim(J, B) \leq n$ we will define an NPC polyhedron $\mathcal{I}(J, B)$ $(= \mathcal{I}_n(J, B))$ with the following properties:

(1) $\mathcal{I}(J, B)$ contains 2^n disjoint copies of B, each of which is embedded as a totally geodesic subcomplex.
(2) $\mathcal{I}(J, B)$ depends only $\mathcal{P}(J, B)$, i.e., it depends only on the first derived neighborhood of B in J.
(3) If σ is a k-cell in $\mathcal{P}(J, B)$, then $\mathcal{I}(\sigma, \sigma \cap B)$ is an NPC piecewise linear k-manifold with boundary.
(4) $(\mathbf{C}_2)^n \curvearrowright \mathcal{I}(J, B)$ as a reflection group. A fundamental chamber $K(J, B)$ is identified with the first derived neighborhood of B in J.
(5) $\mathcal{I}(J, B)$ retracts onto $K(J, B)$ and from there onto B.

In the next few paragraphs we define a sequence of functors $\mathcal{I}_n(J, B)$, essentially by repeating the definition of $\mathcal{I}(J, v)$ after replacing the vertex v by the subcomplex B. The definition is by induction on $\dim(J, B)$ Put $\mathcal{I}_0(J, B) = B$. Assuming by induction that $\mathcal{I}_{n-1}$ has been defined, put $\mathcal{I}_n(J^{n-1} \cup B, B) = \mathcal{I}_{n-1}(J^{n-1} \cup B, B) \times \{\pm 1\}$ and for each n-cell $\sigma \in \mathcal{P}(J, B)$, put $\mathcal{I}_n(\sigma, \sigma \cap B) = \mathcal{I}_{n-1}(\partial\sigma, \partial\sigma \cap B) \times [-1, 1]$. For an arbitrary pair (J, B) with $\dim(J, B) = n$, $\mathcal{I}_n(J, B)$ is defined as the result of gluing a copy of $\mathcal{I}_n(\sigma^n, \sigma^n \cap B)$ onto $\mathcal{I}_n(J^{n-1} \cup B, B)$ via the natural embedding $\mathcal{I}_{n-1}(\partial\sigma^n, \partial\sigma^n \cap B) \times \{\pm 1\} \hookrightarrow \mathcal{I}_{n-1}(\partial\sigma^n \cap B) \times [-1, 1]$. If $\dim(J, B) \leq n$, then put $\mathcal{I}(J, B) = \mathcal{I}_n(J, B)$.

The cell structure on $\mathcal{I}(J, B)$ is a generalization of the one defined previously in the case where B was a single point v. Let $\mathcal{F}(J, B)$ denote the order complex of $\mathcal{P}(J, B)$. For each $f \in \mathcal{F}(J, B)$, the minimum element of f has nonempty intersection with a face $\tau(f)$ of B. For each $f \in \mathcal{F}(J, B)$ and integer i, with $1 \leq i \leq n$, there is at most one cell $\sigma \in f$ with $\dim \sigma = i$. As before, denote this cell by f_i. If there is no such cell, put $f_i = *_i$ and let $A(f) = \{i \mid 1 \leq i \leq n,\ \ f_i \neq *_i\}$.

Let $\square(f)$ to be the $(\mathbf{C}_2)^n$-orbit of the face of $[-1, 1]^n$ defined by the equations, $t_i = \pm 1$, where $i \notin A(f)$. So, $\square(f)$ is a disjoint union of k-cubes, where $k = \operatorname{Card}(A(f))$. Since $\tau(f)$ is a cell in B, it is a euclidean polytope; hence, each component of $\tau(f) \times \square(f)$ is also a euclidean polytope (with the product metric) of dimension $\dim \tau(f) + k$. As in (6.1) put

$$\mathcal{I}(J, B) = \left(\bigsqcup_{f \in \mathcal{F}(J,B)} f \times \tau(f) \times \square(f) \right) / \sim, \tag{6.3}$$

where f is regarded as a point.

Proposition 6.13 *Suppose* $\dim(J, B) = n$ *and* $\mathcal{I}(J, B)$ *denotes the result of the relative product with interval hyperbolization. Then* $\mathcal{I}(J, B)$ *is an NPC piecewise euclidean polyhedron. Moreover, it has properties* (1) *to* (5) *listed above.*

Proof Since each cell of $\mathcal{I}(J, B)$ is the product of a euclidean polytope and a cube, $\mathcal{I}(J, B)$ is a polyhedron with a piecewise euclidean metric. To show the metric on $M^n = \mathcal{I}(J, B)$ is NPC one uses the inductive definition and Reshetnyak's Gluing Lemma (see [35, p. 350], [42, p. 316] or [136, p. 124]). This lemma asserts that when we glue together two NPC spaces via an isometry of two totally geodesic subspaces (or a self- isometry of a single totally geodesic subspace), the result is NPC. In the case at hand, we can assume by induction that $\mathcal{I}_{n-1}(J^{n-1} \cup B, B) \times \{\pm 1\}$ is NPC as is $\mathcal{I}_n(\sigma, \sigma \cap B) = \mathcal{I}_{n-1}(\partial\sigma, \partial\sigma \cap B) \times [-1, 1]$ (with the product metric). So, the result follows from Reshetnyak's Lemma. Note that each $(l + k)$-cell of $\mathcal{I}(J, B)$ is isometric to $\tau^l \times [-1, 1]^k$ for some l-cell τ in B. (Alternatively, when B is an NPC cube complex, one could use Gromov's Lemma and the fact that the geometric realization of $\mathcal{F}(J, B)$ is a flag complex to prove that the cube complex $\mathcal{I}(J, B)$ is NPC.)

As before, $(\mathbf{C}_2)^n \curvearrowright \mathfrak{I}(J, B)$. A strict fundamental domain $K(J, B)$ is defined by the inequalities $t_i \geq 0$, for $1 \leq i \leq n$. The space $K(J, B)$ is a regular neighborhood of B in J. There are some further RACGs in the background. Let $S = \mathcal{P}(J, B)$. There is a RACS (W, S). Its nerve $L(W, S)$ is equal to the flag complex $\mathcal{F}(J, B)$. The set S indexes a mirror structure on $K(J, B)$. As in formulas (3.9) or (4.4) of Sect. 3.1.1 or 4.1.1, the basic construction yields

$$\overline{\mathfrak{I}}(J, B) = W \times K(J, B))\,/\sim,$$

and this can be identified with a covering space of $\mathfrak{I}(J, B)$. (In fact, in [82, §12.8], this description of $\overline{\mathfrak{I}}(J, B)$ used in the first definition of relative hyperbolization.)

Let $\tilde{K}(J, B)$ be the universal cover of $K(J, B)$. Let $\tilde{S}$ denote the inverse image of S in $\tilde{K}(J, B)$ and let $(\tilde{W}, \tilde{S})$ be the corresponding (infinitely generated) RACS. The space $\tilde{\mathfrak{I}}(J, B) = \tilde{W} \times \tilde{K}(J, B))\,/\sim$ is the universal cover of $\overline{\mathfrak{I}}(J, B)$, hence, also the universal cover of $\mathfrak{I}(J, B)$. Put $\pi = \pi_1(K(J, B)) = \pi_1(B)$. Since $\pi \curvearrowright \tilde{K}(J, B)$ by deck transformations, it acts on $(\tilde{W}, \tilde{S})$ by diagram automorphisms. The semidirect product (or wreath-graphproduct) $\tilde{W} \rtimes \pi$ is the group of all lifts of the W-action on $\overline{\mathfrak{I}}(J, B)$; hence, it is also the group of all lifts of the $(\mathbf{C}_2)^n$-action on $\mathfrak{I}(J, B)$. It follows that the fundamental group of $\mathfrak{I}(J, B)$ is given by

$$\pi_1(\mathfrak{I}(J, B)) = \Gamma \rtimes \pi,$$

where Γ is the kernel of the natural epimorphism $\tilde{W} \to (\mathbf{C}_2)^n$. (See [82, pp. 168–170] for more details.) Taking $M = \mathfrak{I}(J, B)$, the next theorem is a corollary to Proposition 6.13.

Theorem 6.14 (A Nonpositively Curved Version of the Reflection Group Trick) *Suppose B is a compact NPC polyhedron that can be embedded in some PL n-manifold. Let J be a regular neighborhood of B in the manifold and suppose (J, B) satisfies* (6.2). *Then there is a closed manifold M^n $(= \mathfrak{I}(J, B))$ with a nonpositively curved polyhedral metric so that B is totally geodesic in M^n and so that M^n retracts onto B. If B is an NPC cube complex, then $\mathfrak{I}(J, B)$ also can be taken to be an NPC cube complex.*

There are two types of applications for this theorem. First, as with the reflection group trick, if the fundamental group of B has a property which is inherited by any group that retracts onto it, then the theorem implies that we can produce a closed NPC manifold with the same property. Secondly, one can show that certain results which are known to hold for NPC manifolds (such as the Farrell–Jones Conjecture) descend to results for a compact NPC polyhedron B. An example of the first type of application is the following.

Corollary 6.15 *For each $n \geq 4$, there is a closed n-manifold with an NPC cubical structure, that retracts onto any one of the Burger-Mozes examples from Sect. 2.4.4. In particular, the fundamental group of such a manifold retracts onto an infinite* *simple group and hence, is not residually finite (since residual finiteness is preserved under retractions).*

Alternatively, one could start from one of Higman's groups from Sect. 2.4.3 to get NPC closed manifolds whose fundamental groups are not residually finite.

For any group π, there is an assembly map $A : H_n(B\pi; \mathbb{L}) \to L_n(\pi)$, where $L_n(\pi)$ is Wall's surgery obstruction group for π and where $\mathbb{L}$ means the Ω-spectrum associated to $\mathbb{Z} \times G/TOP$. The Novikov Conjecture for a torsion-free group π asserts that A is injective. The Borel Conjecture asserts that A is an isomorphism. As an illustration of the second type of application, we have the following.

Corollary 6.16 *Let π be the fundamental group of a compact NPC polyhedron B.*

(i) (Hu [154]).*The Whitehead group, $Wh(\pi)$, is trivial.*
(ii) (Hu [155]). *The Novikov Conjecture holds for π.*
(iii) (Bartels-Lück [13]). *The Borel Conjecture holds for π.*

Proof Farrell and Jones proved that $Wh(\pi) = 0$ for π the fundamental group of a closed Riemannian manifold of nonpositive sectional curvature. They also proved the Assembly Map Conjecture in this context (cf. [116]). Hu [154] showed that the Farrell–Jones argument for the Whitehead group worked for an NPC polyhedral metric on a manifold M and hence, the relative hyperbolization of Theorem 6.14 gives (i). It seems likely that the Farrell–Jones proof of the Assembly Map Conjecture also holds for polyhedral manifolds, in which case Theorem 6.14 also gives (iii). The point is that since π is a retract of $\pi_1(M)$ the abelian groups $Wh(\pi)$, $H_n(B\pi; \mathbb{L})$ and $L_n(\pi)$ are direct summands of the corresponding groups for $\pi_1(M)$. This is enough to conclude (ii), e.g., see [80]. In [13] Bartels and Lück give a different argument that does not use relative hyperbolization.

6.4 Gromov's Hyperbolization Procedure

6.4.1 A General Technique for Hyperbolization

We begin with some general remarks following [91, Section 1].

Let Δ^n denote the simplex with vertex set $\{0, 1, \dots, n\}$. Given a simplicial complex J of dimension $\leq n$, a *nondegenerate simplicial map* $p : J \to \Delta^n$ means the same as a "coloring" $\mathrm{Vert}(J) \to \{0, 1 \dots, n\}$ from Definition 4.121.

Definition 6.17 (cf. [91, Def. (1a.1)]) A *simplicial complex over* Δ^n is a pair (J, p) where J is a simplicial complex and $p : J \to \Delta^n$ is a nondegenerate simplicial map. If (J_1, p_1) and (J_2, p_2) are simplicial complexes over Δ^n, then a simplicial map $f : J_1 \to J_2$ is a *map over* Δ^n if $p_1 = p_2 f$.

Similarly, one can define a *cube complex over* $\Box^n$: it is a pair (H, p) where H is a cube complex and $p : H \to \Box^n$ is a cellular map whose restriction to each cube is a combinatorial isomorphism onto its image (cf. [56, Def. 7.2]).

For example, if X is any convex cell complex of dimension $\leq n$, then its barycentric subdivision bX is a simplicial complex over Δ^n. The map $p : bX \to \Delta^n$ sends the barycenter of an i-cell to the integer i in Vert(Δ^n). A *space over* Δ^n is a pair (X, π), where $\pi : X \to \Delta^n$ is a continuous surjection. If E is a subcomplex of Δ^n, then put $X_E = \pi^{-1}(E)$. Eventually we want to use a *manifold with corners over* Δ^n. This means that X is a smooth n-manifold with corners, that for each face τ of Δ^n of codimension k, X_τ is a codimension k stratum of X, and that the normal bundle of X_τ in X is mapped fiberwise isomorphically to the normal bundle of τ in Δ^n.

Definition 6.18 Suppose (J, p) is a simplicial complex over Δ^n and (X, π) is a space over Δ^n. Define $J\tilde{\Delta}X$ to be the fiber product:

$$\begin{array}{ccc} J\tilde{\Delta}X & \longrightarrow & J \\ \downarrow & & \downarrow p \\ X & \xrightarrow{\pi} & \Delta \end{array} \tag{6.4}$$

In other words,

$$J\tilde{\Delta}X = \{(x, y) \in J \times X \mid p(x) = \pi(y)\}. \tag{6.5}$$

In all that follows we shall only be concerned with the case where the simpicial complex is the barycentric subdivision bJ of an n-dimensional simplicial complex J and $p : bJ \to \Delta^n$ is the natural coloring. In [91] we defined the *Williams functor* on J and (X, π) to be the space:

$$J\Delta X := (bJ)\tilde{\Delta}X = \{(x, y) \in bJ \times X \mid p(x) = \pi(y)\}.$$

Let $c : J\Delta X \to bJ$ denote projection onto the first factor.

The method of [91] for constructing a hyperbolization goes as follows. For each $n > 0$, suppose we are given a "hyperbolized n-simplex," $(\mathcal{H}_n(\Delta^n), \pi)$. It should satisfy the usual list of properties in Sect. 6.1.1, e.g., $\mathcal{H}_n(\Delta^n))$ is a smooth n-manifold with corners, $\mathcal{H}_n(\Delta^n)$ is isometric to a NPC cube complex, and for each subcomplex P of Δ^n, the space $\pi^{-1}(P)$ is a totally geodesic subspace of $\mathcal{H}_n(\Delta^n)$. If J is an n-dimensional simplicial complex, one then defines its *hyperbolization* $\mathcal{H}_n(J)$ by

$$\mathcal{H}_n(J) = J\Delta\mathcal{H}_n(\Delta^n). \tag{6.6}$$

In other words, $\mathcal{H}_n(J)$ is $J\Delta X$ for $X = \mathcal{H}_n(\Delta^n)$. (The reason for including the subscript n in our notation for hyperbolization $\mathcal{H}_n$ is that in practice $\mathcal{H}_n(\Delta^n)$ will be defined by induction, and we will first need to define $\mathcal{H}_{n-1}(\partial\Delta^n)$.) In order to show that $\mathcal{H}_n(J)$ satisfies the axioms in Sect. 6.1.1, we will need to show that if (X, π) is a space over Δ^n satisfying appropriate conditions, then $J\Delta X$ satisfies versions of the axioms. Assume the following:

(1) X is a smooth n-manifold with corners. If τ is a codimension-k face of Δ^n, then X_τ is a codimension-k stratum of X.
(2) X is a piecewise euclidean polyhedron and for each subcomplex P of Δ^n, X_P $(= \pi^{-1}(P))$ is a totally geodesic subpolyhedron of X.
(3) If J is a smooth triangulation of a smooth n-manifold, then the inclusion $J\Delta X \to bJ \times X$ is a smooth embedding with trivial normal bundle. (This is automatic if we choose an appropriate smooth version of the map $p : bJ \to \Delta^n$.)

Definition 6.19 Suppose (X, π) satisfies (6.4.1). Then (X, π) is *degree one* if $\pi : (X, \partial X) \to (\Delta^n, \partial\Delta^n)$ induces an isomorphism on H_n and furthermore, if for any codimension-k face τ of Δ^n, the map $(X_\tau, \partial X_\tau) \to (\tau, \partial\tau)$ induces an isomorphism on H_{n-k}. One says that (X, π) is *tangentially trivial* if the tangent bundle of X is trivial. The polyhedron (X, π) is *nonpositively curved* if X is NPC. (It then follows from (2) that the totally geodesic subspace X_P is NPC for each subcomplex P of Δ^n.)

The proof of the next lemma is left for the reader. Details can be found in [91, §1].

Lemma 6.20 *Suppose J is an n-dimensional simplicial complex and (X, π) is as in statement* (1) *of* (6.4.1).

(i) *If (X, π) is degree one, then $c_* : H_*(J\Delta X) \to H_*(J)$ is a surjection.*
(ii) *If (X, π) is nonpositively curved, then $J\Delta X$ is an NPC polyhedron.*
(iii) *Suppose X is a smooth n-manifold with corners and that J is a smooth triangulation of a smooth n-manifold so that $J\Delta X$ also is a smooth n-manifold. If X is tangentially trivial, then the stable tangent bundle of J pulls back to the stable tangent bundle of $J\Delta X$. In particular, the Stiefel-Whitney classes and the Pontrjagin classes of J pull back to the corresponding classes of $J\Delta X$.*

6.4.2 The Space $\Omega(A, \partial A)$

Suppose $(A, \partial A)$ is a manifold with boundary. Let $D(A, \partial A)$ denote the double of A along ∂A and let r be the reflection of $D(A, \partial A)$ that switches the two copies of A and fixes ∂A. The subspace A is a *fundamental half-space*. The proof of the next lemma is easy.

Lemma 6.21 (cf. [91, Lemma 4c.1]). *The following statements are equivalent.*

(i) *The stable tangent bundle of $D(A, \partial A)$ is trivial.*
(ii) *The stable tangent bundle of A is trivial.*
(iii) *The stable tangent bundle of $D(A, \partial A)$ is $\mathbb{Z}_2$-equivariantly trivial. (The $\mathbb{Z}_2$-action is defined by the reflection r.)*

Remark 6.22 This lemma is only true for the stable tangent bundle. For example, if A is the 2-disk, then it has trivial tangent bundle. Since $D(A, \partial A) = S^2$ which has nonzero Euler characteristic, its tangent bundle cannot be trivial.

Following Gromov [136, p. 116], we define a manifold with boundary $\Omega(A, \partial A)$, its boundary being $D(A, \partial A)$. Start with $D(A, \partial A) \times [-1, 1]$ and then glue $r(A) \times \{-1\}$ to $r(A) \times \{+1\}$ using the identity map on $r(A)$. The result is $\Omega(A, \partial A)$. Another way to think of $\Omega(A, \partial A)$ is to start with $D(A, \partial A) \times S^1$ and then slit it open along the fundamental half-space $A \times 1$. (In [91] $\Omega(A, \partial A)$ is called "the cylinder construction.")

Lemma 6.23 (cf. [91, Prop. 4c.2]) *Write D for $D(A, \partial A)$ and Ω for $\Omega(A, \partial A)$.*

(i) Let $\tau(A)$ and $\tau(\Omega)$ denote the stable tangent bundles of A and Ω, respectively. Suppose the bundle $\tau(A)$ is trivial and $\psi : \tau(A) \to A \times \mathbb{R}^N$ is a trivialization. Then ψ extends to a trivialization $\tau(\Omega) \to \Omega \times \mathbb{R}^N$. In particular, $\tau(\Omega)$ is trivial.

(ii) If D is an NPC cube complex, then so is Ω.

Proof Statement (i) follows from Lemma 6.21. As for (ii), if D is an NPC cube complex, then so is $D \times S^1$ and hence, so is Ω (since it is the result of sliting $D \times S^1$ open along $A \times 1$).

If r is a locally linear reflection on a closed manifold M, then $M = D(A, \partial A)$ where ∂A is the fixed set of r and A is a choice of fundamental half-space. We will write $\Omega(M, r)$ instead of $\Omega(D(A, \partial A))$.

6.4.3 Gromov's Construction

Given a simplicial complex J of dimension $\leq n$, Gromov's hyperbolization $\mathcal{H}_n(J)$ is defined by induction on n. For $n = 1$ put $\mathcal{H}_1(J) = J$. On the n-simplex Δ^n let $r : \Delta^n \to \Delta^n$ be a linear reflection given by transposition of two vertices. The homeomorphism r is not a simplicial map; however, it is a simplicial isomorphism if we pass to the barycentric subdivision $b\Delta^n$. Suppose by induction that $\mathcal{H}_{n-1}(J)$ has been defined for any J of dimension $\leq (n-1)$, so that $\mathcal{H}_{n-1}(b\partial\Delta^n)$ is defined. By functorality, there is an involution $\mathcal{H}_{n-1}(r) : \mathcal{H}_{n-1}(b\partial\Delta^n) \to \mathcal{H}_{n-1}(b\partial\Delta^n)$. Since the fixed set of r separates $b\partial\Delta^n$ into two half-spaces, the fixed set of $\mathcal{H}_{n-1}(r)$ on $\mathcal{H}_{n-1}(b\partial\Delta^n)$ has the same property, i.e., $\mathcal{H}(r)$ is a locally linear reflection. The hyperbolization $\mathcal{H}_n(\Delta^n)$ is defined by:

$$\mathcal{H}_n(\Delta^n) := \Omega(\mathcal{H}_{n-1}(b\partial\Delta^n), \mathcal{H}_{n-1}(r)). \tag{6.7}$$

As before, the hyperbolization $\mathcal{H}_n(J)$ of an n-dimensional complex J is defined as the fiber product (6.6). We simplify notation by writing $\mathcal{H}(J)$ instead of $\mathcal{H}_n(J)$.

Example 6.24 We have that $\mathcal{H}(\Delta^2)$ is $b\partial\Delta^2 \times S^1$ slit open along half of $b\partial\Delta^2$. In other words, $\mathcal{H}(\Delta^2)$ is a 2-torus with the interior of a 2-disk removed and $\partial\mathcal{H}(\Delta^2)$ is identified with $b\partial\Delta^2$. The barycentric subdivision $b\partial\Delta^3$ is a union of twenty four 2-simplices. The surface $\mathcal{H}(\partial\Delta^3)$ is formed by replacing each 2-simplex by the torus with hole, $\mathcal{H}(\Delta^2)$. Thus, $\mathcal{H}(\partial\Delta^3)$ will be a closed surface of genus 24.

Remark 6.25 The problem with the definition in (6.7) is that the reflection $r : \partial\Delta^n \to \partial\Delta^n$ is not canonical, as it requires a choice of two vertices in Δ^n. Furthermore, to get r to be a simplicial automorphism we need to pass to the barycentric subdivision. Hence, there are many barycentric subdivisions of simplices hidden in the definition in (6.7). For example, to define $\mathcal{H}_2(\Delta^2)$ and the reflection $\mathcal{H}_2(r)$ on it, we must barycentrically subdivide Δ^2 in order to define r and then replace each 2-simplex in $b\Delta^2$ by a torus with hole. However, in order to make the attaching map of this torus with hole well-defined we need to choose a reflection on each 2-simplex $b\Delta^2$ and to make this reflection simplicial we need to pass to the second barycentric subdivision.

Theorem 6.26 (cf. [91]) *Suppose J is an n-dimensional simpicial complex. Then the result of Gromov's hyperbolization procedure, $\mathcal{H}(J)$, is a NPC cube complex. Moreover, the functor $J \to \mathcal{H}(J)$ satisfies Axioms* (**H0**) *through* (**H6**) *in Sect. 6.1.1.*

Proof We can assume by induction that $\mathcal{H}(b\partial\Delta^n)$ is a $(n-1)$-dimensional NPC cube complex and that $\mathcal{H}_{n-1}(r)$ is a locally linear reflection on it. Applying Lemma 6.23 with $D = D(\mathcal{H}(b\partial\Delta^n), \mathcal{H}(r))$, we see that $\Omega = \mathcal{H}(\Delta^n)$ is an n-dimensional NPC cube complex. Since we also have by induction that the hyperbolization of the $(n-1)$ skeleton of J is an NPC cube complex, it follows from Reshetnyak's Gluing Lemma that $\mathcal{H}(J)$ is an NPC cube complex. This proves the first sentence of the theorem.

Since the definition of $\mathcal{H}(J)$ uses induction on dimension, the verification of the axioms also uses induction. Verifications of (**H0**) through (**H3**) are routine and are left to the reader. By induction, $\mathcal{H}(b\partial\Delta^n)$ is a smooth manifold with stably trivial tangent bundle (Since $\partial\Delta^n$ is a sphere its stable tangent bundle is trivial. Since $\mathcal{H}(\Delta^n)$ is defined by (6.7), it is a smooth manifold with corners with trivial tangent bundle.) If J is a smooth triangulation of a smooth manifold, then, using (6.5), (6.6) and the Implicit Function Theorem, we see that $\mathcal{H}(J)$ is a smooth submanifold of $bJ \times \mathcal{H}(\Delta^n)$ with trivial normal bundle (cf. [91, Prop. 1f.5]). It follows that the stable tangent bundle of J is the pullback of the stable tangent bundle of J. Axioms (**H4**), (**H4**), (**H6**) follow.

6.5 Strict Hyperbolization

There is a construction of [59] for converting a finite dimensional cube complex J into a piecewise hyperbolic space J_X. The idea is to first construct, for each positive integer n, a hyperbolic n-manifold with corners $X = X^n$ such that the strata of X can be organized to give it the structure of a manifold with faces so that

(a) Each face of X is a totally geodesic submanifold. (A k-dimensional "face" is a union of k-dimensional strata of the manifold with corners X. N.B. A face need not be connected.) If two or more faces have nonempty intersection, then the intersection is orthogonal.
(b) The poset of faces of X is isomorphic to the poset of faces of the n-cube $\square^n$. Moreover, the group of symmetries of $\square^n$ (i.e., the Coxeter group $W(\mathbf{B}_n)$) acts by isometries on X^n.

One can then glue together copies of X and its faces in the same combinatorial pattern as the faces of $\square^n$ are glued together in forming J to form a new space J_X. We also get a strata-preserving map $f : J_X \to J$ unique up to a homotopy through strata-preserving maps (cf. Axiom (**H1**)). If X satisfies a couple of other additional properties (that its faces are orientable and that its set of 0-dimensional faces is isomorphic to $\{\pm 1\}^n$), then $f : J_X \to J$ is degree one (cf. Sect. 6.1.1). Moreover, if the cube complex J has flag complexes as its links, then the new space J_X will be locally CAT(-1). Finally, we can precompose this with one of the hyperbolization procedures of Sect. 6.3 or 6.4, which takes values in cube complexes to get a strict hyperbolization procedure $J \mapsto \mathcal{H}(J) \mapsto \mathcal{H}(J)_X$.

The Hyperbolic Manifold with Faces X In the next theorem we use Example 5.22 to describe some properties of certain arithmetic hyperbolic n-manifolds.

Theorem 6.27 ([59]) *For each $n > 0$ there is a closed, connected, hyperbolic manifold M^n, a system $\{M_1, \dots, M_n\}$ of closed, connected submanifolds of codimension one in M^n, and an isometric action of $W(\mathbf{B}_n)$ (the symmetry group of $\square^n$) on M^n preserving the system of submanifolds such that:*

(1) *M_i is a component of the fixed point set of a reflection $r_i \in W(\mathbf{B}_n)$, where $\langle r_1, \dots, r_n\rangle$ is a standard subgroup isomorphic to $\mathbf{C}_2^n$.*
(2) *Each M_i is totally geodesic in M^n.*
(3) *The M_i s intersect orthogonally.*
(4) *$M_1 \cap \cdots \cap M_n$ is a single point x.*
(5) *$W(\mathbf{B}_n)$ fixes x and its representation on $T_x M^n$ is equivalent to its standard representation as a finite reflection group.*
(6) *M^n, as well as each M_i, is orientable.*

Sketch of Proof Here is a sketch of the proof in [59]. Consider the quadratic form φ on $\mathbb{R}^{n+1}$ given by

$$\varphi(x_0, x_1, \dots, x_n) = -\sqrt{2}\,(x_0)^2 + (x_1)^2 + \cdots (x_n)^2.$$

Let A be the ring of integers in the number field $\mathbb{Q}(\sqrt{2})$. Let $O(\varphi, A)$ be the automorphism group of φ on A^{n+1}. On $\mathbb{R}^{n+1}$, φ is equivalent to the standard indefinite form of signature $(n, 1)$; so, $O(\varphi, \mathbb{R})$ is isomorphic to the Lie group $O(n, 1)$ of isometries of $\mathbb{H}^n$ (thought of as one sheet of the hyperboloid $\{\varphi(x) = -1\}$ in $\mathbb{R}^{n,1}$). The manifold M^n will be $\mathbb{H}^n/\Gamma$ where Γ is an appropriately chosen normal, torsion-free subgroup of finite index in $O(\varphi, A)$. The codimension-one submanifold M_i will be the image of the hyperplane $\{x_i = 0\}$ in M^n. Reflection across the hyperplane $\{x_i = 0\}$ descends to a reflection $r_i : M \to M$ so that M_i is a component of the fixed set of r_i on M. (N.B. Although the fixed set of r_i separates M, the component M_i of this fixed set need not.) Let $O_0(n, 1)$ be the subgroup of $O(n, 1)$ that preserves the two sheets of the hyperboloid, let $SO(n, 1) = O(n, 1) \cap SL(n + 1, \mathbb{R})$ and $SO_0(n, 1) = O_0(n, 1) \cap SO(n, 1)$. We need to show that we can choose Γ so that the properties in the theorem hold. Only two of these properties are tricky, (4) and (6). Property (4) holds whenever Γ is a congruence subgroup, $\Gamma(\mathfrak{I}) := \{\gamma \in O(\varphi, A) \mid \gamma \cong 1 \pmod{I}\}$, defined for some nonzero ideal $\mathfrak{I}$ in A. Property (6) holds whenever $\Gamma < SO_0(n, 1)$. So, we need to show that we can find a congruence subgroup $\Gamma(\mathfrak{I})$ which is simultaneously a subgroup of $SO_0(n, 1)$. This is possible by a lemma of Millson-Raganuthan. Its proof depends on the fact that we can choose $\mathfrak{I}$ so that any $\gamma \in \Gamma(\mathfrak{I})$ has trivial "spinor norm" in $\mathbb{R}^*/(\mathbb{R}^*)^2$ (cf. [59, Lemma 6.7]). (Here $\mathbb{R}^* = \mathbb{R} - \{0\}$.)

For each subset $I < \{1, \dots, n\}$, put $M_I = \bigcap_{i \in I} M_i$. The n-manifold with faces X is constructed by "cutting open" the manifold M^n along the system of codimension-one submanifolds $\{M_1, \dots, M_n\}$. In general, a *system of codimension-one submanifolds* $\{F_1, \dots, F_n\}$ is as described in [59, §5]. It means, in particular, that the submanifolds F_i intersect transversely. This process is essentially the same as the one described in Sect. 3.2.5 for cutting open a manifold along a hierarchy. If F_1 is a hypersurface in M, then $M \odot F_1$ denotes the result of adding to $M - F_1$ the normal S^0-bundle to F_1. If F_1 is two-sided, then $M \odot F_1$ is a manifold with boundary, with two boundary components for each component of F_1. If M is connected and F_1 is non-separating, then $M \odot F_1$ is connected. The manifold with corners $M \odot \{F_1, \dots, F_n\}$ is defined by iterating this procedure. If each F_i is two-sided and non-separating, then $M \odot \{F_1, \dots, F_n\}$ is connected. There is a natural obvious gluing map $M \odot \{F_1, \dots, F_n\} \to M$ and a codimension-k face of the manifold with corners is the inverse image of a component of a k-fold intersection. For each component of such an intersection there are 2^k codimension-k faces.

Example 6.28 (Cutting Open an n-Torus to Get an n-Cube) Suppose $M^n = T^n$ and $\{F_1, \dots, F_n\}$ is the standard system of coordinate subtori passing through a given point. Then $M \odot \{F_1, \dots, F_n\}$ is the n-cube $\square^n$.

Let M be the closed hyperbolic n-manifold and $\{M_1, \dots, M_n\}$ the system of submanifolds described in Theorem 6.27. Put

$$X := M \odot \{M_1 \dots, M_n\}. \tag{6.8}$$

Here is a corollary to Theorem 6.27.

Corollary 6.29 (cf. [59, Cor. 6.2]) *There is a compact, connected, orientable hyperbolic n-manifold with faces X together with an action of* $W(\mathbf{B}_n)$ *(the group of symmetries of* $\square^n$*) with the following properties.*

(i) *The poset of faces of* X *is* $W(\mathbf{B}_n)$*-equivariantly isomorphic to the poset of faces of* $\square^n$.
(ii) *Each component of each face of* X *is totally geodesic in* X. *(N.B. A face of* X *need not be connected.)*
(iii) *If a collection of faces has nonempty intersection, then the intersection is orthogonal.*
(iv) *Each* 0*-dimensional face is a single point (i.e.,* X *has precisely* 2^n *vertices).*
(v) *There is natural map* $q : X \to \square^n$ *taking each face of* X *to the corresponding face of* $\square^n$. *The map* q, *as well as its restriction to each face, has degree one.*

Proof Since each M_i is non-separating, it follows that X is connected. Since M is orientable so is X (cf. [59, Lemma 5.8]). Properties (ii) and (iii) follow immediately from corresponding facts about $\{M_1, \ldots, M_n\}$. Property (iv) follows from property (4) of Theorem 6.27. The proof of (i) uses item (4) as well as [59, Lemma 6]. For the proof of (v), see [59, Lemmas 5.3 and 5.9].

The Piecewise Hyperbolic Space J_X

Theorem 6.30 (cf. [59, Proposition 7.1]). *Suppose* J *is a cube complex of dimension* $\leq n$ *and* X *is the hyperbolic* n*-manifold with faces defined by* (6.8). *Then there is a piecewise hyperbolic space* J_X *and a map* $f : J_X \to J$ *with the following properties.*

(1) *For each* k*-cell* $\square^k$ *in* J, $f^{-1}(\square^k)$ *is a* k*-dimensional face of* X. *Furthermore, if* J' *is any subcomplex of* J, *then* $f^{-1}(J')$ *is isometric to* J'_X.
(2) *The directions in* J_X *that are orthogonal to* $f^{-1}(\square^k)$ *form a piecewise spherical polyhedron* $\mathrm{Lk}(f^{-1}(\square^k), J_X)$ *that is isometric to* $\mathrm{Lk}(\square^k, J)$.
(3) *The construction* $J \mapsto J_X$ *is a functor from the category of cube complexes and cellular immersions to the category of piecewise hyperbolic spaces and isometric immersions.*
(4) *The map* f *induces a surjection on homology.*
(5) *If* J *is a manifold, then so is* J_X. *If* J *has a smooth or PL structure, then the same is true for* J_X.
(6) *If* J *is NPC, then* J_X *is locally* $\mathrm{CAT}(-1)$.

Comments on the Proof A straightforward proof of this theorem can be found in [59, pp. 347–348]. The space J_X is constructed by gluing together faces of X in the same combinatorial fashion as the cubes of J are glued together. One comment that needs to be made is that this process is independent of the combinatorial equivalence of a k-cell in J with the standard k-cube. This is guaranteed by item (i) in Corollary 6.29: every combinatorial symmetry of $\square^k$ corresponds to an isometry of the corresponding face of X. Since each face of X is orientable and since each 0-dimensional face is a point, we get property (4), that f induces a surjection on homology. As for property (6), suppose a point $z \in J_X$ lies in the relative interior of some face of X corresponding to some k-dimensional cube $\square^k$ of J.

A neighborhood of z in J_X is isometric to a neighborhood of the cone point in the hyperbolic cone, $\mathrm{Cone}_\varepsilon^{-1}(\mathrm{Lk}(z, J_X))$, defined by formula (2.3) in Sect. 2.1. By Lemma 2.8 this cone is CAT(−1) if and only if $\mathrm{Lk}(z, J_X)$ is CAT(1). By item (2) above, this link is isometric to $\mathrm{Lk}(f(z), J) \equiv \mathbb{S}^{k-1} * \mathrm{Lk}(\square^k, J)$. By Gromov's Lemma 2.28, J is NPC if and only if $\mathrm{Lk}(\square^k, J)$ is a flag complex for each cell $\square^k$ and by (2) this means that $\mathrm{Lk}(f^{-1}(\square^k), J_X)$ is also a flag complex and hence, that $\mathrm{Cone}_\varepsilon^{-1}(\mathrm{Lk}(z, J_X))$ is CAT(−1).

To establish properties of the stable tangent bundle of J_X we need to know something about the tangent bundle of X and we would like a description of J_X similar to the fiber product construction in Definition 6.18.

Definition 6.31 Suppose J is a cube complex. A *projection from* J *to* $\square^n$ is a cellular map $p : J \to \square^n$ such that the restriction of p to any cell is a combinatorial isomorphism onto a face of $\square^n$. (Compare this to the definition of a "coloring" of a simplicial complex given in Definition 4.121.)

If J admits a projection $p : J \to \square^n$, then, as in Definition 6.18, J_X is a fiber product:

$$\begin{array}{ccc} J_X & \xrightarrow{f} & J \\ \downarrow & & \downarrow p \\ X & \xrightarrow{q} & \square^n \end{array} \tag{6.9}$$

That is to say, $J_X = \{x, y) \in X \times J \mid q(x) = p(y)\}$. This means that if J is a smooth or PL manifold, then J_X is a smooth or PL submanifold of $X \times J$ emdedded with trivial normal bundle. Hence, if we can find an example of a hyperbolic manifold with faces X that has trivial tangent bundle, then the tangent bundle of J_X will be stably equivalent to the pullback of the tangent bundle of J via f.

In order to get the tangent bundle of X to be trivial, the question arises: can we choose the manifold M^n $(= \mathbb{H}^n/\Gamma(\mathfrak{I}))$ to have stably trivial tangent bundle? The answer is yes. First observe that the stable tangent bundle $TM \oplus \mathbb{R}$ is flat, where the associated holonomy $\Gamma \to O(n, 1)$ is the representation that defines M^n. It is then a result of Deligne and Sullivan that the flat bundle $TM \oplus \mathbb{R}$ becomes trivial after passing to a finite cover $M' \to M$ corresponding to some subgroup Γ' of finite index in Γ. To obtain M as in Theorem 6.27 we need Γ to be a congruence subgroup for some ideal $\mathfrak{I}$. Passing to a smaller ideal $\mathfrak{I}'$ we can assume that there is a deeper congruence subgroup $\Gamma(\mathfrak{I}')$ such that $\Gamma(\mathfrak{I}') \le \Gamma'$. Since the tangent bundle of M' is stably trivial so is its pullback to the finite cover $M'' = \mathbb{H}^n/\Gamma(\mathfrak{I})$. So, by taking $M = M''$ we get the desired stably parallelizable hyperbolic manifold. Cutting it open as before we obtain the hyperbolic manifold with faces X with trivial tangent bundle. So, we have proved the following.

Lemma 6.32 *Suppose J is a cube complex projecting to $\square^n$ and that J is a smooth or PL manifold. Choose a hyperbolic manifold with faces X as above so that its* *tangent bundle is trivial. Let* $f : J_X \to J$ *be the map defined in* (6.9). *Then the tangent bundle of* J_X *is stably isomorphic to* $f^*(TJ)$.

This lemma means in particular that the Stiefel-Whitney classes and Pontrjagin classes of $T(J_X)$ are the pullbacks of the corresponding characteristic classes of TJ. (Note, however, that the same is not true for Euler classes since $\chi(X) \neq 1$.)

Converting Hyperbolizations to Strict Hyperbolizations Suppose J is an n-dimensional simplicial complex. As we have seen in Sect. 6.3, the product with interval construction $\mathcal{I}_n(J, v)$ is an NPC cube complex and as in Sect. 6.4.3, the same is true for Gromov's hyperbolization construction $\mathcal{H}_n(J)$. Let $\mathcal{G}(J)$ stand for either one of these constructions. Given a hyperbolic n-manifoold with faces X we can then compose the functor $J \mapsto \mathcal{G}(J)$ with the functor $\mathcal{G}(J) \mapsto \mathcal{G}(J)_X$ to get the desired strict hyperbolization functor $J \mapsto \mathcal{G}(J)_X$ satisfying Axioms (**H0**), (**H1**), (**H3**) and (**H4**) and in the case of $\mathcal{H}(J)_X$, satisfying Axiom (**H2**) as well.

In order to verify that Axioms (**H5**) and (**H6**) hold we need to show (a) that there is a projection map $p : \mathcal{G}_n(J) \to \square^n$ and (b) that X can be chosen so that the tangent bundle of X is trivial. Condition (a) is easily proved by induction on n. Condition (b) is Lemma 6.32 above. So, as promised earlier, we have the following theorem (which we state only in the case of Gromov's hyperbolization so that it is not necessary to exclude (**H2**)).

Theorem 6.33 *There is a strict hyperbolization procedure,* $J \mapsto \mathcal{H}(J)_X$*, defined on any* n*-dimensional simplical complex* J*, that satisfies Axioms* (**H0**)–(**H6**).

Remark 6.34 In [186, 187] Ontaneda makes good use of strict hyperbolization to produce many new examples of smooth, negatively curved riemannian manifolds that satisfy a variety of interesting properties.

Chapter 7
Morse Theory and Bestvina–Brady Groups

Section 7.1 develops "Morse theory" for real-valued functions on convex cell complexes. There are two purposes: (1) to study finiteness properties of groups that are defined as the kernel of a homomorphism from a group to the infinite cyclic group, and (2) for groups of type F to study their fundamental groups at infinity and their cohomology groups with group ring coefficients. In Sect. 7.3 the Bestvina–Brady group BB_L is defined as the kernel of a surjection $A_L \to \mathbb{Z}$, where A_L is the RAAG associated to a flag complex L. Bestvina and Brady proved that if L is acyclic but not simply connected then BB_L is a group that is type FP but not type F. There are various applications. In Sect. 7.4.1 examples are given of PD^n groups that are not the fundamental group of any closed aspherical manifold. In Sect. 7.4.2 we describe the examples of Leary–Nucinkis of virtually torsion-free groups G whose torsion-free subgroup is type F, yet G is not type $\mathcal{V}$F. In Sect. 7.4.3 variations of Bestvina–Brady groups are used to get Leary's construction of uncountably many groups of type FP.

The origin of these ideas lies in work of Stallings [208]. Let F_2 denote the free group on two generators. Stallings showed that the kernel of a natural surjection $F_2 \times F_2 \times F_2 \to \mathbb{Z}$ is a finitely presented group which is not type F_3 (i.e., it does not have a classifying space with finite 3-skeleton). In [25] this was generalized to the kernel of $(F_2)^m \to \mathbb{Z}$. Such a group was type F_{m-1} but not type F_m. (N.B. $(F_2)^m$ is a RAAG.) The proof that the kernel G is not type F_m is that $H_m(G; \mathbb{Z})$ is not finitely generated (and hence, BG cannot have a finite m-skeleton). These examples are usually called the *Bieri–Stallings groups*. So, the Bieri–Stallings groups are examples of Bestvina–Brady groups.

M. W. Davis, *Infinite Group Actions on Polyhedra*, Ergebnisse der Mathematik und ihrer Grenzgebiete. 3. Folge / A Series of Modern Surveys in Mathematics 77,
https://doi.org/10.1007/978-3-031-48443-8_7

7.1 Morse Theory on Convex Cell Complexes

Suppose that $\varphi : Y \to \mathbb{R}$ is a real-valued function on a convex cell complex Y. This section deals with such functions in the case where the homotopy type of the level sets only change as $\varphi^{-1}(t)$ crosses a vertex.

7.1.1 *Functions that Restrict to an Affine Map on Each Cell*

We are primarily interested in the case where the restriction of φ to each cell C is an affine map which is not constant whenever $\dim C > 0$. Further assume that φ takes the 0-skeleton Y^0 to a discrete subset of $\mathbb{R}$. As in [23] such a φ will be called a *Morse function* on Y. For each cell C of Y, $\varphi|_C$ is an affine map with a unique minimum and maximum (which necessarily occur at vertices of C). For any vertex v of C, recall that $\mathrm{Lk}(v, C)$ is the set of inward-pointing unit tangent vectors to C at v. As in (2.1) from Sect. 2.1, the *link* at v in Y is the set of all such directions at v so that $\mathrm{Lk}(v, Y) := \bigcup \mathrm{Lk}(v, C)$, where the union is over all cells C containing v. The *descending link at* v, denoted $\mathrm{Lk}_{\downarrow}(v)$, is the union of the $\mathrm{Lk}(v, C)$, over all cells C containing v, such that $\varphi|_C$ has a maximum at v. Similarly, the *ascending link*, $\mathrm{Lk}_{\uparrow}(v)$ is the corresponding union where $\varphi|_C$ has a minimum at v. The idea behind Morse theory on cell complexes is contained in the following lemma.

Lemma 7.1 ([23, Lemma 2.5]) *Suppose $J < J' \leq \mathbb{R}$ are nonempty closed intervals such that $J' - J$ contains only one number r that is equal to φ(0-cell). If $\inf J = \inf J'$, then $\varphi^{-1}(J')$ is homotopy equivalent to $\varphi^{-1}(J)$ with the copies of $\mathrm{Lk}_{\downarrow}(v)$, $v \in \varphi^{-1}(r)$, coned off.*

(This lemma can also be used whenever the vertices in $\varphi^{-1}(r)$ have disjoint open stars.) A similar lemma holds when $\sup J = \sup J'$ if $\mathrm{Lk}_{\downarrow}(v)$ is replaced by $\mathrm{Lk}_{\uparrow}(v)$.

7.1.2 *Distance Functions*

The discussion in this subsection follows Brady–McCammond–Meiers [33]. Suppose that Y is a CAT(0) polyhedron, that v is a choice of base vertex, and that the function $D_v : Y \to [0, \infty)$ is defined by $D_v(y) = d(v, y)$. The level set $D_v^{-1}(r)$ is then $S_v(r)$, the metric sphere of radius r. The topology of the level set can change whenever $S_v(r)$ first touches a cell. When Y is a cube complex or when Y is a Davis–Moussong complex, such critical points occur only at vertices. (As is explained in [33] if critical points occur at interior points of cells, then we can subdivide Y so that all critical points occur at vertices. Henceforth, we assume that this is the case.)

Define the *ascending link*, $\mathrm{Lk}_{\uparrow}(w)$, for D_v at a vertex $w \in S_v(r)$ as follows: it is the link of w in the induced subcomplex of Y spanned by those vertices w' such that

$D_v(w') > D_v(w)$. (If $w = v$, then $\mathrm{Lk}_\uparrow(w) = \mathrm{Lk}(v)$.) As r decreases as we cross a vertex w, the effect on the level set $S_v(r)$ is to attach a copy of $\mathrm{Cone}(\mathrm{Lk}_\uparrow(w))$ to $S_v(r+\varepsilon)$. So, the ascending links can be used to understand the homology (or cohomology) of Y at infinity. For example, we find the following lemma in [33].

Lemma 7.2 ([33, Lemma 3.5]) *Suppose Y is a locally finite,* CAT(0) *complex and that for each vertex $w \in \mathrm{Vert}(Y)$, the ascending link, $\mathrm{Lk}_\uparrow(w)$, is m-acyclic. Then Y is m-acyclic at infinity. (The notion of "m-acyclic at infinity" is defined in the next subsection.)*

If p is a point in $\mathrm{Lk}(w)$, then the *punctured link*, $P\,\mathrm{Lk}(w, p)$, is the complement of the open ball centered at p of radius $\pi/2$ (with respect to the piecewise spherical metric on $\mathrm{Lk}(v)$).

Lemma 7.3 ([33, Lemma 3.6]) *Let $\mathrm{Lk}_\uparrow(w)$ denote the ascending link for D_v at a vertex $w \in Y$, If $w \neq v$, then $\mathrm{Lk}_\uparrow(w)$ is a deformation retract of $P\,\mathrm{Lk}(w, p)$ where p is the point of $\mathrm{Lk}(v)$ determined by the geodesic from w to v.*

If σ is a simplex of $\mathrm{Lk}(v)$, then $\mathrm{Lk}(v) - \sigma$ denotes the subcomplex of $\mathrm{Lk}(v)$ spanned by the vertices of $\mathrm{Lk}(v)$ other than those that are vertices of σ. In the case of a CAT(0) cube complex (and in many other cases), $\mathrm{Lk}_\uparrow(w) = \mathrm{Lk}(w) - \sigma$, where σ is the simplex of $\mathrm{Lk}(w)$ which contains p in its relative interior. (In [87, 102] all subcomplexes of the form $\mathrm{Lk}(v) - \sigma$ are called "punctured links".)

Remark 7.4 These lemmas provide a different argument for the contractibility of the Davis–Moussong complex. Similarly, these lemmas show that the positive cone in the universal cover of the Salvetti complex (consisting of the chambers corresponding to the positive Artin monoid) is contractible.

7.1.3 Topology at Infinity

We recall some definitions from [130] or [82, Appendix G.2]. Suppose Y is a locally finite CW complex. Then $C_i^{lf}(Y)$, the group of *locally finite cellular i-chains* on Y, is defined to be the group of all integer-valued functions on the set of i-cells of Y, The group of ordinary cellular i-chains, $C_i(Y)$, defined as the free abelian group on the set of i-cells. The group of *i-chains at infinity of Y* is defined by $C_i^\infty(Y) := C_{i+1}^{lf}(Y)/C_{i+1}(Y)$. The *homology at infinity of Y* is the homology of the chain complex $C_*^\infty(Y)$. The space Y is *m-acyclic at infinity* if $H_i^\infty(Y) = 0$ for $i \leq m$. Similarly, $C_c^i(Y)$, the *compactly supported cochains*, means the group of finitely supported cochains, $C_\infty^i(Y) := C^i(Y)/C_c^i(Y)$, and $H_\infty^i(Y) = H^i(C_\infty^*(Y))$.

Suppose a discrete group G acts freely and cellularly on a contractible complex Y with quotient space a finite CW complex BG. In [39, Prop. 7.5] we find the following formula:

$$H^*(G; \mathbb{Z}G) = H_c^*(Y). \tag{7.1}$$

More generally, if G acts properly on Y with only finitely many orbits of cells, then (7.1) continues to hold (see [39, Exercise 4, p. 209]). So, (7.1) holds whenever G acts on $\underline{E}G$ with only finitely many orbits of cells. Applying the method of Lemma 7.3 allows us to compute the cohomology with group ring coefficients of Coxeter groups, as well as, of cocompact lattices in automorphism groups of locally finite buildings.

First consider Coxeter groups. As in Sect. 4.2, suppose (W, S) is a Coxeter system, that $L(W, S)$ is its nerve and that $\Sigma = \Sigma(W, S)$ is its Davis–Moussong complex. Let $K = K(W, S)$ be the fundamental chamber. The link of any vertex of Σ is $L(W, S)$; moreover, the barycentric subdivision of $L(W, S)$ can be identified with K^S (the boundary of K). For any subset J of S, let K^J denote the union of the mirrors K_s indexed by $\{s \mid s \in J\}$. If σ is a simplex of $L(W, S)$, then let T be its vertex set (so, T is a spherical subset of S). One sees that K^{S-T} is homeomorphic to $L(W, S) - \sigma$ when both are regarded as subcomplexes of the barycentric subdivision. In other words,

$$K^{S-T} \sim L(W, S) - \sigma. \tag{7.2}$$

Theorem 7.5 ([78, Thm. A] or [82, Thm. 8.5.1]) *Suppose (W, S) is a Coxeter system and K is the associated chamber. Then $H^*(W; \mathbb{Z}W)$ is a direct sum of terms of the form $H^*(K, K^{S-T})$, where T is a spherical subset of S. To be more precise, for each $w \in W$, let $T(w)$ be the spherical subset defined by $T(w) := \{s \in S \mid \ell(sw) < \ell(w)\}$. Then*

$$H^*(W; \mathbb{Z}W) = H_c^*(\Sigma) = \bigoplus_{w \in W} H^*(K, K^{S-T(w)}).$$

Note that $H^*(K, K^{S-T})$ is isomorphic to the reduced cohomology of the punctured link $L(W, S) - \sigma(T)$ where $\sigma(T)$ is the spherical simplex with vertex set T.

As stated in the following theorem a similar formula holds for any RAB of finite thickness (cf. [88]) or more generally for any building of finite thickness (cf. [87]). However, the proof uses more than just the Morse theoretic method of Lemma 7.3.

Theorem 7.6 ([88] or [87, Main Theorem]) *Let $\mathcal{C}$ be a building of type (W, S) and of finite thickness. Let $D(\mathcal{C}, K)$ be its standard realization (as in Sect. 4.4.2). Then*

$$H_c^*(D(\mathcal{C}, K)) = \bigoplus_{T \in \mathcal{S}} H^*(K, K^{S-T}) \otimes \hat{A}^T,$$

where $\hat{A}^T$ is a certain free abelian group of functions on the T-residues of $\mathcal{C}$.

A corollary gives a generalization of Theorem 7.5 to an arbitrary graph product of finite groups. It gives a calculation of the cohomology with group ring coefficients of the graph product Γ of a family of finite groups $(G_{s\in S})$ (over the 1-skeleton of a flag complex L), to wit:

Corollary 7.7 ([87, 88, Cor. 9.4], or [97, Prop. 9.7]) *Let Γ be a graph product of finite groups with associated* RACS, (W, S). *Then*

$$H^*(\Gamma; \mathbb{Z}\Gamma) = \bigoplus_{T\in\mathcal{S}(W,S)} H^*(K, K^{S-T}) \otimes \hat{A}^T,$$

where $\hat{A}^T$ is a certain free abelian group.

7.2 Finiteness Properties of Groups

A discrete group G is *type* F if BG is modeled by a finite complex CW complex. A homological version of this is that G is *type* FP. This means that the $\mathbb{Z}G$-module $\mathbb{Z}$ (with trivial G-action) admits a resolution of finite length by finitely generated projective $\mathbb{Z}G$-modules. Intermediate between types F and FP are types FH and FL. *Type* FL means that $\mathbb{Z}$ admits a resolution of finite length by finitely generated free $\mathbb{Z}G$-modules. (If G is torsion-free, then, conjecturally, $\tilde{K}(\mathbb{Z}G) = 0$. This conjecture implies that type FP and type FL are the same class of groups.) *Type* FH means that G acts freely and cellularly on an acyclic complex Y. The cellular chains on Y then provide a free resolution of $\mathbb{Z}$. Obviously,

$$F \implies FH \implies FL \implies FP.$$

For $m \in \mathbb{N}$ properties F_m, FH_m, FL_m, FP_m are defined analogously. For example, type F_m means that BG has a model with a finite m-skeleton. Type F_1 and type FP_1 are both equivalent to the group being finitely generated. Type F_2 implies that G is finitely presented; however, as we shall see below, type FP_2 is strictly weaker than being finitely presented. By a theorem of Wall, if G is type FL and is finitely presented, then it is type F. So, the difference between type F and FL lies entirely within finite presentability.

7.2.1 Finiteness Properties for Proper Actions

If G has nontrivial torsion, then BG cannot be finite dimensional. (The proof is due to P.A. Smith: if H is a finite cyclic subgroup of G, then $BH \to BG$ is a covering space. If H is nontrivial, BH has cohomology in arbitrarily high degrees. So, BH and, a fortiori, BG must be an infinite dimensional.)

A group G is *virtually torsion-free* if it has a torsion-free subgroup Γ of finite index. The *virtual cohomological dimension of* G, denoted by $\operatorname{vcd} G$, is the cohomological dimension of such a Γ. The group G is *type* VF if there is a finite index subgroup Γ of type F. The notions of type VFP, VFL, VF_m, etc, are defined similarly. (These notions are independent of the choice of Γ, cf. [39, Ch. VIII].)

When a group G has nontrivial torsion, there is a notion of a "universal space for proper actions," which is sometimes more useful than the universal space EG.

Definition 7.8 For a discrete group G, its *universal proper G-space* is a CW complex $\underline{E}G$ together with a proper cellular action, $G \curvearrowright \underline{E}G$, so that for any finite subgroup $H < G$, the point set of H, $(\underline{E}G)^H$, is a contractible subcomplex. (In particular, if every finite subgroup of G is trivial, then $\underline{E}G = EG$.) The quotient complex $\underline{E}G/G$ is denoted by $\underline{B}G$. (The complex $\underline{B}G$ is an "orbihedron.")

For example, if $G \curvearrowright Y$, with Y a CAT(0) polyhedron, then $Y = \underline{E}G$.

The universal property of $\underline{E}G$ is that for any CW complex Y with a cellular, proper G-action (i.e., with finite isotropy groups), there is an equivariant map $Y \to \underline{E}G$, unique up to an equivariant homotopy.

One says that G is *type* $\mathcal{V}$F if there is a model for $\underline{E}G$ with only finitely many G-orbits of cells, i.e., if $\underline{B}G$ is a finite complex. So, if G acts cocompactly on a CAT(0) polyhedron Y, then G is type $\mathcal{V}$F. Types $\mathcal{V}$FH, $\mathcal{V}$FP, $\mathcal{V}\mathrm{F}_m$, etc., are defined analogously using an appropriate version of projective modules. (See [100, Appendix].)

The notion of type $\mathcal{V}F$ is closely related being type VF, although a group of type $\mathcal{V}F$ need not be virtually torsion-free. If a group G of type $\mathcal{V}F$ is virtually torsion-free, then it is obviously type VF. In Theorem 7.19 below, we shall give examples of Leary–Nucinkis of groups that are VF but not $\mathcal{V}$F.

Note that if G is $\mathcal{V}F$, then for any finite subgroup $P < G$, its normalizer, $N(P)$, is type F_∞. (Proof: $(\underline{E}G)^P$ is contractible and $N(P)$ acts cocompactly on it with finite stabilizers.) This gives a necessary condition for G to be $\mathcal{V}F$. When G is type VF the condition is close to being sufficient: it implies that $\underline{E}G$ is finitely dominated.

7.3 Level Sets and Bestvina–Brady Groups

Let A_L be the RAAG associated to a flag complex L, $\mathbb{T}_L$ its standard classifying space and $Y_L = EA_L$ be the universal cover of $\mathbb{T}_L$. As explained in Theorem 3.14 of Sect. 3.1.1, Y_L has the structure of a CAT(0) cube complex. We recall why the link of each vertex is a flag complex. The link of a vertex $v \in Y_L$ is a polyhedral join of 0-spheres. We denote this polyhedral join by OL,

$$OL := *_L S^0 = \mathrm{Lk}(v, Y_L), \tag{7.3}$$

and call it the *octahedralization of* L. Thus, OL is formed by replacing each vertex of L by a copy of S^0 and each k-simplex by the boundary complex of a $(k+1)$-

dimensional octahedron (= "cross polytope"). it is easy to see that if L is a flag complex, then so is OL.

Let $\phi : A_L \to \mathbb{Z}$ be the homomorphism that sends each Artin generator a_i to the generator of the infinite cyclic group. Then $BB_L := \ker\phi$ is the *Bestvina–Brady group* associated to L. Extend ϕ to a map $\varphi : Y_L \to \mathbb{R}$ which is affine and nonconstant on each positive dimensional cube. The restriction of φ to the 0-skeleton can be identified with $\phi : A_L \to \mathbb{Z}$. The group BB_L acts cocompactly on each level set $\varphi^{-1}(t)$. By Lemma 7.1 the homotopy type of $\varphi^{-1}(t)$ does not change as t moves within any interval disjoint from $\mathbb{Z}$.

Example 7.9 If L is a 1-simplex, then $A_L = \mathbb{Z} \times \mathbb{Z}$ and $Y_L = \mathbb{E}^2$ tiled by squares. One should think of the squares as being placed diagonally in $\mathbb{E}^2$ so that each edge is parallel to one of the vectors $\pm(e_1 + e_2)$ or $\pm(e_1 - e_2)$. The "height function" $\varphi : \mathbb{E}^2 \to \mathbb{R}$ can then be regarded as projection onto the y-axis.

For each vertex v of Y_L, its ascending link $\mathrm{Lk}_\uparrow(v)$ is the set of all directions at v corresponding to edges with bottom vertex v. Hence, the vertex set of $\mathrm{Lk}_\uparrow(v)$ is $\mathrm{Vert}(L)$ and $\mathrm{Lk}_\uparrow(v)$ can be identified with L. Similarly, $\mathrm{Lk}_\downarrow(v) = L$. One can then use Lemma 7.1 to analyze how $\varphi^{-1}(t)$ changes as t crosses an integer point. The main theorem in [23] is the following.

Theorem 7.10 (Bestvina–Brady [23, Main Theorem])

(i) BB_L is type F if and only if L is contractible.
(ii) BB_L is type FH (or FP or FL) if and only if L is acyclic.

Moreover, if L is not simply connected, then BB_L is not finitely presentable.

The proof of Theorem 7.10 uses Morse theory on convex cell complexes developed in Sect. 7.1.1, i.e., in Lemma 7.1. In particular, statement (ii) follows from Lemma 7.1. To prove statements about the fundamental groups of the level sets, the following stronger version of this lemma is proved in the context of RAAGs when $Y = EA_L$.

Theorem 7.11 (Bestvina–Brady [23, Theorem 8.6]) *The level set Y_t is homotopy equivalent to a wedge of copies of L, one for each vertex not in Y_t.*

This is all that is needed to prove the results of [23]. Here are some further corollaries to it.

Corollary 7.12 ([23])

(i) BB_L is finitely generated if and only if L is connected.
(ii) BB_L is finitely presented if and only if L is simply connected.

More generally we have the following.

Corollary 7.13 ([23])

(i) BB_L is type F_m if and only if L is $(m-1)$-connected.
(ii) BB_L is type FH_m if and only if L is $(m-1)$-acyclic.

Remark 7.14 For any commutative ring R one can define corresponding finiteness properties $FH(R)$, $FL(R)$, and $FP(R)$ as well as $FH_m(R)$. The appropriate generalizations of Theorem 7.10 and its corollaries are proved in this context in [23]. An application of this generalization is given in Theorem 7.17 below.

Remark 7.15 (Subgroups of Hyperbolic Groups that Are Not Hyperbolic) I believe that the ideas for the Bestvina–Brady paper [23] originated in work of N. Brady [32]. He wanted to find examples of a finitely presented group H which was the kernel of an epimorphism from a word hyperbolic group G to $\mathbb{Z}$ so that H was not word hyperbolic. According to Theorem 2.22, H cannot be word hyperbolic by virtue of being a quasiconvex subgroup since such kernels in word hyperbolic groups are never quasiconvex. If G is the fundamental group of a hyperbolic 3-manifold M^3 and the homomorphism to $\mathbb{Z}$ comes from a fiber bundle projection $M^3 \to S^1$, then the kernel H is word hyperbolic since it is a surface group (the fundamental group of the fiber). As we saw in Sect. 5.3.5 all hyperbolic 3-manifolds virtually admit such bundle projections onto a circle. In [136, pp. 125-126] Gromov considered a similar situation in higher dimensions. For any odd n, he shows that there are branched covers M^n of T^n which fiber over S^1 and that fundamental group H of the fiber can never be word hyperbolic for $n > 3$. Furthermore, he gave a convincing (although incorrect) condition for when $G = \pi_1(M^n)$ is word hyperbolic. The example of Brady [32] was of a similar nature in dimension 3. In Brady's example, G is not a 3-manifold group; rather it is a word hyperbolic fundamental group of a cube complex obtained as the branched cover of the product of 3 copies of a finite graph and $G \to \mathbb{Z}$ was a standard map. Using the ideas from this chapter Brady proved that the kernel H is finitely presented but not type FP_3, hence, not word hyperbolic (since torsion-free word hyperbolic groups are type F).

7.4 Some Applications

7.4.1 *Poincaré Duality Groups that Are Not Finitely Presentable*

A group G is a *Poincaré duality group of dimension n over a commutative ring R* if G is type $FP(R)$ and

$$H^i(G; RG) = \begin{cases} 0, & \text{for } i \neq n, \\ R, & \text{for } i = n. \end{cases}$$

When $R = \mathbb{Z}$, G is called simply a "Poincaré duality group of dimension n," abbreviated as PD^n-group.

The fundamental group of a closed aspherical n-manifold is a PD^n-group. However, since any closed manifold is homotopy equivalent to a finite CW complex, any such fundamental group is type F, while a PD^n-group need not be. In fact, one can use the Reflection Group Trick together with the Bestvina–Brady construction to find examples of such G that are not finitely presentable. It is still an open question if every finitely presented PD^n-group is the fundamental group of a closed aspherical manifold.

Theorem 7.16 ([78, Ex. 6.7] or [79, Thm. 7.1.5]) *For each integer $n \geq 4$, there is a* PD^n*-group G that is not finitely presentable (so, BG is not homotopy equivalent to a closed manifold).*

Proof Let L be a 2-dimensional, acyclic flag complex that is not simply connected. By Corollary 7.12 (ii), the Bestvina–Brady kernel, BB_L, is not finitely presentable. The level set $\varphi^{-1}(t)$ is a 2-dimensional acyclic CW complex and its quotient $B = \varphi^{-1}(t)/BB_L$ is a compact 2-dimensional CW complex. The complex B can be thickened to a compact n-manifold with boundary X for any $n \geq 4$. (Any compact d-dimensional complex is homotopy equivalent to a $2d$-manifold; this can be seen by attaching handles instead of cells.) Triangulate ∂X as a flag complex and then give X the structure of a manifold with corners with strata the dual cells of the triangulation. Apply the Reflection Group Trick of Sect. 3.2.6. However, instead of the universal cover of X, use the covering space $\widetilde{X}$ corresponding to BB_L and use the RACG, $\widetilde{W}$, corresponding to the triangulation of $\partial \widetilde{X}$. The result is an acyclic n-manifold $\widetilde{M}$ together with a cocompact action of the semidirect product $\widetilde{W} \rtimes BB_L$. Since $\widetilde{M}$ is an n-manifold, it satisfies Poincaré duality and for $G = \widetilde{W} \rtimes BB_L$ we see that:

$$H^i(G; \mathbb{Z}G) = H^i_c(\widetilde{M}) = H_{n-i}(\widetilde{M}),$$

which vanishes for $i \neq n$ since $\widetilde{M}$ is acyclic. Passing to an appropriate torsion-free, finite-index subgroup Γ of $\widetilde{W}$ we obtain the desired PD$_n$-group $\Gamma \rtimes BB_L$ (in fact, it is of type FH). Since $\Gamma \rtimes BB_L$ retracts onto BB_L, it is not finitely presentable. (Further details of this argument can be found in [78, Example 6.7].)

For a given commutative ring R it is natural to ask if every finitely presented PD$^n(R)$-group is the fundamental group of a closed, aspherical R-homology manifold. In [120] Fowler observes that a variation of the above argument gives a negative answer for any R such that tensoring with R kills some nontrivial finite cyclic group in an integral homology group.

Theorem 7.17 (Fowler [120, Theorem 2.1]) *There is a torsion-free, finitely presented* PD$^n(\mathbb{Q})$*-group that is not the fundamental group of any closed, aspherical rational homology manifold.*

Proof Let L be a simply connected, $\mathbb{Q}$-acyclic flag complex that is not $\mathbb{Z}$-acyclic. (For example, L could be the suspension of $\mathbb{R}P^2$.) Then BB_L is finitely presented and type $FH(\mathbb{Q})$; however, if $H_m(L; \mathbb{Z}) \neq 0$, then BB_L is not type F_{m+1}. As in the proof of Theorem 7.16, we get a finite CW complex $B = \varphi^{-1}(t)/BB_L$, with $\pi_1(B) = BB_L$ and with $\mathbb{Q}$-acyclic universal cover. Since BB_L is not F_{m+1}, B is not homotopy equivalent to a finite aspherical complex. Thickening B to a manifold with boundary X and using the Reflection Group Trick as in the previous proof, we obtain a $\mathrm{PD}^n(\mathbb{Q})$-group $\Gamma \rtimes BB_L$ that is not type F_{m+1} (since it retracts onto BB_L).

7.4.2 *Symmetries of* RAAGs *and Bestvina–Brady Groups*

In [170] Leary and Nucinkis use the Bestvina–Brady construction to find groups of type VF that are not $\mathcal{V}F$. These groups have the form $BB_L \rtimes Q$, where Q is a finite group of simplicial automorphisms of L acting without inversions. (A group action on a simplicial complex is *without inversions* if the pointwise stabilizer of any simplex is equal to its setwise stabilizer.) Since BB_L is torsion-free, $BB_L \rtimes Q$ is virtually torsion-free. By Theorem 7.10 (i), $BB_L \rtimes Q$ is type VF if and only L is contractible. The Leary–Nucinkis result (Theorem 7.19 below) is that $BB_L \rtimes Q$ is $\mathcal{V}F$ if and only if for each subgroup $P \leq Q$, L^P is contractible. Before continuing to discuss this result, we make a few general comments about the wreath-graph product of a RAAG and a finite group.

Suppose Q is a group of automorphisms of a finite flag complex L. Since Q acts on the 1-skeleton, L^1, it defines a group of diagram automorphisms of the associated RACS, (W_L, S), and hence, a group of automorphisms of the associated RAAG, A_L. Form the semidirect product, $A_L \rtimes Q$. It acts by isometries on the associated CAT(0) cube complex Y_L. Since Y_L is CAT(0), it is equal to $\underline{E}(A_L \rtimes Q)$.

Suppose the Q-action on L is without inversions and that $M = L^Q$ is its fixed set. Then

- The fixed set of Q on the cube complex Y_L is the subcomplex Y_M.
- The RAAG A_M centralizes Q. The normalizer $N(Q)$ of Q in $A_L \rtimes Q$ is equal to $A_M \times Q$.

The above two facts are consequences of the following.

Theorem 7.18 (Crisp [69], Leary–Nucinkis [170, Theorem 3]) *Suppose a finite group Q acts admissibly on a flag complex L and that $M = L^Q$ is the subcomplex fixed by Q.*

(i) The fixed set of Q on Y_L is the CAT(0) *cube complex Y_M associated to the* RAAG*, A_M.*
(ii) The subgroup of A_L fixed by Q is A_M.

Let $\mathcal{P}$ be the set of finite subgroups of $BB_L \rtimes Q$ that map isomorphically onto Q. For each $P \in \mathcal{P}$, let $N(P)$ denote its normalizer in $BB_L \rtimes Q$.

(iii) If $M = \emptyset$, then $\mathcal{P}$ contains infinitely many conjugacy classes of subgroups of $BB_L \rtimes Q$. For each $P \in \mathcal{P}$, $N(P) = P$.
(iv) If $M \neq \emptyset$, then all elements of $\mathcal{P}$ are conjugate in $BB_L \rtimes Q$ and $N(Q) = A_M \times Q$.

Proof Statements (i) and (ii) are proved by Crisp [69]. (More generally, Crisp describes the subgroup fixed by Q even when Q-action is allowed to have inversions.)

Any $P \in \mathcal{P}$ is the isotropy subgroup of some vertex $y_P \in Y_L$. If $y_P = gy_Q$, for some $g \in A_L$, then $P = gQg^{-1}$. Since A_L acts transitively on the vertices of Y_L, this shows that any two elements of $\mathcal{P}$ are conjugate in $A_L \rtimes Q$. Since the fixed set of Q on Y_L is contractible, when $M = \emptyset$, $(Y_L)^Q$ must consist of a single vertex y_Q. Similarly, $(Y_L)^P$ must be a single vertex y_P. Since the $(BB_L \rtimes Q)$-action preserves the level sets of φ, we see that when $M = \emptyset$, φ must take y_P and y_Q to the same element of $\mathbb{Z}$. Hence, when $M = \emptyset$ there are infinitely many conjugacy classes parameterized by $\varphi(y_P) \in \mathbb{Z}$. This proves (iii). On the other hand, if $M \neq \emptyset$, then Q must fix a vertex i of L and therefore, centralize the corresponding Artin generator $a = a_i$. Since a suitable power of a takes y_P to y_Q, this means that P and Q are conjugate in $BB_L \rtimes Q$. Hence, (iv).

The subgroup $BB_L \rtimes Q$ of $A_L \rtimes Q$ acts cocompactly on each level set $\varphi^{-1}(t)$.

Theorem 7.19 (Leary–Nucinkis [170, Theorem 5]) *The following are equivalent.*

(i) The group $BB_L \rtimes Q$ is $\mathcal{V}F$.
(ii) For each subgroup $P \leq Q$, L^P is contractible.
(iii) For each $t \in \mathbb{R}$, $\varphi^{-1}(t)$ is a model for $\underline{E}(BB_L \rtimes Q)$.

Proof If (ii) holds, then $(\varphi^{-1}(t))^P$ is $N(P)$-equivariantly homotopy equivalent to $(Y_L)^P$, which is contractible (since $Y_L = \underline{E}(A_L \rtimes Q)$). Conversely, suppose (ii) fails. Then, for some subgroup $P \leq Q$ and some natural number m, L^P is $(m-1)$-connected but not m-connected. By Theorem 7.10 (i)$'$, for $M = L^P$, BB_M is type F_m but not F_{m+1}. Therefore, $N(P) = A_M \times P$ is not type F_{m+1}; so, $BB_L \rtimes Q$ cannot be $\mathcal{V}F$.

A classical topic in transformation groups is to produce examples of finite group actions on finite contractible complexes (e.g., disks) such that the fixed point sets are not contractible. For example, for any sufficiently complicated finite group Q there is Q-action on a contractible L with $L^Q = \emptyset$ (see [184]). One can combine such examples with Theorems 7.18 and 7.19 to produce examples of groups $BB_L \rtimes Q$ with exotic finiteness properties.

Corollary 7.20 (Leary–Nucinkis [170, Section 5]) *There are examples of groups G of type VF with either of the following properties.*

(i) G contains infinitely many conjugacy classes of finite subgroups.
(ii) G is $\mathcal{V}FH$, but not $\mathcal{V}F$

Proof If L is contractible, then $BB_L \rtimes Q$ is VF. For (i), take a Q-action on L with empty fixed set. For (ii), take a Q-action on L so that the fixed set of each subgroup of Q is acyclic, but for at least one $P \leq Q$, L^P is not simply connected.

7.4.3 Uncountably Many Groups of Type FP

Although there are uncountably many isomorphism classes of finitely generated groups, only countably many are finitely presentable. Higman showed that a finitely generated group can be embedded in a finitely presented group if and only if it has a recursive presentation. To study such questions for groups of type FP, Ian Leary [169] develops the following interesting variation of the Bestvina–Brady construction.

As usual, L is a connected, finite flag complex. In our application L will be acyclic but not simply connected and, in addition, L will be aspherical. Given a subset $S \leq \mathbb{Z}$, let $Y_L^{(S)} \to Y_L$ be the universal branched cover, where the branch set is the set of 0-cells v such that $\varphi(v) \in \mathbb{Z} - S$ (i.e., the branch set is $\varphi^{-1}(\mathbb{Z} - S)$). In other words, $Y_L^{(S)}$ is the metric completion of the universal cover of $Y_L - \varphi^{-1}(\mathbb{Z} - S)$. The cubical structure of Y_L lifts to $Y_L^{(S)}$ to give it the structure of a CAT(0) cube complex. The link of a 0-cell in $\varphi^{-1}(S)$ is, as before, equal to OL while the link of a 0-cell in $\varphi^{-1}(\mathbb{Z} - S)$ is $O\tilde{L}$, where $\tilde{L}$ means the universal cover of L. The composition $Y_L^{(S)} \to Y_L \to Y_L/BB_L$ is also a regular branched cover. Leary [169] defines $G_L(S)$ to be the group of deck transformations of $Y_L^{(S)} \to Y_L/BB_L$. The group $G_L(\mathbb{Z})$ is obviously equal to BB_L and it turns out that $G_L(\emptyset) \cong BB_{\tilde{L}} \rtimes \pi_1(L)$.

Theorem 7.21 (Leary [169, Theorem A] also compare [168]) *Suppose L is a given connected flag complex that is not simply connected.*

(i) There are uncountably many (in fact, $2^{\aleph_0}$) isomorphism types of groups $G_L(S)$.
(ii) $G_L(S)$ is finitely presentable if and only if S is finite.
(iii) $G_L(S)$ is isomorphic to a subgroup of a finitely presented group if and only if S is recursively enumerable.

Let $\overline{\varphi} : Y_L^{(S)} \to \mathbb{R}$ be the Morse function obtained by precomposing $\varphi : Y_L \to \mathbb{R}$ with the projection $Y_L^{(S)} \to Y_L$. For each $v \in \overline{\varphi}^{-1}(S)$ the descending and ascending links, $\mathrm{Lk}_\downarrow(v, Y_L^{(S)})$ and $\mathrm{Lk}_\uparrow(v, Y_L^{(S)})$, are isomorphic to L, while for $v \in \overline{\varphi}^{-1}(\mathbb{Z} - S)$ they become isomorphic to $\tilde{L}$, the universal cover of L. Suppose that

(a) L is acyclic but not simply connected.
(b) $\tilde{L}$ is contractible (i.e., L is aspherical).
(c) L has no local cut points.

Condition (c) is needed to insure that $G_L(S)$ is isomorphic to $\pi_1(Y_L/BB_L - V(S))$, where $V(S) = \{v \in \mathrm{Vert}(Y_L/BB_L) \mid \varphi(v) \notin S\}$ (cf. [169, Prop. 31]). As before, one can study the level sets $\overline{\varphi}^{-1}(t)$, for $t \in \mathbb{R} - \mathbb{Z}$. The hypotheses imply that each ascending or descending link, being isomorphic to either L or its universal cover $\tilde{L}$, is acyclic. Hence, the proof of Theorem 7.10 in Sect. 7.3, gives the following result (the point being that under conditions (a) and (b) all ascending and descending links are acyclic).

Theorem 7.22 (Leary [169, Theorem B]) *With hypotheses as above, for any $S \leq \mathbb{Z}$, $G_L(S)$ is type FH. When S is infinite, $G_L(S)$ is not finitely presentable, hence, not type F.*

The conclusion on finite presentability follows from an analog of Theorem 7.10—since when S is infinite, there are infinitely many vertices with links of type L.

It can happen that $G_L(S)$ and $G_L(S')$ are isomorphic when $S \neq S'$. (Indeed, this is the case whenever S and S' differ by an affine automorphism λ of $\mathbb{R}$ such that $\lambda(\mathbb{Z}) \leq \mathbb{Z}$.) The heart of [169] lies in the proof of statement (i) of Theorem 7.21—when $\pi_1(L) \neq 1$, the groups $G_L(S)$ run through uncountably many isomorphism types as S varies. To prove this, an invariant is needed. For a countable group G and finite sequence $\mathbf{g} = (g_1, \dots, g_l)$ in G, Leary [169, §15] introduces an invariant $\mathcal{R}(G, \mathbf{g})$ which takes values in subsets of $\mathbb{Z}$. (In fact, $\mathcal{R}(G, \mathbf{g}) = \{n \in \mathbb{Z} \mid g_1^n \cdots g_l^n = 1\}$.) He proves that for any fixed isomorphism type of G, $\mathcal{R}(G, \mathbf{g})$ can take on only countably many values. When $G = G_L(S)$, with $0 \in S$, and $\mathbf{a} = (a_1, \dots, a_l)$ is a sequence corresponding to a homotopically nontrivial edge loop in L, it is proved [169, Lemma 48] that $\mathcal{R}(G_L(S), \mathbf{a}) = S$. Since after a translation of $\mathbb{Z}$, one can assume $0 \in \mathbb{Z}$, the uncountability of the set of isomorphism types follows.

Correction to: Infinite Group Actions on Polyhedra

Correction to:
M. W. Davis, *Infinite Group Actions on Polyhedra*,
Ergebnisse der Mathematik und ihrer Grenzgebiete.
3. Folge / A Series of Modern Surveys in Mathematics 77,
https://doi.org/10.1007/978-3-031-48443-8

The original version of the book was inadvertently published with errors in Chapters 2 and 4.

- **In Chapter 2: Polyhedral Preliminaries**
  In Section 2.4.3, on page 34, line 5: G2 * G4 should be H2 * H4.
- **In Chapter 4: Coxeter Groups, Artin Groups, Buildings**
  - In Sections 4.2–4.4: On pages 95–163, the odd numbered pages: in these sections, the running title Outline of This Section should never appear as a running title on the right-hand page. Instead, these running titles should be the appropriate section number: 4.2 Coxeter Groups, 4.3 Artin Groups, and 4.4 Buildings.
  - On p. 95, p. 122, p. 137: On these pages, Outline of Section is not the title of a new section or subsection, it is merely the title of a paragraph in Sections 4.2, 4.3, and 4.4. So, the font should be somewhat smaller.
  - On p. 112, line 4: "the 120" should be "the 120-cell".
  - On p. 112, line 8 of Example 4.36: "is $\pi/2$" should be "of $\pi/2$".
  - On p. 112, 5 lines before Remark 4.37: "right-angled" polytope should be convex polytope.
  - On p. 113, first paragraph after Remark 4.37: "Poincar'e" should be Poincaré.
  - On p. 137, line 11: should be "always holds".

The updated version of these chapters can be found at
https://doi.org/10.1007/978-3-031-48443-8_2
https://doi.org/10.1007/978-3-031-48443-8_4

M. W. Davis, *Infinite Group Actions on Polyhedra*, Ergebnisse der Mathematik und ihrer Grenzgebiete. 3. Folge / A Series of Modern Surveys in Mathematics 77,
https://doi.org/10.1007/978-3-031-48443-8_8

Appendix A
Complexes of Groups

The theory of complexes of groups is developed in [35, 65, 138]. Most of the time we will only use the easier notion of a "simple complex of groups." This notion is also developed in [35], as well as in other papers such as [56]. The idea of a complex of groups arises from the action of a discrete group G acting on a polyhedron. Another word for a complex of groups is "orbihedron," which conveys the idea of the orbit space of such a group action together with some data about isomorphism types of isotropy subgroups. We get a simple complex of groups when the G-action has a "strict" fundamental domain (see Definition 3.5). The most basic examples of complexes of groups are graphs of groups. When the underlying graph is a tree, we get a "tree of groups," the basic example of a simple complex of groups. When the underlying graph contains a loop we are forced to consider general complexes of groups. Given a simple complex of groups, the group action on a simply connected polyhedron can be recovered by a simple gluing procedure. This procedure is the "basic construction," described in Sect. 3.1.1 and in Sect. A.2 below. For a general complex of groups the process of recovering the simply connected polyhedron involves covering space theory as well as this pasting procedure. This is explained in Sect. A.3. In Sect. A.4 we discuss the notion of an "aspherical realization" of a complex of groups and we explain the $K(\pi, 1)$-Question for a simple complex of groups.

Suppose G acts on a space X. A *strict fundamental domain* for G on X is a closed subspace $C \leq X$ such that C intersects each orbit in exactly one point. It follows that the restriction of the orbit projection $p : X \to X/G$ to the subspace C is a homeomorphism. Hence, C provides a section $s : X/G \to X$ of p defined by $s = (p|_C)^{-1}$. Usually we require X to be a cell complex and furthermore, that the G-action permutes the cells, in which case one can require C to be a subcomplex of X. Two points of C belong to the same *pure stratum* if their isotropy subgroups are equal. More precisely, a *pure stratum* is a connected component of an equivalence class in C defined by equality of isotropy subgroups. A *stratum* of C is the closure of a pure stratum. So, C is a stratified space as in [35, II.12.1]. The set of strata of

M. W. Davis, *Infinite Group Actions on Polyhedra*, Ergebnisse der Mathematik und ihrer Grenzgebiete. 3. Folge / A Series of Modern Surveys in Mathematics 77, https://doi.org/10.1007/978-3-031-48443-8

C is naturally a poset $\mathcal{Q}$ ordered by inclusion. Similarly, if G acts on a poset $\mathcal{P}$ by order-preserving automorphisms, then a *strict fundamental domain* for G on $\mathcal{P}$ is a subposet $\mathcal{Q}$ which intersects each G-orbit in exactly one element.

A.1 Simple Complexes of Groups: Definitions

The basic reference for the material in this appendix is [35]. Other references include [56, 99, 115, 191] and [82, Appendix E].

Definition A.1 (cf. Bridson-Haefliger [35, II.12.11, p. 375]) A *simple complex of groups* $G\mathcal{Q}$ over a poset $\mathcal{Q}$ is a collection of groups $\{G_\sigma\}_{\sigma\in\mathcal{Q}}$ and monomorphisms $\phi_{\tau\sigma} : G_\sigma \to G_\tau$ defined whenever $\tau < \sigma$. The G_σ are the *local groups*. Furthermore, $G\mathcal{Q}$ must be a cofunctor from $\mathcal{Q}$ to the category of groups and monomorphisms in the sense that $\phi_{\mu\tau}\phi_{\tau\sigma} = \phi_{\mu\sigma}$ whenever $\mu < \tau < \sigma$. (The term *cofunctor* means a contravariant functor, i.e., a functor on the dual poset, $\mathcal{Q}^{\mathrm{op}}$, where the order relation is reversed.)

Often $\mathcal{Q}$ will be the set of cells in a cell complex, partially ordered by inclusion.

Suppose G acts on a cell complex X with strict fundamental domain C. The set of strata of C is a poset $\mathcal{Q}$ and the isotropy subgroups of the pure strata define a simple complex of groups $G\mathcal{Q} = \{G_\sigma\}_{\sigma\in\mathcal{Q}}$. Here the monomorphisms $\varphi_{\tau\sigma} : G_\sigma \to G_\tau$ for $\tau < \sigma$ are given by the natural inclusion of the isotropy subgroup at a large stratum into the isotropy subgroup at a smaller stratum. A simple complex of groups $G\mathcal{Q}$ is said to be *developable* if it arises in this way from an action, $G \curvearrowright X$, and the G-space X is called a *development* of $G\mathcal{Q}$. Similarly, if $G \curvearrowright \mathcal{P}$, where $\mathcal{P}$ is a poset with a poset $\mathcal{Q}$ as a strict fundamental domain, then we again get a simple complex of groups, $G\mathcal{Q} = \{G_\sigma\}$. The G-poset $\mathcal{P}$ is a *development*. N.B. Not every simple complex of groups is developable. We shall give a necessary and sufficient condition below for $G\mathcal{Q}$ to be developable.

A *simple morphism* $\psi = (\psi_\sigma)$ from a simple complex of groups $G\mathcal{Q}$ to a group H is a function which assigns to each $\sigma \in \mathcal{Q}$ a homomorphism $\psi_\sigma : G_\sigma \to H$ such that $\psi_\sigma = \psi_\tau\phi_{\tau\sigma}$ whenever $\tau < \sigma$. The simple morphism ψ is *injective on local groups* if each ψ_σ is injective. If $H \curvearrowright \mathcal{P}$ with strict fundamental domain $\mathcal{Q}$ and $G\mathcal{Q}$ is the corresponding simple complex of groups, then by taking isotropy subgroups we get a simple morphism $\psi : G\mathcal{Q} \to H$ that is injective on local groups. Conversely, we shall see below in Theorem A.6 that if there is a simple morphism $\psi : G\mathcal{Q} \to H$ that is injective on local groups, then $G\mathcal{Q}$ is developable. (In fact, we can take H to being the direct limit of the system of groups $G\mathcal{Q}$, as defined below.)

Definition A.2 (This Definition Is Also Given in Sect. 3.1.2) The *direct limit* of $G\mathcal{Q}$ is a group G (also denoted by $\lim G\mathcal{Q}$) together with a simple morphism ψ : $G\mathcal{Q} \to G$ (called the *canonical simple morphism*) with the following universal property: if $\psi' : G\mathcal{Q} \to H$ is any simple morphism, then there exists a unique homomorphism $\theta : G \to H$ so that ψ' factors through G, that is to say, $\theta\psi_\sigma = \psi'_\sigma : G_\sigma \to H$ for all $\sigma \in \mathcal{Q}$. It follows that the direct limit is unique up to canonical isomorphism. To establish its existence we note that the direct limit can be constructed in a standard fashion by taking the free product of the G_τ and then quotienting by relations to insure that G_σ is identified with its image under $\phi_{\tau\sigma}$ in G_τ and that $\phi_{\mu\tau}\phi_{\tau\sigma}(g) = \phi_{\mu\sigma}(g)$ whenever $\mu < \tau < \sigma$, for all $g \in G_\sigma$ (cf. [207, p. 1]).

Example A.3 (Graphs of Groups) Suppose Y is a connected graph without loops. This means that each edge of Y has two distinct end points. In other words, Y is a 1-dimensional regular CW complex. (To say that CW complex is *regular* means that each cell is embedded.) Its vertex set and edge set are denoted $\operatorname{Vert} Y$ and $\operatorname{Edge} Y$, respectively. Let $\mathcal{Q} = \operatorname{Vert} Y \bigsqcup \operatorname{Edge} Y$ be its set of cells, partially ordered by inclusion. A *graph of groups* on Y is a simple complex of groups over $\mathcal{Q}$ defined by specifying a *vertex group* G_τ for each $\tau \in \operatorname{Vert} \Omega$, an edge group G_σ for each $\sigma \in \operatorname{Edge} \Omega$, and a monomorphism $G_\sigma \hookrightarrow G_\tau$ whenever $\tau < \sigma$. If Y is a tree, then $G\mathcal{Q}$ is a *tree of groups*. The direct limit G should be thought of as an amalgam. For example, if Ω is an interval, then G is the amalgam, $G_1 *_H G_2$ where G_1, G_2 are the vertex stabilizers and H is the edge stabilizer. The standard definition of a graph of groups in [207, Section 4.4] allows the possibility of loops: one must use directed edges and for each edge σ one must give monomorphisms from G_σ to the local groups at its initial and terminal vertices. When Y has loops, instead of considering the poset of cells in Y one should let $\mathcal{Q}$ be the "scwol" (= a small category without loops) where if an edge σ has only one vertex τ, then there are two arrows from σ to τ (while in a poset there is only one such arrow). For the corresponding notion of a graph of groups $G\mathcal{Q}$ over $\mathcal{Q}$ there should be two injections $G_\sigma \to G_\tau$ between the local groups. (For the definition of "scwol" see Sect. A.3.1 below.)

Example A.4 (Cell Complexes of Groups) Suppose C is a regular CW complex and let $\mathcal{Q}$ be its poset of cells. A simple complex of groups $G\mathcal{Q} := \{G_\sigma\}_{\sigma\in\mathcal{Q}}$ is then called a *cell complex of groups*. For example, if C is a simplex we get a *simplex of groups* generalizing the notion of an interval of groups in the previous example. More generally, if C is a polyhedral cell complex, where the cells are not necessarily embedded, then instead of the poset of cells of C it is better to consider an appropriately defined scwol.

Definition A.5 (Polytopes of Groups) Suppose P is a simple convex polytope and that $\mathcal{F}(P)$ denotes its poset of faces. Put $\mathcal{Q} = \mathcal{F}(P)$. A simple complex of groups $G\mathcal{Q}$ over $\mathcal{Q}$ is a *polytope of groups*. If $\dim P = 2$, then $G\mathcal{Q}$ is a *polygon of groups*.

A.2 The Basic Construction for a Simple Complex of Groups

Suppose $G\mathcal{Q}$ is a simple complex of groups over a poset $\mathcal{Q}$ and that $\psi' = (\psi'_\sigma)$ is a simple morphism from $G\mathcal{Q}$ to a group H that is injective on local groups. Define a poset of cosets by

$$D(\psi', \mathcal{Q}) := \coprod_{\sigma \in \mathcal{Q}} H/G_\sigma . \tag{A.1}$$

The partial order is the natural one, defined by $(hG_\tau, \tau) < (gG_\sigma, \sigma) \iff \tau < \sigma$ and $h^{-1}g \in G_\tau$. There is a natural inclusion $\mathcal{Q} \hookrightarrow D(\psi', \mathcal{Q})$; moreover, $H \curvearrowright D(\psi', \mathcal{Q})$ with quotient poset $\mathcal{Q}$. The poset $D(\psi', \mathcal{Q})$ is the *development* of $G\mathcal{Q}$ with respect to ψ. Examples of (A.1) are given by the poset of spherical cosets in a Coxeter group (4.7) or an Artin group (4.8) discusssed in Sect. 4.1.2.

Next suppose as in [35, pp. 368–370] that C is a stratified space with strata indexed by the poset $\mathcal{Q}$. We take this simply to mean that $\{C_\sigma\}_{\sigma \in \mathcal{Q}}$ is a collection of closed subspaces of C whose union is C so that $C_\sigma < C_\tau$ whenever $\sigma < \tau$. (We might want to add the hypothesis that C is a CW complex and that each C_σ is a subcomplex.) Put $\partial C_\sigma := \bigcup_{\tau < \sigma}(C_\sigma - C_\tau)$ and define the *pure stratum* indexed by σ to be $C^\circ_\sigma := C_\sigma - \partial C_\sigma$. For each $x \in C$, let $\sigma(x) \in \mathcal{Q}$ be the index of the pure stratum which contains x. The *basic construction*, $D(\psi', C)$, is defined by

$$D(\psi', C) := (H \times C)/\sim, \tag{A.2}$$

where, as in (3.9), the equivalence relation $\sim$ is defined by $(g, x) \sim (g', x') \iff x = x'$and $gG_{\sigma(x)} = g'G_{\sigma(x)}$. Then H acts on $D(\psi', C)$ with strict fundamental domain C (cf. Definition 3.5, Lemmas 3.6 and 3.7).

Our principal example of a stratified space is $|\mathcal{Q}|$, the geometric realization of the order complex of the poset $\mathcal{Q}$. Its strata are defined by

$$|\mathcal{Q}|_\sigma := |\mathcal{Q}_{\leq\sigma}|, \tag{A.3}$$

where, as usual, $\mathcal{Q}_{\leq\sigma} = \{\tau \in \mathcal{Q} \mid \tau \leq \sigma\}$.

Theorem A.6 (Some Properties of the Basic Construction, cf. [35, pp 381–387], [82, §5.1] or [227]) *Suppose $G\mathcal{Q}$ is a simple complex of groups over a poset $\mathcal{Q}$ and that $\psi' = (\psi'_\sigma)$ is a simple morphism from $G\mathcal{Q}$ to a group H that is injective on local groups. Let $G = \lim G\mathcal{Q}$ and let $\varphi : G \to H$ be the homomorphism defined by the universal property of a direct limit. Let C be a stratified space over $\mathcal{Q}$ as above.*

(i) There is an H-action on the poset $D(\psi', \mathcal{Q})$ defined by (A.1) *with $\mathcal{Q}$ as a strict fundamental poset.*
(ii) There is an H-action on the space $D(\psi', C)$ with strict fundamental domain C.
(iii) (Compare [227].) The H-action on $D(\psi', C)$ has the following universal property: suppose $H \curvearrowright Z$ and $f : C \to Z$ is a map such that for all $\sigma \in \mathcal{Q}$,

$f(C_\sigma)$ is contained in $\mathrm{Fix}(G_\sigma, Z)$, *(the fixed point set of G_σ on Z). Then there is a unique extension of f to a H-equivariant map $\tilde{f} : D(\psi', \mathcal{Q}) \to Z$.*

(iv) The orbit space of $D(\psi', C)$ is C. The orbit projection $p : D(\psi', C) \to C$ is a retraction.

(v) If $D(\psi', C)$ is path connected, then so is C; moreover, the collection of subgroups $\{G_\sigma\}_{\sigma \in \mathcal{Q}}$ generates H. (This implies that the canonical homomorphism $\varphi : G \to H$ is onto).

(vi) If $D(\psi', C)$ is simply connected, then so is C. Moreover, this implies that the canonical homomorphism $\varphi : G \to H$ is an isomorphism.

Proof Since ψ' is injective on local groups, we may identify each local group G_σ with its image $\psi'_\sigma(G_\sigma)$ in H. Statements (i) and (ii) are immediate from the definitions in Eqs. (A.1) and (A.2).

(iii) Denote by $[g, x]$ the image of $(g, x) \in H \times C$ in $(H \times C)/\sim$. Then $\tilde{f} : D(\psi', C) \to Z$ is defined by $\tilde{f}([g, x]) = gf(x)$.

(iv) Part (iv) is precisely what is meant by saying that C is a strict fundamental domain.

(v) Since $p : D(\psi', C) \to C$ is a retraction, it induces a surjection on π_0 (in fact, on all homotopy groups π_i). So, if $D(\psi', C)$ is path connected, then so is C.

(vi) Suppose $D(\psi', C)$ is simply connected. Since p is a retraction, C is also simply connected. The universal property of the direct limit gives a homomorphism $\varphi : G \to H$. Put $D = D(\psi, C)$ and $D' = D(\psi', C)$, where $\psi = (\psi_\sigma)$ is the canonical simple morphism from $G\mathcal{Q}$ to G. Since C is path connected and the direct limit G is generated by local groups, it follows that D is path connected and that φ is onto. Let $N = \ker(\varphi : G \to H)$. By (iii) there is a φ-equivariant map $\pi : D \to D'$ which is a covering projection with fiber N. Since D' is simply connected, N is the trivial subgroup of G. □

In the next theorem we prove converses to parts (v) and (vi) of the previous theorem.

Theorem A.7 ([35]) *Suppose $G\mathcal{Q}$ is a simple complex of groups over a poset $\mathcal{Q}$ and that $\psi' = (\psi'_\sigma)$ is a simple morphism from $G\mathcal{Q}$ to a group H that is injective on local groups. Let $G = \lim G\mathcal{Q}$ and let $\varphi : G \to H$ be the homomorphsim defined by the universal property of a direct limit. Then H acts on the poset $D(\psi', \mathcal{Q})$ defined by* (A.1). *Let C be a stratified space over $\mathcal{Q}$. Let H' be the subgroup of H generated by $\{\varphi(G_\sigma)\}_{\sigma \in \mathcal{Q}}$. The simple morphism ψ' from $G\mathcal{Q}$ to H factors through a simple morphism $\psi'' : G\mathcal{Q} \to H'$. The following converses to statements* (v) *and* (vi) *of Theorem A.6 hold.*

(v)$'$ *If C is path connected and $\varphi : G \to H$ is onto, then $D(\psi', C)$ is path connected. This shows that if C is path connected, then so is $D(\psi'', C)$.*

(vi)$'$ *If C is simply connected and $\varphi : G \to H$ is an isomorphism, then $D(\psi', C)$ is simply connected.*

Proof of Theorem A.7

(v)′ The simple morphism ψ'' gives an H'-equivariant embedding $D(\psi'', C) \hookrightarrow D(\psi', C)$. Moreover, $D(\psi'', C)$ is both open and closed in $D(\psi', C)$. (This uses the fact that if U is a open neighborhood of x in C, then $(G_\sigma \times U)/\sim$, the basic construction on U, is an open neighborhood of $[1, x]$ in $D(\psi', C)$.) So, if C is connected, then $D(\psi'', C)$ is connected and $D(\psi', C)$ is a disjoint union of copies of $D(\psi'', C)$, one for each coset of H' in H. Hence, $H' = H$ if and only if $D(\psi', |\mathcal{Q}|)$ is connected.

(vi)′ Suppose C is simply connected and $H = G$. Put $D = D(\psi, C)$. To show D is simply connected, we must show that D admits only the trivial covering space. Let $E \to D$ be a connected covering space. Since C is simply connected, the inclusion $C \hookrightarrow D$ lifts to E so that the image of C is a fundamental domain for the group $\tilde{G}$ of all lifts to E of the G-action on D (this is the key point). Each local group G_σ lifts isomorphically to the isotropy subgroup of $\tilde{G}$ at a point of C. So, the $\tilde{G}$-action defines the same simple complex of groups $G\mathcal{Q}$. By part (iii) of Theorem A.6, the lift of the fundamental domain extends to a section $D \to E$ of the covering projection $E \to D$. Therefore, $E \to D$ is a homeomorphism. □

When $H = G$ and ψ' is the canonical simple morphism $\psi : G\mathcal{Q} \to G$ we will write $D(G, C)$ instead of $D(\psi, C)$.

Suppose $G\mathcal{Q}$ is developable and $|\mathcal{Q}|$ is simply connected. Let $G = \lim G\mathcal{Q}$. By Theorem A.7 (vi), $D(G, |\mathcal{Q}|)$ is simply connected. Hence, $D(G, |\mathcal{Q}|)$ is the universal cover of the simple complex of groups $G\mathcal{Q}$ in the sense of Sect. A.3 below. Moreover, G is the "fundamental group" of $G\mathcal{Q}$. (We shall also say that the poset $D(G, \mathcal{Q})$ is the "universal cover" of $G\mathcal{Q}$.) We will see in Sect. A.2.1 below how to describe the universal cover, in the case when $|\mathcal{Q}|$ is not simply connected.

A.2.1 The Universal Cover When $|\mathcal{Q}|$ Is Not Simply Connected

The purpose of this subsection is to describe the universal cover of a simple complex of groups when the order complex of the underlying poset $\mathcal{Q}$ is not simply connected. The construction uses the semidirect product construction.

So, suppose $\mathcal{Q}$ is a poset whose order complex $|\mathcal{Q}|$ is connected but not necessarily simply connected. The elements of $\mathcal{Q}$ are identified with Vert $|\mathcal{Q}|$ and the edges of $|\mathcal{Q}|$ define the order relation. Let X denote the universal cover of the Δ-complex $|\mathcal{Q}|$. The simplicial structure on $|\mathcal{Q}|$ lifts to a simplicial structure on X. Let $\tilde{\mathcal{Q}}$ be the poset with underlying set Vert X and with the partial order defined by Edge X. There is an induced simple complex of groups $G\tilde{\mathcal{Q}}$ over $\tilde{\mathcal{Q}}$ defined by $G_{\tilde{\sigma}} := G_\sigma$, where $\tilde{\sigma}$ is a vertex lying σ. Similarly, if $\{\tilde{\tau}, \tilde{\sigma}\}$ is an edge of X with $\tilde{\tau} < \tilde{\sigma}$, then put $\phi_{\tilde{\sigma}\tilde{\tau}} := \phi_{\sigma\tau}$. Let $\tilde{G} = \lim G\tilde{\mathcal{Q}}$. Form the basic construction $D(\tilde{G}, |\tilde{\mathcal{Q}}|)$. Since $|\tilde{\mathcal{Q}}|$ is simply connected, it follows from Theorem A.7 (vi)′ that $D(\tilde{G}, |\tilde{\mathcal{Q}}|)$ is simply connected.

Put $\pi = \pi_1(|\mathcal{Q}|)$. Then $\pi \curvearrowright |\tilde{\mathcal{Q}}|$ by deck transformations. It also acts on the poset $\tilde{\mathcal{Q}}$, as well as, on the cofunctor $G\tilde{\mathcal{Q}}$ as a group of natural transformations. The action of π on $G\tilde{\mathcal{Q}}$ defines a (nonsimple) complex of groups $G\tilde{\mathcal{Q}} \rtimes \pi$. Similarly, since π acts on the direct limit $\tilde{G}$ as a group of automorphisms, we can form the semidirect product $\tilde{G} \rtimes \pi$. An action of this semidirect product on $D(\tilde{G}, |\tilde{\mathcal{Q}}|)$ is defined by the formula:

$$(h, a) \cdot [g, x] = [ha(g), ax] \tag{A.4}$$

Here $(h, a) \in \tilde{G} \rtimes \pi$, $[g, x] \in D(\tilde{G}, |\tilde{\mathcal{Q}}|)$ $(= \tilde{G} \times C)/\sim)$, and $a(g)$ means the automorphism corresponding to $a \in \pi$ applied to element $g \in \tilde{G}$. It is a simple exercise to check that formula (A.4) gives a well-defined action on $D(\tilde{G}, |\tilde{\mathcal{Q}}|)$ (cf. [82, p. 169]). By Theorem A.7 (vi), $D(\tilde{G}, |\tilde{\mathcal{Q}}|)$ is simply-connected. Hence, $D(\tilde{G}, |\tilde{\mathcal{Q}}|)$ is the universal cover of $G\mathcal{Q}$ and $\pi_1(G\mathcal{Q}) = \tilde{G} \rtimes \pi$. In other words, for any developable simple complex of groups, $G\mathcal{Q}$, with $|\mathcal{Q}|$ connected, its universal cover and fundamental group come from the semidirect product construction. In summary, this gives the following.

Proposition A.8 (The Universal Cover of a Simple Complex of Groups) *Suppose $G\mathcal{Q}$ is a developable simple complex of groups over a poset $\mathcal{Q}$ whose order complex has fundamental group $\pi_1(|\mathcal{Q}|) = \pi$ and universal cover $|\tilde{\mathcal{Q}}|$. Let $\tilde{\mathcal{Q}}$ be the poset of lifted simplices in $|\tilde{\mathcal{Q}}|$ and $G\tilde{\mathcal{Q}}$ the induced simple complex of groups. Let $\tilde{G} = \lim G\tilde{\mathcal{Q}}$. Then the fundamental group of $G\mathcal{Q}$ is the semidirect product, $\tilde{G} \rtimes \pi$, and its universal cover is $D(\tilde{G}, |\tilde{\mathcal{Q}}|)$.*

Example A.9 (The HNN Construction, cf. [207, Prop. 5, p. 8]) Here we are considering the case of a graph of groups where the graph consists of a single 0-cell and a single 1-cell (i.e., a *loop of groups*). A loop of groups is an example of a complex of groups which is not a simple complex of groups. Suppose A is a subgroup of a group G and that $\theta : A \to G$ is an injective homomorphism. For each $n \in \mathbb{Z}$, put $A_n = A$, $G_n = G$ and let $\iota : A \to G$ denote the inclusion.The injections $\theta : A_n \to G_n$ and $\iota : A_n \to G_{n+1}$ define a tree of groups. Let H be the associated direct limit described by the diagram:

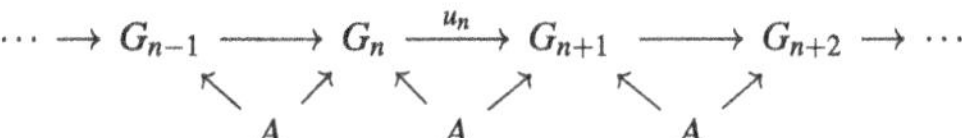

Let $u_n : G_n \to G_{n+1}$ be the canonical isomorphism. Then u_n extends to a automorphism u of H. Use u to form the semidirect product $G' = H \rtimes \mathbb{Z}$, where a generator of $\mathbb{Z}$ is denoted t. Identify G with G_0; so, if $a \in A$ the element $u(a) \in G_1$ is identified with $\theta(a) \in G_0$. The group G' is denoted by $*_A G$ and is called the HNN construction on G, A and θ. So, the graph gives a scwol rather than a poset. It is not useful to form the direct limit over the scwol; however, H is the direct limit of the tree of groups (it is the group $\tilde{G}$ from the previous paragraph) and the semidirect product construction agrees with (A.4). Thus, the universal cover of the loop of groups is given by the basic construction on the tree of groups. The fundamental group of the loop of groups is the HNN construction $*_A G$.

A.3 General Complexes of Groups

A.3.1 Overview

In the first part of this appendix we define the notions of a scwol and of a complex of groups over a scwol.

Scwols A small category $\mathcal{C}$ defines a directed graph whose vertex set is the set of objects in $\mathcal{C}$ and whose edge set is the set of morphisms that are not equal to the identity morphism from an object to itself. Morphisms are also called "arrows." If a is a morphism, then the initial vertex $i(\alpha)$ is the source of a and its terminal vertex $t(\alpha)$ is its target. We also regard an object $\sigma \in \mathrm{Ob}(\mathcal{C})$ as being the identity morphism in $\mathrm{Hom}(\sigma, \sigma)$.

A poset is a small category in which there is a single morphism from σ to τ whenever $\sigma \leq \tau$. A *small category without loops* (abbreviated *scwol*) is similarly defined except that there can be more than one morphism from σ to τ. In particular, there is a poset structure on the objects of a scwol: write $\sigma < \tau$ whenever $\sigma \neq \tau$ and $\mathrm{Hom}(\sigma, \tau)$ is nonempty. The geometric realization of a poset is a simplicial complex (the order complex), while the geometric realization of a scwol is an ordered simplicial cell complex. For example, if X is a polyhedral cell complex as classically defined (i.e., the intersection of two distinct cells is a common face), then the geometric realization of its face poset is the barycentric subdivision of X. On the other hand, if X is a more general polyhedral cell complex (for example a Δ-complex), then its set of cells is the set of objects in a scwol, the geometric realization of which is, as before, the barycentric subdivision of X .

Suppose $\mathcal{R}$ is a small category without loops. The *geometric realization* of $\mathcal{R}$ is the ordered simplicial cell complex $|\mathcal{R}|$ defined as follows. Its vertex set, $\mathrm{Vert}(\mathcal{R})$, is the set of objects of $\mathcal{R}$. Its *edge set*, $\mathrm{Edge}(\mathcal{R})$, is the set of non-identity morphisms between two vertices. If $a \in \mathrm{Hom}(\sigma, \tau)$ is an edge, we have its *initial vertex*, $i(a) = \sigma$, as well as, its *terminal vertex*, $t(a) = \sigma$. A k-simplex of $|\mathcal{R}|$ is a k-tuple $(a_1, \dots, a_k)$ of composable edges, i.e., $t(a_j) = i(a_{j+1})$. (N.B. $|\mathcal{R}|$ is not necessarily a simplicial complex, rather, as on the first page of Sect. 2.1, it is a "Δ-complex," cf. [146].) In particular, a poset $\mathcal{Q}$ is a scwol; its geometric realization $|\mathcal{Q}|$ is an ordinary simplicial complex (the *order complex*). A k-simplex in $|\mathcal{Q}|$ is a chain $\{\sigma_0 < \cdots < \sigma_k\}$.

What is the scwol associated to a graph Ω? We have $\text{Vert}\,\mathcal{R} = \text{Vert}\,\Omega \bigsqcup \text{Edge}\,\Omega$. For each vertex $\tau \in \Omega$ which is an endpoint of an edge of $\sigma \in \Omega$, there is an edge of $\mathcal{R}$ running from $\sigma \in \text{Vert}\,\mathcal{R}$ to $\tau \in \text{Vert}\,\mathcal{R}$. So, the geometric realization of the scwol is the barycentric subdivision of the graph. For example, if Ω is a loop consisting of a single vertex τ and a single 1-cell σ, then there are two edges in $|\mathcal{R}|$ connecting the vertices corresponding to σ and τ. The general theory of scwols is developed in [35, III.$\mathcal{C}$.1].

There is a natural generalization of the notion of a simple complex of groups $G\mathcal{Q}$ to a "complex of groups." There are two aspects to this generalization. First, we can replace the poset $\mathcal{Q}$ by a scwol $\mathcal{R}$ and consider a cofunctor $G\mathcal{R}$ from $\mathcal{R}$ to the category of groups and monomorphisms. A loop of groups as in Example A.9 gives a complex of groups of this type. A second aspect of this generalization is to relax the requirement that $G\mathcal{R}$ be a cofunctor. To wit, if a and b are composable morphisms in $\mathcal{R}$, then one only requires the homomorphisms ϕ_{ab} and $\phi_a\phi_b$ from $G_{t(a)}$ to $G_{i(b)}$ agree up to conjugation by an element $g_{a,b}$ in the target group $G_{i(b)}$, i.e.,

$$\text{Ad}(g_{a,b})\phi_{ab} = \phi_b\phi_a. \tag{A.5}$$

Moreover, these elements $g_{a,b}$ are required to satisfy an appropriate cocycle condition for each triple of composable morphisms (a, b, c). (See [35, III.$\mathcal{C}$.2.1, p.535].) A *complex of groups* over a scwol $\mathcal{R}$ is then a "lax cofunctor," as defined above, from $\mathcal{R}$ to the category of groups and monomorphisms.

There are layers of complications built into the definitions of a complex of groups, its fundamental group and its universal cover. These complications are spelled out in detail in the book of Bridson-Haefliger [35]. We will only indicate these complications by giving definitions that are sometimes incomplete and then referring to [35] for the full details. Here is the official definition of a "complex of groups."

Definition A.10 (Bridson-Haefliger [35, III.$\mathcal{C}$.2, p. 535]) A *complex of groups* $G\mathcal{R} = \{G_\sigma, \psi_a, g_{a,b}\}$ over a scwol $\mathcal{R}$ consists of the following data:

(i) for each $\sigma \in \text{Vert}(\mathcal{R})$, a group G_σ, called the *local group* at σ,
(ii) for each $a \in \text{Edge}(\mathcal{R})$ an injective homomorphism $\psi_a : G_{t(a)} \to G_{i(a)}$,
(iii) for each pair of composable edges $(a, b) \in \text{Edge}(\mathcal{R}) \times \text{Edge}(\mathcal{R})$, a *twisting element* $g_{a,b} \in G_{i(b)}$ so that

$$\text{Ad}(g_{a,b})\psi_{ab} = \psi_b\psi_a\,. \tag{A.6}$$

Moreover, the $\{g_{a,b}\}$ must satisfy a certain cocycle condition [35, p. 535].

The complex of groups is *simple* if each $g_{a,b}$ is equal to the trivial element of $G_{i(b)}$.

Remark A.11 For a simple complex of groups over a poset the same functoriality condition is used in [35] as in Sect. A.1: it is a contravariant functor to the category of groups and monomorphisms. In Definition A.10 we maintain the condition of being contravariant. For example, given a loop of groups as in Example A.9, there are two injections from the edge stabilizer to the vertex stabilizer. However, in Bridson-Haefliger [35] a complex of groups over a scwol becomes covariant. At some point we to need return to the covariance of [35, III.$\mathcal{C}$.2] by replacing the scwol $\mathcal{R}$ by the dual scwol $\mathcal{R}^{\mathrm{op}}$ and we shall do so when we define the category $CG\mathcal{R}$ below.

There is a natural notion of a morphism from a complex of groups $G\mathcal{R}$ to a group G. (The reader can either supply her own definition or else look at [35, III.$\mathcal{C}$.2.4, p. 536].)

Definition A.12 Suppose $G\mathcal{R}$ is a complex of groups over a scwol $\mathcal{R}$ and $\psi : G\mathcal{R} \to G$ is a morphism to a group G, which is injective on local groups. A *development* of $G\mathcal{R}$ with respect to ψ is a scwol $D(\psi, \mathcal{R})$ together with an action of G on it so that the quotient scwol is $\mathcal{R}$ and so that the associated complex of groups is $G\mathcal{R}$.

There is the following generalization of Theorem A.6.

Theorem A.13 (Bridson-Haefliger [35, III.$\mathcal{C}$.2, Thm. 2.13]) *For any morphism $\psi : G\mathcal{R} \to G$, there is a development $D(\psi, \mathcal{R})$. If ψ is injective on local groups, then $G\mathcal{R}$ is the complex of groups associated to the G-action on $D(\psi, \mathcal{R})$.*

The construction of the development $D(\psi, \mathcal{R})$ is similar to the case of a simple complex of groups over a poset $\mathcal{Q}$, which was done in Eq. (A.1). To wit, the objects of the scwol $D(\psi, \mathcal{R})$ are cosets,

$$\mathrm{Vert}(D(\psi, \mathcal{R})) := \bigsqcup_{\sigma \in \mathrm{Vert}(\mathcal{R})} G/\psi_\sigma(G_\sigma), \tag{A.7}$$

and the edges (= morphisms) of $D(\psi, \mathcal{R})$ are given by

$$\mathrm{Edge}(D(\psi, \mathcal{R})) := \{(gG_{i(a)}, a) \mid a \in \mathrm{Edge}(\mathcal{R}) \text{ and } gG_{i(a)} \in G/G_{i(a)}\}, \tag{A.8}$$

where to simplify notation we have written $gG_{i(a)}$ instead of $g\psi_{i(a)}(G_{i(a)})$.

A.3.2 The Category $CG\mathcal{R}$

In a scwol such as $\mathcal{R}$, each object σ corresponds to a unique morphism $\mathbb{I}_\sigma$ (the identity) from σ to itself; hence, Vert($\mathcal{R}$) is identified as a subset of the set of all morphisms. So, we can think of $\mathcal{R}$ as the set of morphisms (i.e., the set of arrows).

The information contained in a complex of groups $G\mathcal{R}$ yields a small category $CG\mathcal{R}$ which we will now define. The discussion follows [35, III.$\mathcal{C}$.2.8]. The set of objects of $CG\mathcal{R}$ is Vert($\mathcal{R}$) (= Vert($\mathcal{R}^{\mathrm{op}}$), where $\mathcal{R}^{\mathrm{op}}$ means the dual scwol). As we indicated in Remark A.11, it is now time to reverse the directions of the arrows. The elements of $CG\mathcal{R}$ are pairs (g, a) where $a \in \mathcal{R}^{\mathrm{op}}$ and $g \in G_{t(a}$. If $i(a) = t(b)$, then the composition $(g, a)(h, b)$ is defined by

$$(g, a)(h, b) = (g\psi_a(h)g_{a,b}, ab).$$

Note that the map $CG\mathcal{R} \mapsto \mathcal{R}^{\mathrm{op}}$ is a functor.

Associated to any category C there is a CW complex BC called its *nerve* (or its *classifying space*). In the case C is a poset or a scwol $\mathcal{R}$, BC is its *geometric realizattion*. When C is a group G, then $BC = BG$, the usual $K(G, 1)$-complex. The functor $CG\mathcal{R} \to \mathcal{R}^{\mathrm{op}}$ induces a map $BCG\mathcal{R} \to B\mathcal{R}^{\mathrm{op}} = |\mathcal{R}^{\mathrm{op}}| = |\mathcal{R}|$. As we shall see in Sect. A.4, this exhibits $BCG\mathcal{R}$ as an "aspherical realization" of the complex of groups $G\mathcal{R}$.

A.3.3 The Fundamental Group and Universal Cover of a Complex of Groups

There are at least two general methods for defining the fundamental group of a complex of groups $G\mathcal{R}$. The first uses the category $CG\mathcal{R}$ defined in the previous subsection. The procedure is to first take the nerve $BCG\mathcal{R}$ of the category and then to take the ordinary fundamental group of this topological space:

$$\pi_1(G\mathcal{R}) := \pi_1(BCG\mathcal{R}). \tag{A.9}$$

There is a second method, developed in [35, III.$\mathcal{C}$.3], which is more concrete and easier to use. It is based on the construction of Bass-Serre in [207]. One defines the fundamental group by using a combination of the local groups and homotopy classes of edge paths in the geometric realization of the underlying scwol (cf. [35, III.$\mathcal{C}$.3.3]). First one defines the *universal group* $F(G\mathcal{R})$. The generators are the elements of G_σ, $\sigma \in \mathrm{Vert}(\mathcal{R})$, and the set of signed edges, $E^{\pm}(\mathcal{R})$ where $E^{\pm}(\mathcal{R})$ means that we take two copies a^+ and a^- for each edge $a \in \mathrm{Edge}(\mathcal{R})$. By introducing the obvious relations (cf. [35, p. 546]) we get a group $F(G\mathcal{R})$ which is much larger than the fundamental group we wish to define. There is a natural morphism $\iota = (\iota_\sigma, \iota_a) : G\mathcal{R} \to F(G\mathcal{R})$, where $\iota_\sigma : G_\sigma \to F(G\mathcal{R})$ is the natural homomorphism and where $\iota(a) = a^+$. A *$G\mathcal{R}$-path* issuing from $\sigma_0 \in \mathrm{Vert}(\mathcal{R})$ is a sequence $c = (g_0, e_1, g_1, \dots, e_k, g_k)$, where $(e_1, \dots, e_k)$ is an edge path in the geometric realization of $\mathcal{R}$, $g_0 \in G_{\sigma_0}$ and g_i belongs to the local group at the appropriate end point of e_i. Associated to a $G\mathcal{R}$ path $c = (g_0, e_1, g_1, \dots, e_k, g_k)$, there is an element $\pi(c) \in F(G\mathcal{R})$ represented by the word $g_0e_1g_1 \cdots e_kg_k$. Two $G\mathcal{R}$-paths c and c' are *homotopic* if $\pi(c) = \pi(c')$. Given $\sigma_0 \in \mathrm{Vert}(\mathcal{R})$, let $\pi_1(G\mathcal{R}, \sigma_0)$ denote the set of homotopy classes of $G\mathcal{R}$-loops based at σ_0. It has a group structure induced by concatenation of paths (cf. [35, III.$\mathcal{C}$.3.3]). The group $\pi_1(G\mathcal{R}, \sigma_0)$ is the second definition of the *fundamental group of* $G\mathcal{R}$.

The complex of groups $G\mathcal{R}$ is *developable* if for each $\sigma \in \text{Vert}(\mathcal{R})$ the natural homomorphism $G_\sigma \to \pi_1(G\mathcal{R}, \sigma_0)$ induced by $\iota_\sigma : G_\sigma \to F(G\mathcal{R})$ is injective.

The universal cover. Suppose $G\mathcal{R}$ is a developable complex of groups over a connected scwol $\mathcal{R}$. The *universal cover* $G\mathcal{R}$ is the development $D(\psi, \mathcal{R})$ (cf. Definition A.12), where $\psi : G\mathcal{R} \to \pi_1(G\mathcal{R}, T)$ is the canonical morphism. We shall often confuse the scwol $D(\psi, \mathcal{R})$ with its geometric realization. Also, we shall write π for $\pi_1(G\mathcal{R}, T)$ and use the notation $D(\pi, \mathcal{R})$ for $D(\psi, \mathcal{R})$. Finally, we will sometimes denote the universal cover by

$$UG\mathcal{R} := D(\pi, \mathcal{R}). \tag{A.10}$$

A.4 Aspherical Realizations and the $K(\pi, 1)$-Question

A.4.1 Complexes of Spaces and van Kampen's Theorem

There is a natural notion of a "complex of spaces" over a poset or scwol $\mathcal{R}$. (See Definition A.14 below.) It consists of a collection of path connected subspaces $\{X_\sigma\}_{\sigma\in\mathcal{R}}$ of some given space X, together with some other data. Each complex of spaces has an associated complex of groups $G\mathcal{R} = \{G_\sigma\}_{\sigma\in\mathcal{R}}$, where $G_\sigma = \pi_1(X_\sigma)$. Van Kampen's Theorem tells us that, for $X = \bigcup X_\sigma$, the fundamental group of X can be computed from the complex of groups as: $\pi_1(X) = \pi_1(G\mathcal{R})$. (See Theorem A.15 below.) The theory is easier to understand when $\mathcal{R}$ is a poset and $G\mathcal{R}$ is a simple complex of groups. This is the case of primary interest to us.

It is necessary to introduce more complications in order to define the local structure of the geometric realization of a general scwol $\mathcal{R}$ as opposed to the case when $\mathcal{R}$ is a poset $\mathcal{R}$. For example, if $\sigma \in \text{Vert}(\mathcal{R})$ and $\mathcal{R}$ is a poset, then its *upper star* is simply the cone, $|\mathcal{R}_{\geq\sigma}|$. The corresponding notion for a scwol $\mathcal{R}$ is the *dual cone* D_σ, the precise definition of which can be found in [138, 1.3]. For example, suppose $\mathcal{R}$ is the scwol associated to a loop with a single 1-cell τ and a single 0-cell σ. There are two edges in $|\mathcal{R}|$ emanating from the vertex τ. These two edges have distinct end points in D_σ. In other words, D_σ is the cone on the *upper link*, $\text{Lk}(\sigma)$, which in this example is isomorphic to S^0.

Definition A.14 (Haefliger [138, Def. 3.3.2]) Suppose $\mathcal{R}$ is a scwol. A *complex of spaces* over a scwol $\mathcal{R}$ is a space X together with a projection map $p : X \to |\mathcal{R}|$ satisfying the following additional properties. We are given a collection of path connected spaces $\{X_\sigma\}_{\sigma\in\text{Vert}(\mathcal{R})}$ and for each $a \in \text{Edge}(\mathcal{R})$ a π_1-injective map $f_a : X_{t(a)} \to X_{i(a)}$. For each vertex $\sigma \in \text{Vert}(\mathcal{R})$, $p^{-1}(\sigma) = X_\sigma$ and so that $p^{-1}(D_\sigma)$ deformation retracts onto $p^{-1}(\sigma)$ relative to $p^{-1}(\sigma)$. For each $a \in \text{Edge}(\mathcal{R})$ with $t(a) = \sigma$ and $i(a) = \tau$, we are given an inclusion $\sigma \hookrightarrow D_\sigma$ and a projection $D_\sigma \to \tau$. These are covered by a homotopy equivalence $p^{-1}(\sigma) \hookrightarrow p^{-1}(D_\sigma)$

and a projection $p^{-1}(D_\sigma) \to p^{-1}(\tau)$. Their composition gives a map $f_a : X_\sigma \to X_\tau$, which is required to be π_1-injective. The *complex of groups associated to* X is defined by setting $G_\sigma = \pi_1(X_\sigma)$ for each $\sigma \in \text{Vert}(\mathcal{R})$. For each $a \in \text{Edge}(\mathcal{R})$, the monomorphism $\psi_a : G_{t(a)} \to G_{i(a)}$ is defined by $\psi_a = (f_a)_* : \pi_1(X(t(a)) \to \pi_1(X(i(a))$. A complex of spaces is a *complex of CW complexes* if each X_σ is a CW complex and each f_a is a cellular map.

Suppose $G\mathcal{R} = \{G_\sigma\}$ is a complex of groups and that $X = \bigcup_{\sigma \in \text{Vert}\,\mathcal{R}} X_\sigma$ is a complex of CW complexes over $\mathcal{R}$, with $p : X \to |\mathcal{R}|$ so that the associated complex of groups is $G\mathcal{R}$. Then X is called a (cellular) *realization* of $G\mathcal{R}$.

Of course, the choice of base points is important in the previous definition and must be made explicit. This is conveniently done by choosing a section of $p : X \to |\mathcal{R}|$ defined over the 1-skeleton of $\mathcal{R}$. In other words, for each $\sigma \in \text{Vert}(\mathcal{R})$ we have a base point $s(\sigma) \in X_\sigma$ and for each $a \in \text{Edge}(\mathcal{R})$, a lift of the segment in $|\mathcal{R}|^1$ corresponding to a to a path $s(a)$ from $s(i(a))$ to $s(t(a))$. According to the definition of a complex of groups (cf. Definition A.10), the correspondence is required to be functorial only up to conjugation by certain elements $g_{a,b} \in G_{i(b)}$ (cf. (A.5)). Note that the path $s(a)s(b)\overline{s(ab)}$ is a closed loop in $X_{i(a)}$ based at $s(i(a))$. Hence, it defines an element $g_{a,b} \in G_{i(a)}$ and we have $\psi_{ab} = \text{Ad}(g_{a,b})\psi_a\psi_b$. It follows that $G\mathcal{R}$ is a complex of groups where the twisting element $g_{a,b}$ is the homotopy class of $s(a)s(b)\overline{s(ab)}$ in $\pi_1(X_{i(a)}, s(i(a)))$. The theory of complexes of groups gives the following generalization of van Kampen's Theorem.

Theorem A.15 (cf. Haefliger [138, Prop. 3.3.3]) *Suppose $p : X \to |\mathcal{R}|$ is a complex of CW complexes and that $G\mathcal{R}$ is given as above. Then $G\mathcal{R}$ is a complex of groups over $\mathcal{R}$. Moreover, $\pi_1(X) = \pi_1(G\mathcal{R}, s)$, where s is a section defined over the* 1*-skeleton of* $|\mathcal{R}|$.

Example A.16 (Graphs of Spaces) Suppose Y is a graph without loops and that $\mathcal{Q}$ denotes its poset of cells. The elements of $\mathcal{Q}$ are 0-cells and 1-cells. A 1-cell gives a vertex σ of $\mathcal{Q}$, namely, the midpoint of the 1-cell. If τ is a 0-cell, then D_τ is the star of τ in the barycentric subdivision of Y. A *graph of spaces* is then a space X and a map $p : X \to Y$ giving X a decomposition into *vertex spaces* $p^{-1}(D(\tau))$ and *edge spaces* X_σ. The edge space X_σ is the intersection of the two vertex spaces corresponding to the endpoints of σ. The statement that $\pi_1(X)$ is the fundamental group of the graph of groups is a well-known version of van Kampen's Theorem.

A.4.2 Aspherical Realizations

Definition A.17 Suppose $p : X \to |\mathcal{R}|$ is a cellular realization of a complex of groups $G\mathcal{R}$. Then $p : X \to |\mathcal{R}|$ is called an *aspherical realization* of $G\mathcal{R}$ if for each $\sigma \in \text{Vert}(\mathcal{R})$, the cell complex $X_\sigma := p^{-1}(\sigma)$ is aspherical. In other words, if for each $\sigma \in \text{Vert}(\mathcal{R})$, X_σ is homotopy equivalent to BG_σ.

The nerve of the category $CG\mathcal{R}$ from Sect. A.3.2 is easily seen to be an aspherical realization of $G\mathcal{R}$. The projection map $p : BCG\mathcal{R} \to |\mathcal{R}|$ is the map induced by the functor $CG\mathcal{R} \to \mathcal{R}^{\mathrm{op}}$.

An aspherical realization $p : X \to |\mathcal{R}|$ has the following universal property: if $p' : X' \to |\mathcal{R}|$ is another complex of CW complexes over $\mathcal{R}$ associated to the same complex of groups $G\mathcal{R}$, then there is a cellular map $f : X' \to X$ covering the identity on $|\mathcal{R}|$ so that the following diagram commutes:

$$\begin{array}{ccc} X' & \xrightarrow{f} & X \\ {\scriptstyle p'}\downarrow & & \downarrow{\scriptstyle p} \\ |\mathcal{R}| & \xrightarrow[id]{} & |\mathcal{R}| \end{array} . \tag{A.11}$$

Of course, part of the data for an aspherical realization should include a section $s : |\mathcal{R}|^1 \to X$ over the 1-skeleton and the map f should take the section s' for X' to the section s. When this is the case, for $\sigma \in \operatorname{Vert} \mathcal{R}$, the induced map $\pi_1(X', s'(\sigma)) \to \pi_1(X, s(\sigma))$ should be the identity, i.e., f should induce the identity homomorphism on the G_σ. In what follows we will usually suppress all mention of these sections and choices of base points.

By a standard argument, it follows from the universal property of an aspherical realization that if X and X' are both aspherical realizations of a complex of groups $G\mathcal{R}$, then there is a homotopy equivalence $f : X' \to X$ covering the identity on $|\mathcal{R}|$ and restricting to a homotopy equivalence $X'_\sigma \to X_\sigma$ for each $\sigma \in \operatorname{Vert}(\mathcal{R})$. When we are not choosing the specific model $BCG\mathcal{R}$, we will denote an aspherical realization of $G\mathcal{R}$ by $BG\mathcal{R}$.

Theorem A.18 (Haefliger [138, §3.4]) *Any complex of groups $G\mathcal{R}$ has an aspherical realization $p : BG\mathcal{R} \to |\mathcal{R}|$. Moreover, $BG\mathcal{R}$ is unique up to homotopy equivalence.*

Discussion of proof The existence of an aspherical realization can be established without invoking the category $CG\mathcal{R}$. The details of the proof can be found [138, §3.4]. To simplify the discussion first suppose $G\mathcal{Q}$ is a simple complex of groups over a poset $\mathcal{Q}$, that is, suppose we are given a collection of groups $\{G_\sigma\}_{\sigma\in\mathcal{Q}}$ and monomorphisms $\phi_{\tau\sigma} : G_\tau \to G_\sigma$ whenever $\sigma < \tau$. The idea is to glue together the collection of classifying spaces $\{BG_\sigma\}_{\sigma\in\mathcal{Q}}$ by using the maps $f_{\tau\sigma} : BG_\tau \to BG_\sigma$ corresponding to the $\phi_{\tau\sigma}$. To further simplify the discussion suppose $G\mathcal{Q}$ is a graph of groups where $\mathcal{Q}$ is the poset of cells in the underlying graph. Let e_τ be an edge and v_σ one of its endpoints. Then τ and σ are vertices of $|\mathcal{Q}|$, the barycentric subdivision of the underlying graph; τ is the midpoint of the edge e_τ; the 1-simplex $[\tau, \sigma]$ is one half of the edge e_τ. Moreover, $[\tau, \sigma]$ is contained in D_σ (the upper star of σ in $|\mathcal{Q}|$). The mapping cylinder of $f_{\tau\sigma}$ lies above $[\tau, \sigma]$ and gives a gluing of $p^{-1}(\tau) \sim BG_\tau$ to $p^{-1}([\tau, \sigma]) \subset p^{-1}D_\sigma \sim BG_\sigma$. In the case where the order complex $|\mathcal{Q}|$ has dimension > 1, to construct the preimage of a simplex $[\tau_0 < \cdots < \tau_k]$ in $|\mathcal{R}|$ we use iterated mapping cylinders of the maps $f_{\tau\sigma}$, where τ and σ are vertices of the simplex. The uniqueness of aspherical realizations follows immediately from their universal property. □

Theorem A.19 *Suppose that $G\mathcal{R}$ is a developable complex of groups, that $\pi = \pi_1(G\mathcal{R})$ and that $UG\mathcal{R}$ is its universal cover. Then $BG\mathcal{R}$ is homotopy equivalent to the Borel construction on $UG\mathcal{R}$:*

$$BG\mathcal{R} = UG\mathcal{R} \times_\pi E\pi.$$

Proof Projection on the first factor induces the projection map $p : UG\mathcal{R} \times_\pi E\pi \to UG\mathcal{R}/\pi = |\mathcal{R}|$, fulfilling one of the requirements in the definition of an aspherical realization. The fiber over σ is $\pi/G_\sigma \times_\pi E\pi$ which is homotopy equivalent to $E\pi/G_\sigma = BG_\sigma$. The theorem follows. □

The $K(\pi, 1)$-Question A primary reason for introducing the notion of a complex of groups was to address the following.

Question A.20 (The $K(\pi, 1)$-Question) Suppose $G\mathcal{R}$ is a complex of groups. Is $BG\mathcal{R}$ a $K(\pi, 1)$-complex? In other words, is the aspherical realization of a complex of groups an aspherical complex?

In symbols, is $BG\mathcal{R} = B\pi$ for $\pi = \pi_1(G\mathcal{R})$?

The following is a corollary of Theorem A.19.

Theorem A.21 (Haefliger [138, Prop. 3.2.3]) *The $K(\pi, 1)$-Question for a developable complex of groups $G\mathcal{R}$ has a positive answer if and only if the universal cover $UG\mathcal{R}$ of $G\mathcal{R}$ is contractible. In other words, the aspherical realization $BG\mathcal{R}$ is homotopy equivalent to $B\pi$ if and only if $UG\mathcal{R}$ is contractible.*

Proof Put $U = UG\mathcal{R}$. By Theorem A.19, $BG\mathcal{R}$ is homotopy equivalent to $U \times_\pi E\pi$. The universal cover of $U \times_\pi E\pi$ is $U \times E\pi$ which is homotopy equivalent to U (since $E\pi$ is contractible). The proposition follows. □

Here are some corollaries to Theorem A.21.

Corollary A.22 *Let $G\mathcal{R}$ be a graph of groups. Then the $K(\pi, 1)$-Question for $G\mathcal{R}$ has an affirmative answer, i.e., $BG\mathcal{R}$ is aspherical.*

Proof For a graph of groups, its universal cover $UG\mathcal{R}$ is a tree. □

A special case of this corollary is the following.

Corollary A.23 (Whitehead's Lemma)

(i) Suppose $G\mathcal{Q}$ is a segment of groups, with vertex groups G_1 and G_2 and with edge group A. Then the result of gluing BG_1 to BG_2 along BA is homotopy equivalent to the classifying space of the amalgamated product $G_1 *_A G_2$.

(ii) Suppose the data for a loop of groups (or HNN construction) is given by as in Example A.9. Then the result of gluing $BA \times [0, 1]$ to BG via the inclusion on $BA \times 0$ and the map induced by θ on $BA \times 1$ is the classifying space for the HNN construction $*_A G$.

A direct proof of Whitehead's Lemma can be found in [146, Thm. 1B.11].

Another criterion which allows us to apply Theorem A.21 is the nonpositive curvature property. A central result in the book of Bridson-Haefliger [35] is the following.

Theorem A.24 (Bridson-Haefliger [35, Def. 4.16 and Thm. 4.17 in III.$\mathcal{C}$.4]) *Suppose $G\mathcal{R}$ is a nonpositively curved complex of groups. (This means that each local development is* CAT(0).*) Then $G\mathcal{R}$ is developable and its universal cover $UG\mathcal{R}$ is* CAT(0). *Hence, the $K(\pi, 1)$-Question for $G\mathcal{R}$ has a positive answer.*

This theorem is the method we used in Chap. 4 to answer the $K(\pi, 1)$-Question for the Coxeter complexes $W\mathcal{S}^{\mathrm{op}}$ (see Theorem 4.12) and for Artin complexes $A\mathcal{S}^{\mathrm{op}}$ when A is type FC (see Theorem 4.76 in Sect. 4.3.4). It also shows that the answer is positive for simple complexes of groups associated to a building (see Theorem 4.107 in Sect. 4.4.2). In the case of right-angled buildings we recover the result that the answer is positive for the graph product complex (cf. Proposition 3.27). In other words, that if $\Gamma = \prod_{L^1}(G_i, H_i)$ is a graph product (or relative graph product), then $B\Gamma$ is $\{(BG_i, BH_i)\}^L$, the polyhedral product of the classifying spaces of the factors.

Of course, the $K(\pi, 1)$-Question is a question about a complex of groups and not just a question about its fundamental group. For example, it can easily happen that we have a simple complex of groups $G\mathcal{Q}$ a simple subcomplex of groups $G\mathcal{Q}'$ with $\mathcal{Q}'$ a subcomplex of $\mathcal{Q}$ so that $\pi_1(G\mathcal{Q}) = \pi_1(G\mathcal{Q}')$ and $UG\mathcal{Q}$ is contractible but $UG\mathcal{Q}'$ is not.

Example A.25 (Some Triangles of Groups, cf. Example 2.59 in Sect. 2.4.5) A reflection group with fundamental chamber a triangle defines a triangle of groups. The local group of each 2-cell is trivial, each edge stabilizer is cyclic of order 2 and the vertex stabilizers are dihedral groups $I_2(p)$, $I_2(q)$ and $I_2(r)$. The direct limit is a Coxeter group denoted by $W(p, q, r)$. It can be represented as a reflection group on $\mathbb{S}^2$, $\mathbb{E}^2$, or $\mathbb{H}^2$ as the sum of the reciprocals $1/p + 1/q + 1/r$ is, respectively, $>$, $=$, or < 1. So, the universal cover of the triangle of groups is contractible if and only if the sum of reciprocals is ≤ 1. (Geometrically, this means that the sum of the angles in the triangle is $\leq \pi$.)

References

1. P. Abramenko and K. S. Brown, *Buildings*, Graduate Texts in Mathematics, vol. 248, Springer, New York, 2008. Theory and applications. MR2439729
2. A. Abrams, *Configuration spaces of colored graphs*, Geom. Dedicata **92** (2002), 185–194, https://doi.org/10.1023/A:1019662529807. Dedicated to John Stallings on the occasion of his 65th birthday. MR1934018
3. I. Agol, *The virtual Haken conjecture*, Doc. Math. **18** (2013), 1045–1087. With an appendix by Agol, Daniel Groves, and Jason Manning. MR3104553
4. S. Alexander, K. Vitali, and A. Petrunin, *Invitation to Alexandrov geometry:* CAT(0) *spaces*, Springer-Verlag, Berlin.
5. D. Allcock, *Completions, branched covers, Artin groups, and singularity theory*, Duke Math. J. **162** (2013), no. 14, 2645–2689, https://doi.org/10.1215/00127094-2380977. MR3127810
6. J. M. Alonso, T. Brady, D. Cooper, V. Ferlini, M. Lustig, M. Mihalik, M. Shapiro, and H. Short, *Notes on word hyperbolic groups*, Group theory from a geometrical viewpoint (Trieste, 1990), World Sci. Publ., River Edge, NJ, 1991, pp. 3–63. Edited by Short. MR1170363
7. F. D. Ancel, M. W. Davis, and C. R. Guilbault, CAT(0) *reflection manifolds*, Geometric topology (Athens, GA, 1993), AMS/IP Stud. Adv. Math., vol. 2, Amer. Math. Soc., Providence, RI, 1997, pp. 441–445. MR1470741
8. E. M. Andreev, *Convex polyhedra in Lobačevskiĭ spaces*, Mat. Sb. (N.S.) **81 (123)** (1970), 445–478 (Russian). MR0259734
9. ——, *Convex polyhedra of finite volume in Lobačevskiĭ space*, Mat. Sb. (N.S.) **83 (125)** (1970), 256–260 (Russian). MR0273510
10. K. I. Appel and P. E. Schupp, *Artin groups and infinite Coxeter groups*, Invent. Math. **72** (1983), no. 2, 201–220, https://doi.org/10.1007/BF01389320. MR700768
11. E. Artin, *Theory of braids*, Ann. of Math. (2) **48** (1947), 101–126, https://doi.org/10.2307/1969218. MR0019087
12. G. Avramidi, M. W. Davis, B. Okun, and K. Schreve, *The action dimension of right-angled Artin groups*, Bull. Lond. Math. Soc. **48** (2016), no. 1, 115–126, https://doi.org/10.1112/blms/bdv083. MR3455755
13. A. Bartels and W. Lück, *The Borel conjecture for hyperbolic and* CAT(0)*-groups*, Ann. of Math. (2) **175**(2012), no. 2, 631–689, https://doi.org/10.4007/annals.2012.175.2.5. MR2993750
14. J. Behrstock, M. F. Hagen, and A. Sisto, *Hierarchically hyperbolic spaces, I: Curve complexes for cubical groups*, Geom.Topol. **21** (2017), no. 3, 1731–1804, https://doi.org/10.2140/gt.2017.21.1731. MR3650081

M. W. Davis, *Infinite Group Actions on Polyhedra*, Ergebnisse der Mathematik und ihrer Grenzgebiete. 3. Folge / A Series of Modern Surveys in Mathematics 77, https://doi.org/10.1007/978-3-031-48443-8

15. I. Belegradek, *Aspherical manifolds with relatively hyperbolic fundamental groups*, Geom. Dedicata **129** (2007), 119–144, https://doi.org/10.1007/s10711-007-9199-8. MR2353987
16. Y. Benoist, *Convexes hyperboliques et quasiisométries*, Geom. Dedicata **122** (2006), 109–134, https://doi.org/10.1007/s10711-006-9066-z (French, with English summary). MR2295544
17. N. Bergeron, F. Haglund, and D. T. Wise, *Hyperplane sections in arithmetic hyperbolic manifolds*, J. Lond. Math. Soc. (2) **83** (2011), no. 2, 431–448, https://doi.org/10.1112/jlms/jdq082. MR2776645
18. N. Bergeron and D. T. Wise, *A boundary criterion for cubulation*, Amer. J. Math. **134** (2012), no. 3, 843–859, https://doi.org/10.1353/ajm.2012.0020. MR2931226
19. E. A. Bering and D. Futer, *Explicit special covers of alternating links*. 1909.12266.
20. D. Bessis, *Finite complex reflection arrangements are $K(\pi,1)$*, Ann. of Math. (2) **181** (2015), no. 3, 809–904, https://doi.org/10.4007/annals.2015.181.3.1. MR3296817
21. M. Bestvina, *Geometric group theory and 3-manifolds hand in hand: the fulfillment of Thurston's vision*, Bull. Amer. Math. Soc. (N.S.) **51** (2014), no. 1, 53–70. MR3119822
22. ——, *Non-positively curved aspects of Artin groups of finite type*, Geom. Topol. **3** (1999), 269–302, https://doi.org/10.2140/gt.1999.3.269. MR1714913
23. M. Bestvina and N. Brady, *Morse theory and finiteness properties of groups*, Invent. Math. **129** (1997), no. 3, 445–470.
24. M. Bhattacharjee, *Constructing finitely presented infinite nearly simple groups*, Comm. Algebra **22** (1994), no. 11, 4561–4589, https://doi.org/10.1080/00927879408825087. MR1284345
25. R. Bieri, *Homological dimension of discrete groups*, Mathematics Department, Queen Mary College, London, 1976. Queen Mary College Mathematics Notes. MR0466344
26. A. Borel, *Compact Clifford-Klein forms of symmetric spaces*, Topology **2** (1963), 111–122, https://doi.org/10.1016/0040-9383(63)90026-0. MR146301
27. F. Bosio and L. Meersseman, *Real quadrics in $\mathbf{C}^n$, complex manifolds and convex polytopes*, Acta Math. **197**(2006), no. 1, 53–127, https://doi.org/10.1007/s11511-006-0008-2. MR2285318
28. N. Bourbaki, *Lie groups and Lie algebras. Chapters 4–6*, Elements of Mathematics (Berlin), Springer-Verlag, Berlin, 2002. Translated from the 1968 French original by Andrew Pressley. MR1890629
29. M. Bourdon, *Immeubles hyperboliques, dimension conforme et rigidité de Mostow*, Geom. Funct. Anal. **7** (1997), no. 2, 245–268 (French, with English and French summaries). MR1445387
30. M. Bourdon, *Sur les immeubles fuchsiens et leur type de quasi-isométrie*, Ergodic Theory Dynam. Systems **20** (2000), no. 2, 343–364, https://doi.org/10.1017/S014338570000016X (French, with English and French summaries). MR1756974
31. B. H. Bowditch, *Relatively hyperbolic groups*, Internat. J. Algebra Comput. **22** (2012), no. 3, 1250016, 66, https://doi.org/10.1142/S0218196712500166. MR2922380
32. N. Brady, *Branched coverings of cubical complexes and subgroups of hyperbolic groups*, J. London Math. Soc. (2) **60** (1999), no. 2, 461–480, https://doi.org/10.1112/S0024610799007644. MR1724853
33. N. Brady, J. McCammond, and J. Meier, *Local-to-asymptotic topology for cocompact* CAT(0) *complexes*, Topology Appl. **131** (2003), no. 2, 177–188, https://doi.org/10.1016/S0166-8641(02)00338-3. MR1981872
34. N. Brady and J. Meier, *Connectivity at infinity for right angled Artin groups*, Trans. Amer. Math. Soc. **353** (2001), no. 1, 117–132, https://doi.org/10.1090/S0002-9947-00-02506-X. MR1675166
35. M. R. Bridson and A. Haefliger, *Metric spaces of non-positive curvature*, Grundlehren der Mathematischen Wissenschaften [Fundamental Principles of Mathematical Sciences], vol. 319, Springer-Verlag, Berlin, 1999. MR1744486
36. E. Brieskorn, *Sur les groupes de tresses [d'après V. I. Arnol'd]*, Séminaire Bourbaki, 24ème année (1971/1972), Exp. No.401, Springer, Berlin, 1973, pp. 21–44. Lecture Notes in Math., Vol. 317 (French). MR0422674

37. E. Brieskorn and K. Saito, *Artin-Gruppen und Coxeter-Gruppen*, Invent. Math. **17** (1972), 245–271, https://doi.org/10.1007/BF01406235 (German). MR323910
38. ——, *Artin groups and Coxeter groups*, English translation of previous entry
39. K. S. Brown, *Cohomology of groups*, Graduate Texts in Mathematics, vol. 87, Springer-Verlag, New York-Berlin, 1982. MR672956
40. ——, *Buildings*, Springer-Verlag, New York, 1989. MR969123
41. V. M. Buchstaber and T. E. Panov, *Toric topology*, Mathematical Surveys and Monographs, vol. 204, American Mathematical Society, Providence, RI, 2015. MR3363157
42. D. Burago, Y. Burago, and S. Ivanov, *A course in metric geometry*, Graduate Studies in Mathematics, vol. 33, American Mathematical Society, Providence, RI, 2001. MR1835418
43. M. Burger and S. Mozes, *Finitely presented simple groups and products of trees*, C. R. Acad. Sci. Paris Sér. I Math. **324** (1997), no. 7, 747–752, https://doi.org/10.1016/S0764-4442(97)86938-8 (English, with English and French summaries). MR1446574
44. ——, *Groups acting on trees: from local to global structure*, Inst. Hautes Études Sci. Publ. Math. **92** (2000), 113–150 (2001). MR1839488
45. ——, *Lattices in product of trees*, Inst. Hautes Études Sci. Publ. Math. **92** (2000), 151–194 (2001). MR1839489
46. P.-E. Caprace, *Conjugacy of 2-spherical subgroups of Coxeter groups and parallel walls*, Algebr. Geom. Topol. **6** (2006), 1987–2029, https://doi.org/10.2140/agt.2006.6.1987. MR2263057
47. ——, *Buildings with isolated subspaces and relatively hyperbolic Coxeter groups*, Innov. Incidence Geom. **10** (2009), 15–31. MR2665193
48. ——, *Erratum to "Buildings with isolated subspaces and relatively hyperbolic Coxeter groups" [MR2665193]*, Innov. Incidence Geom. **14** (2015), 77–79, https://doi.org/10.2140/iig.2015.14.77. MR3450952
49. ——, *Automorphism groups of right-angled buildings: simplicity and local splittings*, Fund. Math. **224** (2014), no. 1, 17–51, https://doi.org/10.4064/fm224-1-2. MR3164745
50. ——, *Finite and infinite quotients of discrete and indiscrete groups*, Groups St Andrews 2017 in Birmingham, London Math. Soc. Lecture Note Ser., vol. 455, Cambridge Univ. Press, Cambridge, 2019, pp. 16–69. MR3931408
51. P.-E. Caprace and F. Haglund, *On geometric flats in the CAT(0) realization of Coxeter groups and Tits buildings*, Canad. J. Math. **61** (2009), no. 4, 740–761, https://doi.org/10.4153/CJM-2009-040-8. MR2541383
52. P.-E. Caprace and B. Rémy, *Simplicity and superrigidity of twin building lattices*, Invent. Math. **176** (2009), no. 1, 169–221, https://doi.org/10.1007/s00222-008-0162-6. MR2485882
53. R. W. Carter, *Simple groups of Lie type*, John Wiley & Sons, London-New York-Sydney, 1972. Pure and Applied Mathematics, Vol. 28. MR0407163
54. R. Charney, *An introduction to right-angled Artin groups*, Geom. Dedicata **125** (2007), 141–158, https://doi.org/10.1007/s10711-007-9148-6. MR2322545
55. R. Charney and M. Davis, *Singular metrics of nonpositive curvature on branched covers of Riemannian manifolds*, Amer. J. Math. **115** (1993), no. 5, 929–1009, https://doi.org/10.2307/2375063. MR1246182
56. R. Charney and M. W. Davis, *The $K(\pi,1)$-problem for hyperplane complements associated to infinite reflection groups*, J. Amer. Math. Soc. **8** (1995), no. 3, 597–627. MR1303028
57. ——, *Finite $K(\pi,1)$s for Artin groups*, Prospects in topology (Princeton, NJ, 1994), Ann. of Math. Stud., vol. 138, Princeton Univ. Press, Princeton, NJ, 1995, pp. 110–124. MR1368655
58. R. Charney and M. Davis, *The Euler characteristic of a nonpositively curved, piecewise Euclidean manifold*, Pacific J. Math. **171** (1995), no. 1, 117–137. MR1362980
59. R. Charney and M. W. Davis, *Strict hyperbolization*, Topology **34** (1995), no. 2, 329–350. MR1318879
60. R. Charney and A. Lytchak, *Metric characterizations of spherical and Euclidean buildings*, Geom. Topol. **5** (2001), 521–550, https://doi.org/10.2140/gt.2001.5.521. MR1833752

61. R. Charney, J. Meier, and K. Whittlesey, *Bestvina's normal form complex and the homology of Garside groups*, Geom. Dedicata **105** (2004), 171–188, https://doi.org/10.1023/B:GEOM.0000024696.69357.73. MR2057250
62. I. Chatterji and G. Niblo, *From wall spaces to* CAT(0) *cube complexes*, Internat. J. Algebra Comput. **15** (2005), no. 5-6, 875–885, https://doi.org/10.1142/S0218196705002669. MR2197811
63. A. Clay and D. Rolfsen, *Ordered groups and topology*, Graduate Studies in Mathematics, vol. 176, American Mathematical Society, Providence, RI, 2016. MR3560661
64. A. M. Cohen and D. B. Wales, *Linearity of Artin groups of finite type*, Israel J. Math. **131** (2002), 101–123, https://doi.org/10.1007/BF02785852. MR1942303
65. J. M. Corson, *Complexes of groups*, Proc. London Math. Soc. (3) **65** (1992), no. 1, 199–224, https://doi.org/10.1112/plms/s3-65.1.199. MR1162493
66. H. S. M. Coxeter, *Discrete groups generated by reflections*, Ann. of Math. (2) **35** (1934), no. 3, 588–621, https://doi.org/10.2307/1968753. MR1503182
67. ——, *Regular polytopes*, 3rd ed., Dover Publications, Inc., New York, 1973. MR370327
68. ——, *Regular skew polyhedra in three and four dimensions and their topological analogues*, Proc. London Math. Soc. **35** (1938), no. 1, 37–62.
69. J. Crisp, *Symmetrical subgroups of Artin groups*, Adv. Math. **152** (2000), no. 1, 159–177, https://doi.org/10.1006/aima.1999.1895. MR1762124
70. F. Dahmani, V. Guirardel, and D. Osin, *Hyperbolically embedded subgroups and rotating families in groups acting on hyperbolic spaces*, Mem. Amer. Math. Soc. **245** (2017), no. 1156, v+152, https://doi.org/10.1090/memo/1156. MR3589159
71. J. Danciger, F. Guéritaud, F. Kassel, G.-S. Lee, and L. Marquis, *Convex cocompactness for Coxeter groups*. arxiv:2102.02757v1.
72. P. Dani, *The large-scale geometry of right-angled Coxeter groups*, Handbook of group actions. V, Adv. Lect. Math. (ALM), vol. 48, Int. Press, Somerville, MA, [2020] ©2020, pp. 107–141. MR4237891
73. M. W. Davis, *Groups generated by reflections and aspherical manifolds not covered by Euclidean space*, Ann. of Math. (2) **117** (1983), no. 2, 293–324, https://doi.org/10.2307/2007079. MR690848
74. ——, *Coxeter groups and aspherical manifolds*, Algebraic topology, Aarhus 1982 (Aarhus, 1982), Lecture Notes in Math., vol. 1051, Springer, Berlin, 1984, pp. 197–221, https://doi.org/10.1007/BFb0075568. MR764580
75. ——, *A hyperbolic* 4*-manifold*, Proc. Amer. Math. Soc. **93** (1985), no. 2, 325–328, https://doi.org/10.2307/2044771. MR770546
76. ——, *Some aspherical manifolds*, Duke Math. J. **55** (1987), no. 1, 105–139, https://doi.org/10.1215/S0012-7094-87-05507-4. MR883666
77. ——, *Buildings are* CAT(0), Geometry and cohomology in group theory (Durham, 1994), London Math. Soc. Lecture Note Ser., vol. 252, Cambridge Univ. Press, Cambridge, 1998, pp. 108–123. MR1709955
78. ——, *The cohomology of a Coxeter group with group ring coefficients*, Duke Math. J. **91** (1998), no. 2, 297–314, https://doi.org/10.1215/S0012-7094-98-09113-X. MR1600586
79. ——, *Poincaré duality groups*, Surveys on surgery theory, Vol. 1, Ann. of Math. Stud., vol. 145, Princeton Univ. Press, Princeton, NJ, 2000, pp. 167–193. MR1747535
80. ——, *Exotic aspherical manifolds*, Topology of high-dimensional manifolds, No. 1, 2 (Trieste, 2001), ICTP Lect. Notes, vol. 9, Abdus Salam Int. Cent. Theoret. Phys., Trieste, 2002, pp. 371–404. MR1937019
81. ——, *Nonpositive curvature and reflection groups*, Handbook of geometric topology, North-Holland, Amsterdam, 2002, pp. 373–422. MR1886674
82. ——, *The geometry and topology of Coxeter groups*, London Mathematical Society Monographs Series, vol. 32, Princeton University Press, Princeton, NJ, 2008. MR2360474
83. ——, *Examples of buildings constructed via covering spaces*, Groups Geom. Dyn. **3** (2009), no. 2, 279–298. MR2486800

84. ——, *Lectures on orbifolds and reflection groups*, Transformation groups and moduli spaces of curves, Adv. Lect. Math. (ALM), vol. 16, Int. Press, Somerville, MA, 2011, pp. 63–93. MR2883685
85. ——, *Right-angularity, flag complexes, asphericity*, Geom. Dedicata **159** (2012), 239–262, https://doi.org/10.1007/s10711-011-9654-4. MR2944529
86. ——, *When are two Coxeter orbifolds diffeomorphic?*, Michigan Math. J. **63** (2014), no. 2, 401–421, https://doi.org/10.1307/mmj/1401973057. MR3215556
87. M. W. Davis, J. Dymara, T. Januszkiewicz, J. Meier, and B. Okun, *Compactly supported cohomology of buildings*, Comment. Math. Helv. **85** (2010), no. 3, 551–582, https://doi.org/10.4171/CMH/205. MR2653692
88. M. W. Davis, J. Dymara, T. Januszkiewicz, and B. Okun, *Cohomology of Coxeter groups with group ring coefficients. II*, Algebr. Geom. Topol. **6** (2006), 1289–1318, https://doi.org/10.2140/agt.2006.6.1289. MR2253447
89. M. W. Davis and A. L. Edmonds, *Euler characteristics of generalized Haken manifolds*, Algebr. Geom. Topol. **14** (2014), no. 6, 3701–3716, https://doi.org/10.2140/agt.2014.14.3701. MR3302976
90. M. W. Davis, J. Fowler, and J.-F. Lafont, *Aspherical manifolds that cannot be triangulated*, Algebr. Geom. Topol. **14** (2014), no. 2, 795–803, https://doi.org/10.2140/agt.2014.14.795. MR3159970
91. M. W. Davis and T. Januszkiewicz, *Hyperbolization of polyhedra*, J. Differential Geom. **34**(1991), no. 2, 347–388. MR1131435
92. ——, *Convex polytopes, Coxeter orbifolds and torus actions*, Duke Math. J. **62** (1991), no. 2, 417–451, https://doi.org/10.1215/S0012-7094-91-06217-4. MR1104531
93. ——, *Right-angled Artin groups are commensurable with right-angled Coxeter groups*, J. Pure Appl. Algebra **153** (2000), no. 3, 229–235.
94. M. W. Davis, T. Januszkiewicz, and R. Scott, *Nonpositive curvature of blow-ups*, Selecta Math. (N.S.) **4** (1998), no. 4, 491–547. MR1668119
95. ——, *Fundamental groups of blow-ups*, Adv. Math. **177** (2003), no. 1, 115–179. MR1985196
96. M. W. Davis, T. Januszkiewicz, and S. Weinberger, *Relative hyperbolization and aspherical bordisms: an addendum to "Hyperbolization of polyhedra"*, J. Differential Geom. **58** (2001), no. 3, 535–541. MR1906785
97. M. W. Davis and M. Kahle, *Random graph products of finite groups are rational duality groups*, J. Topol. **7** (2014), no. 2, 589–606, https://doi.org/10.1112/jtopol/jtt047. MR3217632
98. M. W. Davis and P. H. Kropholler, *Criteria for asphericity of polyhedral products: corrigenda to "right-angularity, flag complexes, asphericity"*, Geom. Dedicata **179** (2015), 39–44, https://doi.org/10.1007/s10711-015-0066-8. MR3424656
99. M. W. Davis, G. Le, and K. Schreve, *Action dimensions of simple complexes of groups*, J. of Topology **12** (2019), 1266-1314.
100. M. W. Davis and I. J. Leary, *Some examples of discrete group actions on aspherical manifolds*, High-dimensional manifold topology, World Sci. Publ., River Edge, NJ, 2003, pp. 139–150. MR2048719
101. M. W. Davis and J. Meier, *The topology at infinity of Coxeter groups and buildings*, Comment. Math. Helv. **77** (2002), no. 4, 746–766, https://doi.org/10.1007/PL00012440. MR1949112
102. ——, *Reflection groups and* CAT(0) *complexes with exotic local structures*, High-dimensional manifold topology, World Sci. Publ., River Edge, NJ, 2003, pp. 151–158. MR2048720
103. M. W. Davis and G. Moussong, *Notes on nonpositively curved polyhedra*, Low dimensional topology (Eger, 1996/Budapest, 1998), Bolyai Soc. Math. Stud., vol. 8, János Bolyai Math. Soc., Budapest, 1999, pp. 11–94. MR1747268
104. M. W. Davis and B. Okun, *Vanishing theorems and conjectures for the ℓ^2-homology of right-angled Coxeter groups*, Geom. Topol. **5** (2001), 7–74, https://doi.org/10.2140/gt.2001.5.7. MR1812434

105. ——, *Cohomology computations for Artin groups, Bestvina-Brady groups, and graph products*, Groups Geom. Dyn. **6** (2012), no. 3, 485–531, https://doi.org/10.4171/GGD/164. MR2961283
106. C. De Concini and C. Procesi, *Wonderful models of subspace arrangements*, Selecta Math. (N.S.) **1** (1995), no. 3, 459–494, https://doi.org/10.1007/BF01589496. MR1366622
107. P. Deligne, *Les immeubles des groupes de tresses généralisés*, Invent. Math. **17** (1972), 273–302 (French). MR0422673
108. M. Desgroseilliers and F. Haglund, *On some convex cocompact groups in real hyperbolic space*, Geom. Topol. **17** (2013), no. 4, 2431–2484, https://doi.org/10.2140/gt.2013.17.2431. MR3110583
109. C. Droms, *Graph groups, coherence, and three-manifolds*, J. Algebra **106** (1987), no. 2, 484–489.
110. M. J. Dunwoody, *The accessibility of finitely presented groups*, Invent. Math. **81** (1985), no. 3, 449–457, https://doi.org/10.1007/BF01388581. MR807066
111. R. D. Edwards, *The topology of manifolds and cell-like maps*, Proceedings of the International Congress of Mathematicians (Helsinki, 1978), Acad. Sci. Fennica, Helsinki, 1980, pp. 111–127. MR562601
112. G. Ellis and E. Sköldberg, *The $K(\pi, 1)$ conjecture for a class of Artin groups*, Comment. Math. Helv. **85** (2010), no. 2, 409–415, https://doi.org/10.4171/CMH/200. MR2595184
113. P. Etingof, A. Henriques, J. Kamnitzer, and E. M. Rains, *The cohomology ring of the real locus of the moduli space of stable curves of genus 0 with marked points*, Ann. of Math. (2) **171** (2010), no. 2, 731–777, https://doi.org/10.4007/annals.2010.171.731. MR2630055
114. B. Farb, *Relatively hyperbolic groups*, Geom. Funct. Anal. **8** (1998), no. 5, 810–840, https://doi.org/10.1007/s000390050075. MR1650094
115. B. Farb, C. Hruska, and A. Thomas, *Problems on automorphism groups of nonpositively curved polyhedral complexes and their lattices*, Geometry, rigidity, and group actions, Chicago Lectures in Math., Univ. Chicago Press, Chicago, IL, 2011, pp. 515–560, https://doi.org/10.7208/chicago/9780226237909.001.0001. MR2807842
116. F. T. Farrell and L. E. Jones, *Topological rigidity for compact non-positively curved manifolds*, Differential geometry: Riemannian geometry (Los Angeles, CA, 1990), Proc. Sympos. Pure Math., vol. 54, Amer. Math. Soc., Providence, RI, 1993, pp. 229–274, https://doi.org/10.1090/pspum/054.3/1216623. MR1216623
117. W. Feit and G. Higman, *The nonexistence of certain generalized polygons*, J. Algebra **1** (1964), 114–131, https://doi.org/10.1016/0021-8693(64)90028-6. MR170955
118. B. Foozwell, *Haken n-manifolds*, ProQuest LLC, Ann Arbor, MI, 2007. Thesis (Ph.D.)–University of Melbourne.
119. B. Foozwell and H. Rubinstein, *Introduction to the theory of Haken n-manifolds*, Topology and geometry in dimension three, Contemp. Math., vol. 560, Amer. Math. Soc., Providence, RI, 2011, pp. 71–84, https://doi.org/10.1090/conm/560/11092. MR2866924
120. J. Fowler, *Finiteness properties for some rational Poincaré duality groups*, Illinois J. Math. **56** (2012), no. 2, 281–299. MR3161324
121. M. H. Freedman, *The topology of four-dimensional manifolds*, J. Differential Geometry **17** (1982), no. 3, 357–453. MR679066
122. D. Gaboriau and F. Paulin, *Sur les immeubles hyperboliques*, Geom. Dedicata **88** (2001), no. 1-3, 153–197 (French, with English summary). MR1877215
123. A. A. Gaĭfullin, *The manifold of isospectral symmetric tridiagonal matrices and the realization of cycles by aspherical manifolds*, Tr. Mat. Inst. Steklova **263** (2008), no. Geometriya, Topologiya i Matematicheskaya Fizika. I, 44–63, https://doi.org/10.1134/S0081543808040044 (Russian, with Russian summary); English transl., Proc. Steklov Inst. Math. **263** (2008), no. 1, 38–56. MR2599370
124. A. Gaifullin, *Universal realisators for homology classes*, Geom. Topol. **17** (2013), no. 3, 1745–1772, https://doi.org/10.2140/gt.2013.17.1745. MR3073934
125. Ś. R. Gal, *Euler characteristic of the configuration space of a complex*, Colloq. Math. **89** (2001), no. 1, 61–67, https://doi.org/10.4064/cm89-1-4. MR1853415

126. D. E. Galewski and R. J. Stern, *Classification of simplicial triangulations of topological manifolds*, Ann. of Math. (2) **111** (1980), no. 1, 1–34, https://doi.org/10.2307/1971215. MR558395
127. D. Galewski and R. Stern, *A universal 5-manifold with respect to simplicial triangulations*, Geometric topology (Proc. Georgia Topology Conf., Athens, Ga., 1977), Academic Press, New York-London, 1979, pp. 345–350. MR537740
128. A. Genevois, *Quasi-isometrically rigid subgroups in right-angled Coxeter groups*, Algebr. Geom. Topol. **22** (2022), no. 2, 657–708, https://doi.org/10.2140/agt.2022.22.657. MR4464462
129. ——, *Cactus groups from the viewpoint of geometric group theory*. arXiv:2212.03494v1.
130. R. Geoghegan, *Topological methods in group theory*, Graduate Texts in Mathematics, vol. 243, Springer, New York, 2008. MR2365352
131. R. Gitik, M. Mitra, E. Rips, and M. Sageev, *Widths of subgroups*, Trans. Amer. Math. Soc. **350** (1998), no. 1, 321–329, https://doi.org/10.1090/S0002-9947-98-01792-9. MR1389776
132. S. Gitler and S. López de Medrano, *Intersections of quadrics, moment-angle manifolds and connected sums*, Geom. Topol. **17** (2013), no. 3, 1497–1534, https://doi.org/10.2140/gt.2013.17.1497. MR3073929
133. E. Godelle and L. Paris, *$K(\pi, 1)$ and word problems for infinite type Artin-Tits groups, and applications to virtual braid groups*, Math. Z. **272** (2012), no. 3-4, 1339–1364, https://doi.org/10.1007/s00209-012-0989-9. MR2995171
134. K. Goldman, *The $K(\pi, 1)$ Conjecture and acylindrical hyperbolicity for relatively extra-large Artin groups*, Algbr. Geom. Topol.
135. M. Gromov, *Groups of polynomial growth and expanding maps*, Inst. Hautes Études Sci. Publ. Math. **53** (1981), 53–73. MR623534
136. M. Gromov, *Hyperbolic groups*, Essays in group theory, Math. Sci. Res. Inst. Publ., vol. 8, Springer, New York, 1987, pp. 75–263. MR919829
137. B. Grünbaum, *Convex polytopes*, 2nd ed., Graduate Texts in Mathematics, vol. 221, Springer-Verlag, New York, 2003. Prepared and with a preface by Volker Kaibel, Victor Klee and Günter M. Ziegler. MR1976856
138. A. Haefliger, *Extension of complexes of groups*, Ann. Inst. Fourier (Grenoble) **42** (1992), no. 1-2, 275–311 (English, with French summary). MR1162563
139. T. Haettel, *Virtually cocompactly cubulated Artin-Tits groups*, Int. Math. Res. Not. IMRN **4** (2021), 2919–2961, https://doi.org/10.1093/imrn/rnaa013. MR4218342
140. F. Haglund, *Finite index subgroups of graph products*, Geom. Dedicata **135** (2008), 167–209. MR2413337
141. F. Haglund and F. Paulin, *Simplicité de groupes d'automorphismes d'espaces à courbure négative*, The Epstein birthday schrift, Geom. Topol. Monogr., vol. 1, Geom. Topol. Publ., Coventry, 1998, pp. 181–248, https://doi.org/10.2140/gtm.1998.1.181 (French, with English and French summaries). MR1668359
142. ——, *Constructions arborescentes d'immeubles*, Math. Ann. **325** (2003), no. 1, 137–164, https://doi.org/10.1007/s00208-002-0373-x (French, with English summary). MR1957268
143. F. Haglund and D. T. Wise, *Special cube complexes*, Geom. Funct. Anal. **17** (2008), no. 5, 1551–1620, https://doi.org/10.1007/s00039-007-0629-4. MR2377497
144. F. Haglund and D. T. Wise, *Coxeter groups are virtually special*, Adv. Math. **224** (2010), no. 5, 1890–1903, https://doi.org/10.1016/j.aim.2010.01.011. MR2646113
145. ——, *A combination theorem for special cube complexes*, Ann. of Math. (2) **176** (2012), no. 3, 1427–1482, https://doi.org/10.4007/annals.2012.176.3.2. MR2979855
146. A. Hatcher, *Algebraic topology*, Cambridge University Press, Cambridge, 2002. MR1867354
147. H. Hendriks, *Hyperplane complements of large type*, Invent. Math. **79** (1985), no. 2, 375–381, https://doi.org/10.1007/BF01388979. MR778133
148. G. Higman, *A finitely generated infinite simple group*, J. London Math. Soc. **26** (1951), 61–64, https://doi.org/10.1112/jlms/s1-26.1.61. MR38348
149. C. Horbez and J. Huang, *Measure equivalence rigidity among Higman groups*. arXiv:2206.00884v1.

150. G. C. Hruska and B. Kleiner, *Hadamard spaces with isolated flats*, Geom. Topol. **9** (2005), 1501–1538, https://doi.org/10.2140/gt.2005.9.1501. With an appendix by the authors and Mohamad Hindawi. MR2175151
151. G. C. Hruska and D. T. Wise, *Finiteness properties of cubulated groups*, Compos. Math. **150** (2014), no. 3, 453–506, https://doi.org/10.1112/S0010437X13007112. MR3187627
152. T. Hsu and D. T. Wise, *On linear and residual properties of graph products*, Michigan Math. J. **46** (1999), no. 2, 251–259.
153. T. Hsu and D. Wise, *A non-residually finite square of finite groups*, Groups St. Andrews 1997 in Bath, I, London Math. Soc. Lecture Note Ser., vol. 260, Cambridge Univ. Press, Cambridge, 1999, pp. 368–378. MR1676633
154. B. Z. Hu, *Whitehead groups of finite polyhedra with nonpositive curvature*, J. Differential Geom. **38** (1993), no. 3, 501–517. MR1243784
155. ——, *Retractions of closed manifolds with nonpositive curvature*, Geometric group theory (Columbus, OH, 1992), Ohio State Univ. Math. Res. Inst. Publ., vol. 3, de Gruyter, Berlin, 1995, pp. 135–147. MR1355108
156. J. Huang, *Labeled four cycles and the $K(\pi,1)$ problem for Artin groups*. arxiv:2305.16847v1.
157. J. Huang and D. Osajda, *Helly meets Garside and Artin*, Invent. Math. **225** (2021), no. 2, 395–426, https://doi.org/10.1007/s00222-021-01030-8. MR4285138
158. T. Januszkiewicz and J. Świątkowski, *Commensurability of graph products*, Algebr. Geom. Topol. **1** (2001), 587–603, https://doi.org/10.2140/agt.2001.1.587. MR1875609
159. ——, *Hyperbolic Coxeter groups of large dimension*, Comment. Math. Helv. **78** (2003), no. 3, 555–583, https://doi.org/10.1007/s00014-003-0763-z. MR1998394
160. ——, *Simplicial nonpositive curvature*, Publ. Math. Inst. Hautes Études Sci. **104** (2006), 1–85, https://doi.org/10.1007/s10240-006-0038-5. MR2264834
161. V. G. Kac and D. H. Peterson, *Defining relations of certain infinite-dimensional groups*, Astérisque **Numéro Hors Série** (1985), 165–208. The mathematical heritage of Élie Cartan (Lyon, 1984). MR837201
162. J. Kahn and V. Markovic, *Immersing almost geodesic surfaces in a closed hyperbolic three manifold*, Ann. of Math. (2) **175** (2012), no. 3, 1127–1190, https://doi.org/10.4007/annals.2012.175.3.4. MR2912704
163. D. M. Kan and W. P. Thurston, *Every connected space has the homology of a $K(\pi,1)$*, Topology **15** (1976), no. 3, 253–258, https://doi.org/10.1016/0040-9383(76)90040-9. MR413089
164. M. Kapovich, *Representations of polygons of finite groups*, Geom. Topol. **9** (2005), 1915–1951, https://doi.org/10.2140/gt.2005.9.1915. MR2175160
165. M. M. Kapranov, *Chow quotients of Grassmannians. I*, I. M. Gel'fand Seminar, Adv. Soviet Math., vol. 16, Amer. Math. Soc., Providence, RI, 1993, pp. 29–110. MR1237834
166. R. C. Kirby and L. C. Siebenmann, *Foundational essays on topological manifolds, smoothings, and triangulations*, Annals of Mathematics Studies, No. 88, Princeton University Press, Princeton, N.J.; University of Tokyo Press, Tokyo, 1977. With notes by John Milnor and Michael Atiyah. MR0645390
167. P. H. Kropholler and A. Martino, *Graph-wreath products and finiteness conditions*, J. Pure Appl. Algebra **220** (2016), no. 1, 422–434, https://doi.org/10.1016/j.jpaa.2015.07.001. MR3393469
168. R. P. Kropholler, I. J. Leary, and I. Soroko, *Uncountably many quasi-isometry classes of groups of type FP*, Amer. J. Math. **142** (2020), no. 6, 1931–1944, https://doi.org/10.1353/ajm.2020.0048. MR4176549
169. I. J. Leary, *Uncountably many groups of type FP*, Proc. Lond. Math. Soc. (3) **117** (2018), no. 2, 246–276, https://doi.org/10.1112/plms.12135. MR3851323
170. I. J. Leary and B. E. A. Nucinkis, *Some groups of type VF*, Invent. Math. **151**(2003), no. 1, 135–165, https://doi.org/10.1007/s00222-002-0254-7. MR1943744
171. H. van der Lek, *The homotopy type of complex hyperplane complements*, 1988. Thesis (Ph.D.)–Nijmegan University.

172. S. López de Medrano, *Topology of the intersection of quadrics in* $\mathbf{R}^n$, Algebraic topology (Arcata, CA, 1986), Lecture Notes in Math., vol. 1370, Springer, Berlin, 1989, pp. 280–292, https://doi.org/10.1007/BFb0085235. MR1000384
173. P. Linnell, B. Okun, and T. Schick, *The strong Atiyah conjecture for right-angled Artin and Coxeter groups*, Geom. Dedicata **158** (2012), 261–266, https://doi.org/10.1007/s10711-011-9631-y. MR2922714
174. C. Manolescu, *Pin(2)-equivariant Seiberg-Witten Floer homology and the triangulation conjecture*, J. Amer. Math. Soc. **29** (2016), no. 1, 147–176, https://doi.org/10.1090/jams829. MR3402697
175. G. A. Margulis, *Discrete subgroups of semisimple Lie groups*, Ergebnisse der Mathematik und ihrer Grenzgebiete (3) [Results in Mathematics and Related Areas (3)], vol. 17, Springer-Verlag, Berlin, 1991. MR1090825
176. A. Martin, *On the cubical geometry of Higman's group*, Duke Math. J. **166** (2017), no. 4, 707–738, https://doi.org/10.1215/00127094-3715913. MR3619304
177. T. Matumoto, *Triangulation of manifolds*, Algebraic and geometric topology (Proc. Sympos. Pure Math., Stanford Univ., Stanford, Calif., 1976), Proc. Sympos. Pure Math., XXXII, Amer. Math. Soc., Providence, R.I., 1978, pp. 3–6. MR520517
178. G. Mess, *Examples of Poincaré duality groups*, Proc. Amer. Math. Soc. **110** (1990), no. 4, 1145–1146, https://doi.org/10.2307/2047770. MR1019274
179. G. Moussong, *Hyperbolic Coxeter groups*, ProQuest LLC, Ann Arbor, MI, 1988. Thesis (Ph.D.)–The Ohio State University. MR2636665
180. G. Moussong, *Some nonsymmetric manifolds*, Differential geometry and its applications (Eger, 1989), Colloq. Math. Soc. János Bolyai, vol. 56, North-Holland, Amsterdam, 1992, pp. 535–546. MR1211680
181. G. Niblo and L. Reeves, *Groups acting on* CAT(0) *cube complexes*, Geom. Topol. **1** (1997), 1–7, https://doi.org/10.2140/gt.1997.1.1. MR1432323
182. G. A. Niblo and L. D. Reeves, *The geometry of cube complexes and the complexity of their fundamental groups*, Topology **37** (1998), no. 3, 621–633, https://doi.org/10.1016/S0040-9383(97)00018-9. MR1604899
183. ——, *Coxeter groups act on* CAT(0) *cube complexes*, J. Group Theory **6** (2003), no. 3, 399–413, https://doi.org/10.1515/jgth.2003.028. MR1983376
184. R. Oliver, *Fixed-point sets of group actions on finite acyclic complexes*, Comment. Math. Helv. **50** (1975), 155–177, https://doi.org/10.1007/BF02565743. MR0375361
185. P. Ontaneda, *The hyperbolization process and its Riemannian version*, Handbook of group actions. Vol. III, Adv. Lect. Math. (ALM), vol. 40, Int. Press, Somerville, MA, 2018, pp. 59–86. MR3888616
186. ——, *Normal smoothings for Charney-Davis strict hyperbolizations*, J. Topol. Anal. **9** (2017), no. 1, 127–165, https://doi.org/10.1142/S1793525317500054. MR3594608
187. ——, *Riemannian hyperbolization*, Publ. Math. Inst. Hautes Études Sci. **131** (2020), 1–72, https://doi.org/10.1007/s10240-020-00113-1. MR4106793
188. G. Paolini and M. Salvetti, *Proof of the* $K(\pi,1)$ *conjecture for affine Artin groups*, Invent. Math. **224**(2021), no. 2, 487–572, https://doi.org/10.1007/s00222-020-01016-y. MR4243019
189. L. Paris, *Parabolic subgroups of Artin groups*, Journal of Algebra **196** (1997), no. 2, 369–399.
190. ——, $K(\pi,1)$ *conjecture for Artin groups*, Ann. Fac. Sci. Toulouse Math. (6) **23** (2014), no. 2, 361–415, https://doi.org/10.5802/afst.1411 (English, with English and French summaries). MR3205598
191. N. Petrosyan and Tomasz Prytula, *Bestvina complex for group actions with a strict fundamental domain*, Groups, Geom. Dynamics.
192. P. Przytycki and J. Świątkowski, *Flag-no-square triangulations and Gromov boundaries in dimension 3*, Groups Geom. Dyn. **3** (2009), no. 3, 453–468, https://doi.org/10.4171/GGD/66. MR2516175
193. G. A. Reisner, *Cohen-Macaulay quotients of polynomial rings*, Advances in Math. **21** (1976), no. 1, 30–49, https://doi.org/10.1016/0001-8708(76)90114-6. MR407036

194. B. Remy, *Kac-Moody groups: split and relative theories. Lattices*, Groups: topological, combinatorial and arithmetic aspects, London Math. Soc. Lecture Note Ser., vol. 311, Cambridge Univ. Press, Cambridge, 2004, pp. 487–541, https://doi.org/10.1017/CBO9780511550706.016. MR2073357
195. B. Rémy, *Groupes de Kac-Moody déployés et presque déployés*, Astérisque **277** (2002), viii+348 (French, with English and French summaries). MR1909671
196. ——, *Kac-Moody groups as discrete groups*, Essays in geometric group theory, Ramanujan Math. Soc. Lect. Notes Ser., vol. 9, Ramanujan Math. Soc., Mysore, 2009, pp. 105–124. MR2605357
197. M. A. Roller, *Pocsets,median algebras and group actions*, Univ. of Southampton, 1998.
198. M. Ronan, *Lectures on buildings*, Perspectives in Mathematics, vol. 7, Academic Press, Inc., Boston, MA, 1989. MR1005533
199. J. H. Rubinstein and S. Wang, *π_1-injective surfaces in graph manifolds*, Comment. Math. Helv. **73**(1998), no. 4, 499–515, https://doi.org/10.1007/s000140050066. MR1639876
200. M. Sageev, *Ends of group pairs and non-positively curved cube complexes*, Proc. London Math. Soc. (3) **71** (1995), no. 3, 585–617. MR1347406
201. ——, *Codimension-1 subgroups and splittings of groups*, J. Algebra **189** (1997), no. 2, 377–389. MR1438181
202. ——, CAT(0) *cube complexes and groups*, Geometric group theory, IAS/Park City Math. Ser., vol. 21, Amer. Math. Soc., Providence, RI, 2014, pp. 7–54. MR3329724
203. M. Salvetti, *Topology of the complement of real hyperplanes in* C^n, Invent. Math. **88** (1987), 603-618.
204. M. Sapir, *A Higman embedding preserving asphericity*, J. Amer. Math. Soc. **27** (2014), no. 1, 1–42, https://doi.org/10.1090/S0894-0347-2013-00776-6. MR3110794
205. P. Scott, *Subgroups of surface groups are almost geometric*, J. London Math. Soc. (2) **17** (1978), no. 3, 555–565, https://doi.org/10.1112/jlms/s2-17.3.555. MR494062
206. R. Scott, *Right-angled mock reflection and mock Artin groups*, Trans. Amer. Math. Soc. **360** (2008), no. 8, 4189–4210, https://doi.org/10.1090/S0002-9947-08-04452-8. MR2395169
207. J.-P. Serre, *Trees*, Springer-Verlag, Berlin-New York, 1980. Translated from the French by John Stillwell. MR607504
208. J. Stallings, *A finitely presented group whose 3-dimensional integral homology is not finitely generated*, Amer. J. Math. **85** (1963), 541–543, https://doi.org/10.2307/2373106. MR0158917
209. J. R. Stallings, *Non-positively curved triangles of groups*, Group theory from a geometrical viewpoint (Trieste, 1990), World Sci. Publ., River Edge, NJ, 1991, pp. 491–503. MR1170374
210. R. P. Stanley, *Combinatorics and commutative algebra*, 2nd ed., Progress in Mathematics, vol. 41, Birkhäuser Boston, Inc., Boston, MA, 1996. MR1453579
211. D. A. Stone, *Geodesics in piecewise linear manifolds*, Trans. Amer. Math. Soc. **215** (1976), 1–44, https://doi.org/10.2307/1999713. MR402648
212. J. Świątkowski, *Trivalent polygonal complexes of nonpositive curvature and Platonic symmetry*, Geom. Dedicata **70** (1998), no. 1, 87–110, https://doi.org/10.1023/A:1004954009167. MR1612350
213. ——, *Estimates for homological dimension of configuration spaces of graphs*, Colloq. Math. **89** (2001), no. 1, 69–79, https://doi.org/10.4064/cm89-1-5. MR1853416
214. W. P. Thurston, *Three-dimensional geometry and topology. Vol. 1*, Princeton Mathematical Series, vol. 35, Princeton University Press, Princeton, NJ, 1997. Edited by Silvio Levy. MR1435975
215. ——, *Three-dimensional manifolds, Kleinian groups and hyperbolic geometry*, Bull. Amer. Math. Soc. (N.S.) **6** (1982), no. 3, 357–381, https://doi.org/10.1090/S0273-0979-1982-15003-0. MR648524
216. J. Tits, *Structures et groupes de Weyl*, Séminaire Bourbaki, Vol. 9, Soc. Math. France, Paris, 1995, pp. Exp. No. 288, 169–183 (French). MR1608796
217. ——, *Sur le groupe des automorphismes d'un arbre*, Essays on topology and related topics (Mémoires dédiés à Georges de Rham), Springer, New York, 1970, pp. 188–211 (French). MR0299534

218. ——, *Buildings of spherical type and finite BN-pairs*, Lecture Notes in Mathematics, Vol. 386, Springer-Verlag, Berlin-New York, 1974. MR0470099
219. J. Tits, *On buildings and their applications*, Proceedings of the International Congress of Mathematicians (Vancouver, B. C., 1974), Canad. Math. Congress, Montreal, Que., 1975, pp. 209–220. MR0439945
220. ——, *A local approach to buildings*, The geometric vein, Springer, New York-Berlin, 1981, pp. 519–547. MR661801
221. J. Tits, *Buildings and group amalgamations*, Proceedings of groups—St. Andrews 1985, London Math. Soc. Lecture Note
Ser., vol. 121, Cambridge Univ. Press, Cambridge, 1986, pp. 110–127. MR896503
222. ——, *Immeubles de type affine*, Buildings and the geometry of diagrams (Como, 1984), Lecture Notes in Math., vol. 1181, Springer, Berlin, 1986, pp. 159–190, https://doi.org/10.1007/BFb0075514 (French). MR843391
223. ——, *Uniqueness and presentation of Kac-Moody groups over fields*, J. Algebra **105** (1987), no. 2, 542–573, https://doi.org/10.1016/0021-8693(87)90214-6. MR873684
224. ——, *Twin buildings and groups of Kac-Moody type*, Groups, combinatorics & geometry (Durham, 1990), London Math. Soc. Lecture Note Ser., vol. 165, Cambridge Univ. Press, Cambridge, 1992, pp. 249–286, https://doi.org/10.1017/CBO9780511629259.023. MR1200265
225. J. Tits and R. M. Weiss, *Moufang polygons*, Springer Monographs in Mathematics, Springer-Verlag, Berlin, 2002. MR1938841
226. C. Tomei, *The topology of isospectral manifolds of tridiagonal matrices*, Duke Math. J. **51** (1984), no. 4, 981–996, https://doi.org/10.1215/S0012-7094-84-05144-5. MR771391
227. È. B. Vinberg, *Discrete linear groups that are generated by reflections*, Izv. Akad. Nauk SSSR Ser. Mat. **35** (1971), 1072–1112 (Russian). MR0302779
228. ——, *Absence of crystallographic groups of reflections in Lobachevskiĭ spaces of large dimension*, Trudy Moskov. Mat. Obshch. **47** (1984), 68–102, 246 (Russian). MR774946
229. ——, *Hyperbolic groups of reflections*, Uspekhi Mat. Nauk **40** (1985), no. 1(241), 29–66, 255 (Russian). MR783604
230. È. B. Vinberg and V. G. Kac, *Quasi-homogeneous cones*, Mat. Zametki **1** (1967), 347–354 (Russian). MR208470
231. S. Weinberger, *Computers, rigidity, and moduli*, M. B. Porter Lectures, Princeton University Press, Princeton, NJ, 2005. The large-scale fractal geometry of Riemannian moduli space. MR2109177
232. D. T. Wise, *Non-positively curved squared complexes: Aperiodic tilings and non-residually finite groups*, ProQuest LLC, Ann Arbor, MI, 1996. Thesis (Ph.D.)–Princeton University. MR2694733
233. ——, *Complete square complexes*, Comment. Math. Helv. **82** (2007), no. 4, 683–724, https://doi.org/10.4171/CMH/107. MR2341837
234. ——, *From riches to raags: 3-manifolds, right-angled Artin groups, and cubical geometry*, CBMS Regional Conference Series in Mathematics, vol. 117, Published for the Conference Board of the Mathematical Sciences, Washington, DC; by the American Mathematical Society, Providence, RI, 2012. MR2986461
235. G. M. Ziegler, *Lectures on polytopes*, Graduate Texts in Mathematics, vol. 152, Springer-Verlag, New York, 1995. MR1311028

Index

M. W. Davis, *Infinite Group Actions on Polyhedra*, Ergebnisse der Mathematik und ihrer Grenzgebiete. 3. Folge / A Series of Modern Surveys in Mathematics 77,
https://doi.org/10.1007/978-3-031-48443-8

The manufacturer's authorised representative in the EU is Springer Nature Customer Service Centre GmbH, Europaplatz 3, 69115 Heidelberg, Germany. If you have any concerns regarding our products, please contact ProductSafety@springernature.com

Printed and bound by CPI Group (UK) Ltd, Croydon, CR0 4YY

07/07/2026

02160914-0002